ANALYTICAL METHODS FOR PESTICIDES AND PLANT GROWTH REGULATORS

VOLUME VIII

GOVERNMENT REGULATIONS, PHEROMONE ANALYSIS,
ADDITIONAL PESTICIDES

Volume I

Principles, Methods, and General Applications

Volume II

Insecticides

Volume III

Fungicides, Nematicides and Soil Fumigants, Rodenticides, and Food and Feed Additives

Volume IV

Herbicides

Volume V

Additional Principles and Methods of Analysis

Volume VI

Gas Chromatographic Analysis

Volume VII

Thin-Layer and Liquid Chromatography and Analyses of Pesticides of International Importance

Volume VIII

Government Regulations, Pheromone Analysis, Additional Pesticides

Analytical Methods for

PESTICIDES AND PLANT GROWTH REGULATORS

Edited by

GUNTER ZWEIG

Office of Pesticide Programs, U.S. Environmental Protection Agency
Washington, D.C.

Volume VIII

GOVERNMENT REGULATIONS, PHEROMONE ANALYSIS, ADDITIONAL PESTICIDES

Edited by

GUNTER ZWEIG *and* JOSEPH SHERMA

Office of Pesticide Programs
U. S. Environmental Protection Agency
Washington, D. C.

Department of Chemistry
Lafayette College
Easton, Pennsylvania

ACADEMIC PRESS New York San Francisco London 1976

A SUBSIDIARY OF HARCOURT BRACE JOVANOVICH, PUBLISHERS

ACADEMIC PRESS, INC.
111 Fifth Avenue, New York, New York 10003

United Kingdom Edition published by
ACADEMIC PRESS, INC. (LONDON) LTD.
24/28 Oval Road, London NW1

Library of Congress Cataloging in Publication Data

Zweig, Gunter.
Analytical methods for pesticides, plant growth regulators, and food additives.

Vols. 6- have title: Analytical methods for pesticides and plant growth regulators.
Includes bibliographies.
CONTENTS.– v. 1. Principles, methods, and general applications. v. 2. Insecticides.–v. 3. Fungicides, nematocides and soil fumigants, rodenticides, and food and feed additives. [etc.]
1. Pesticides–Analysis–Collected works. 2. Plant regulators–Analysis–Collected works. 3. Food additives–Analysis–Collected works. 4. Feed additives–Analysis–Collected works. 5. Chemistry, Analytic–Collected works. I. Sherman, Joseph. II. Title.
SB960.Z9 632'.95 63-16560
ISBN 0-12-784308-6 (v. 8)

PRINTED IN THE UNITED STATES OF AMERICA

Contents

10. Promacyl

D. S. GUNEW

11. Prothoate

B. BAZZI

12. Pyrethrum

JOSEPH SHERMA

13. Thiometon

M. WISSON, C. VAN HOEK, AND H. H. SAUER

PART III. FUNGICIDES

14. 2-Aminobutane

E. W. DAY, JR., AND J. R. KOONS

15. Chlorothalonil

D. L. BALLEE, W. C. DUANE, D. E. STALLARD, AND A. L. WOLFE

16. Dinobuton

17. Ethirimol

18. Mertect®

19. Validacin®

20. Vitavax®

Part IV. Herbicides

21. Balan®

22. Devrinol®

G. G. Patchett, D. J. Brookman, and J. E. Rodebush

23. Dinitramine

H. C. Newsom

24. Isopropalin

W. S. Johnson and O. D. Decker

25. Karbutilate

R. F. Cook

26. MCPB

A. Guardigli, J. Vernetti, and M. A. Gallo

27. Metoxuron

M. Wisson, C. van Hoek, and H. H. Sauer

List of Contributors

Numbers in parentheses refer to the pages on which the authors' contributions begin.

I. L. ADLER (443), *Research Laboratories, Rohm and Haas Co., Spring House, Pennsylvania*

C. A. ANDERSON (453), *Chemagro Agricultural Division, Mobay Chemical Corporation, Kansas City, Missouri*

D. L. BALLEE (263), *T. R. Evans Research Center, Diamond-Shamrock Corporation, Painesville, Ohio*

B. BAZZI (159, 213), *Montedison S.p. A. Centro Ricerche Antiparassitari, Milan, Italy*

MORTON BEROZA (31), *Agricultural Environmental Quality Institute, ARS, USDA, Beltsville, Maryland*

D. J. BROOKMAN (347), *Stauffer Chemical Company, Richmond, California*

D. J. W. BULLOCK (171, 185), *ICI Plant Protection Division, Jealotts Hill Research Station, Bracknell, Berkshire, England*

R. F. COOK (381), *Agricultural Chemical Division, FMC Corporation, Middleport, New York*

E. W. DAY, JR. (251, 475), *Lilly Research Laboratories, Division of Eli Lilly and Company, Greenfield, Indiana*

O. D. DECKER (369, 433), *Lilly Research Laboratories, Division of Eli Lilly and Company, Greenfield, Indiana*

W. C. DUANE (263), *T. R. Evans Research Center, Diamond-Shamrock Corporation, Painesville, Ohio*

D. O. EBERLE (141), *Ciba-Geigy, Basel, Switzerland*

M. J. EDWARDS (285), *ICI Plant Protection Division, Jealotts Hill Research Station, Bracknell, Berkshire, England*

R. FRANK (335, 475), *Lilly Research Laboratories, Division of Eli Lilly and Company, Greenfield, Indiana*

M. A. GALLO (397), *Rhodia Inc., Agricultural Division, New Brunswick, New Jersey*

A. GUARDIGLI (397), *Rhodia, Inc., Agricultural Division, New Brunswick, New Jersey*

D. S. GUNEW (207), *ICI Australia Ltd. Ascot Vale, Victoria, Australia*

L. D. HAINES (443), *Research Laboratories, Rohm and Haas Co., Bristol, Pennsylvania*

M. N. INSCOE (31), *Agricultural Environmental Quality Institute, ARS, USDA, Beltsville, Maryland*

W. S. JOHNSON (335, 369, 433), *Lilly Research Laboratories, Division of Eli Lilly and Company, Greenfield, Indiana*

K. KONISHI (309), *Agricultural Chemicals Division, Takeda Chemical Industries, Ltd., Osaka, Japan*

J. R. KOONS (251), *Lilly Research Laboratories, Division of Eli Lilly and Company, Greenfield, Indiana*

V. P. LYNCH (135, 275, 491), *Murphy Chemical Ltd., Wheathampstead, St. Albans, Hertfordshire, England*

H. C. NEWSOM (359), *U. S. Borax Research Corporation, Anaheim, California*

K. NISHI (309), *Agricultural Chemicals Division, Takeda Chemical Industries, Ltd., Osaka, Japan*

G. G. PATCHETT (347), *Stauffer Chemical Company, Richmond, California*

J. R. RIDEN (483), *Gulf Oil Chemicals Company, Merriam, Kansas*

J. E. RODEBUSH (347), *Stauffer Chemical Company, Richmond, California*

H. H. SAUER (123, 239, 417), *Agrochemical Division, Sandoz Ltd., Basel, Switzerland*

JOSEPH SHERMA (117, 225), *Department of Chemistry, Lafayette College, Easton, Pennsylvania*

D. E. STALLARD (263), *T. R. Evans Research Center, Diamond-Shamrock Corporation, Painesville, Ohio*

G. M. STONE (319), *Uniroyal Chemical, Division of Uniroyal, Inc., Naugatuck, Connecticut*

R. SUTER (141), *Ciba-Geigy, Ltd., Basel, Switzerland*

C. VAN HOEK (123, 239, 417), *Agrochemical Division, Sandoz Ltd., Basel, Switzerland*

J. VERNETTI (397), *Rhodia Inc., Agricultural Division, New Brunswick, New Jersey*

J. P. WARGO, JR. (443), *Research Laboratories, Rohm and Haas Co., Bristol, Pennsylvania*

M. WISSON (123, 239, 417), *Agrochemical Division, Sandoz Ltd., Basel Switzerland*

A. L. WOLFE (263), *T. R. Evans Research Center, Diamond-Shamrock Corporation, Painesville, Ohio*

JACOB WOLFF (3), *Formerly with: Office of Pesticide Programs Criteria and Evaluation Division, Environmental Protection Agency, Washington, D. C.*

J. S. WOOD, JR. (299), *Research Laboratories, Merck, Sharp & Dohme, Rahway, New Jersey*

Preface

In 1972, a new pesticide law was enacted in the United States, and it appeared appropriate to begin Volume VIII of this treatise with a discussion of this law as it pertains to regulations for the registration of pesticides in the United States. A brief discussion of this subject appeared in 1963 in Chapter 1 of Volume I. Since that time a number of new requirements requiring the skills of the analytical pesticide chemist have become mandatory.

Chapter 2 deals with the analysis of naturally occurring chemicals that control the behavior of insects. They may prove to be useful as alternatives to the synthetic chemicals normally used in pest control. Since these compounds (also called pheromones) do not possess characteristic functional atoms (e.g., halides, sulfur, phosphorus, or nitrogen), which simplify the more conventional methods of residue analysis, the analytical techniques differ from those discussed previously in this treatise.

In addition to these two introductory chapters, the reader will find methods for formulation and residue analyses for over twenty-five commercial pesticides not described in prior volumes. In some cases these pesticides are used internationally, but have not been registered in the United States. Two new plant growth regulators are included in this volume, and information on other compounds has been updated, e.g., allethrin, pyrethrum, and succinic acid 2,2-dimethylhydrazide.

As in previous volumes, the editors again request that readers make suggestions for other subjects to be included in subsequent volumes and to submit corrections and omissions.

Gunter Zweig
Joseph Sherma

Part I
General

1

U.S. Government Regulations for Pesticide Uses with Special Emphasis on Analytical Chemical Aspects

JACOB WOLFF

I. INTRODUCTION

Various aspects of the federal regulations pertaining to pesticide uses were discussed in Volume I (1963) of this treatise, particularly in Chapter 1, and parts of Chapters 2 through 4. During the ensuing years, the uses of pesticides in this country have burgeoned. Over 30,000 pesticide products, formulated from 900 chemical compounds, are currently registered with the Environmental Protection Agency (EPA). Some 300 of the chemical compounds have tolerances for residues in food. (About 15,000 pesticide products have state registrations; but some of these are similar to products with federal registrations.) Over 900 million pounds of pesticide active ingredients are used annually. About half of this is by agriculture with the remainder by industry, householders, and government. For 1972, the domestic production of pesticide-type products totaled slightly over one billion pounds with sales running slightly over one billion dollars. This is equivalent to almost five pounds of pesticide product at five dollars for every man, woman, and child in the United States.

In many ways, pesticides have been beneficial to mankind. They have helped assure an adequate supply of food and fiber at a cost that would have been much higher in their absence. They have saved millions of lives that might have been lost due to insect-borne diseases. They have protected forests, parklands, and households from insect damage. However, the use of pesticides has produced a potential hazard to man and to the environment. There is a risk in introducing toxicants into the food supply that must be carefully weighed against the benefits derived for each individual pesticide use. In addition, a number of pesticides used on a large scale until recently have been found to be long-lived in the environment with a tendency to build up in the food web. The eventual cost of those uses in terms of endangered wildlife species and detriment to human health has yet to be determined. These developments have inevitably led to dynamic changes in the regulation of pesticide uses by the federal government. The regulatory agencies have entered new areas and established more stringent requirements in areas already controlled. Thus, it is now appropriate to update the information that was presented on this subject in Volume I.

The most important changes in the regulation of pesticide uses were the establishment of EPA and the passage of the amended pesticide law (PL 92-516). This presentation is prefaced with a discussion of the registration and tolerance-setting processes because these basic control mechanisms must be understood for a proper evaluation of the changes involved. With thousands of pesticide products and hundreds of tolerances involved, there are exceptions to almost every conceivable statement one could make on registrations and tolerances. Nevertheless, the key features, which apply to the vast majority of cases, will be presented.

Following the discussion of the relation between EPA and FDA and the effect of the enactment of the Federal Environmental Pesticide Control Act, monitoring and other regulatory activities of federal agencies will be examined. Finally, the author will examine his somewhat cloudy crystal ball and look into the future of pesticide uses and the government regulation thereof.

II. THE REGISTRATION PROCESS

Registration is an established mechanism for regulating the use of pesticides in the United States. While tolerances are required only for those products that are used on food or feed crops, every pesticide product must be registered with the EPA before it can be shipped in interstate commerce.

In the course of time, the U.S. Department of Agriculture and subsequently the EPA have required increasing amounts of data and information prior to granting registration. At present, these requirements are in a state of flux as EPA develops policy for implementing the amended pesticide law. Important and relatively permanent requirements are issued as regulations and published in the *Federal Register*. Interim requirements or policy statements are issued as PR (pesticide regulation) notices and usually sent directly to registrants. Applications for registrations are reviewed by EPA to determine if the proposed use

(a) can reasonably be expected to yield residues in food or feed (direct uses on food crops are discussed in the next section on tolerance setting; the concern here is for contaminative residues);

(b) is effective for the claims made (the use directions also are evaluated for adequacy and practicality);

(c) bears suitable precautionary statements that, if complied with, would prevent unreasonable injury to man, beneficial animal species, and the environment.

All applications for new registrations must include EPA Form 8570-1, "Confidential Statement of Formula," EPA Form 8570-4 (see Figs. 1 and 2), five copies of the proposed labeling, and the supporting data. (The last may be replaced in part or totally by "offer-to-pay" and "method-of-support" statements.)

Most of the entries for EPA Form 8570-1 are self-explanatory. However, item 11, "DATA SUBMITTED WITH THIS APPLICATION" requires amplification. The supporting data and information are conventionally presented in the format of Sections A–J.

1. SECTION A

Section A characterizes the pesticide product. It includes a description of the basic manufacturing process, information on the purity of starting and intermediate materials, and in some cases, quality control procedures.

Information on active ingredients usually includes the common, trade, and chemical names (*Chemical Abstracts* nomenclature), the molecular weight, structural formula and identifying spectra (usually IR, and in some cases NMR or UV). Chemical properties such as stability and hydrolysis rate are given here, as well as physical properties like physical state, melting and boiling points, vapor pressure, and density or specific gravity. Finally, the composition of the technical material is given along with the names and percentages (and ranges) of impurities.

Information on inert ingredients usually is limited to the chemical name, preferably one that is used in Section 180.1001 of the *Pesticide Regulations*. (The last is a list of substances that have been exempted from tolerance requirements when used in pesticide formulations.) In some cases, the exact nature of an inert ingredient is considered to be a trade secret. As indicated in Fig. 2, the manufacturer arranges for the supplier to provide the confidential information directly to EPA. Where an ingredient is inert with regard to pesticidal effects but presents possible residue or environmental problems, the data requirements for the inert may be as stringent as those for the active ingredient.

Section A also includes similar types of information for the pesticide formulations, as described above, with the following additions:

1. Miscibility with various solvents.
2. The boiling point, specific gravity, viscosity, and vapor pressure and the weight of active ingredient per gallon.
3. The flash point and explosive characteristics.
4. The pH of solutions and corrosion hazards.

FORM APPROVED: OMB NO. 158-R0066

U.S. ENVIRONMENTAL PROTECTION AGENCY
OFFICE OF PESTICIDES PROGRAMS
REGISTRATION DIVISION
WASHINGTON, D.C. 20460

APPLICATION FOR NEW REGISTRATION OF A PESTICIDE

(Under the Federal Insecticide, Fungicide, and Rodenticide Act)

1. DATE OF APPLICATION

2. NAME OF PESTICIDE *(Must be same product name as on label-do not list active ingredients)*

IMPORTANT: READ INSTRUCTIONS ON REVERSE

3. TYPE OF PESTICIDE *(Check each applicable item for combination products)*
INSECTICIDE ☐ FUNGICIDE ☐ HERBICIDE ☐
RODENTICIDE ☐ GERMICIDE–DISINFECTANT ☐
OTHER *(Specify)*

4. NAME & MAILING ADDRESS OF FIRM TO WHOM REGISTRATION IS TO BE ISSUED *(Include Zip Code)*

5. IS THE REGISTRANT SHOWN IN ITEM 4 THE MANUFACTURER?
YES ☐ NO ☐
(If "No", see instruction 5 on reverse)

6. TYPE OF FORMULATION
DUST ☐ WETTABLE POWDER ☐ PRESSURIZED PRODUCT ☐
GRANULAR ☐ EMULSIFIABLE LIQUID ☐ BAIT ☐
OTHER *(Specify)*

7. TYPE OF CONTAINER
METAL ☐ GLASS ☐
PLASTIC ☐ PAPER ☐
OTHER *(Specify)*

8. NET CONTENTS OR CONTAINER SIZES

9. MANNER IN WHICH LABEL IS AFFIXED TO PRODUCT
LITHOGRAPHED ☐ PAPER, GLUED ☐ STENCILED ☐
OTHER *(Specify)*

10. PLACE WHERE DIRECTIONS FOR USE APPEAR
ON LABEL ☐ IN PRINTED MATTER ACCOMPANYING PRODUCT ☐

11. DATA SUBMITTED WITH THIS APPLICATION *(Identify and submit in triplicate)*
EFFICACY DATA ☐ TOXICOLOGY DATA ☐ RESIDUE DATA ☐ PETITION FOR TOLERANCE ☐
OTHER *(Specify)*:

12. ANY ADDITIONAL PERTINENT INFORMATION *(Do not enter confidential formula here-see item 13, below)*

ACKNOWLEDGMENT

13. THE FOLLOWING MUST BE SUBMITTED WITH APPLICATION
- Five (5) copies of proposed labeling, including all printed or graphic matter which may accompany the sale of this product. Copies must be clearly legible and identical.
- Five (5) copies of the complete formula, showing the precise name and percentage of each active and each inert ingredient. *(This information is treated confidentially.)*

14. SIGNATURE OF AUTHORIZED FIRM REPRESENTATIVE

15. TITLE

16. DATE SIGNED

17. RECEIVED BY EPA-ABP/REGISTRATION DIVISION, WASHINGTON, D.C. 20250

IN ANY CORRESPONDENCE ON THIS PRODUCT, REFER TO THIS FILE SYMBOL NO.:

EPA Form 8570-1 (6-73) REPLACES PR FORM 9-199 WHICH IS OBSOLETE.

FIG. 1. EPA Form 8570-1, "Application for New Registration of a Pesticide."

5. Capability of acting as an oxidizing or reducing agent.

6. The results of stability tests (periodic chemical analyses for 1 year).

The "Confidential Statement of Formula" (see Fig. 2) is important

ENVIRONMENTAL PROTECTION AGENCY
OFFICE OF PESTICIDES PROGRAMS, REGISTRATION DIVISION
WASHINGTON, D.C. 20250

CONFIDENTIAL STATEMENT OF FORMULA
(Under the Federal Insecticide, Fungicide and Rodenticide Act)

Form Approved
OMB No. 158-R0066

1. PAGE OF PAGES

IMPORTANT: Read instructions on the reverse before completing form.

2. DATE OF APPLICATION

3. REASON FOR SUBMISSION *(Check one)*

- APPLICATION FOR A NEW REGISTRATION
- APPLICATION FOR AMENDED OR RENEWAL REGISTRATION
- SUBMITTED BY BASIC SUPPLIER *(See instruction C on reverse)*

4. NAME AND ADDRESS OF APPLICANT OR REGISTRANT *(Include ZIP Code)*

5. NAME OF PESTICIDE *(Must be same product name as on label. Do not list active ingredients)*

6. EPA REGISTRATION NO./FILE SYMBOL *(If known)*

7. NAME AND MAILING ADDRESS OF BASIC SUPPLIER *(If applicable) (Include ZIP Code)*

8. NAME OF INGREDIENT *(Product)* PROVIDED BY BASIC SUPPLIER FOR WHICH FORMULATION IS UNKNOWN *(If any)*

9. IS THE INGREDIENT *(Product)* IN ITEM 8 REGISTERED WITH EPA?
YES NO

10. IF YES TO ITEM 9, GIVE REGISTRATION NO.

11. NAME OF EACH ACTIVE AND INERT INGREDIENT USED IN THE FORMULATION *(List both the common name and the precise chemical name of each.)*	12. PERCENT OF EACH BY WEIGHT
CHECK IF CONTINUED ON ATTACHMENT(S) — ALL INGREDIENTS LISTED IN ITEM 11 ABOVE AND ON ADDITIONAL PAGES MUST TOTAL	100%

13. SIGNATURE OF AUTHORIZED FIRM REPRESENTATIVE

14. TITLE AND FIRM NAME

15. DATE SIGNED

16. RECEIVED BY EPA-OPP, REGISTRATION DIVISION, WASHINGTON, D.C. 20259

EPA FORM 8570-4 (1-73) REPLACES PR FORM 9-196 WHICH MAY NOT BE USED.

FIG. 2. EPA Form 8570-4, "Confidential statement of Formula."

in the enforcement of the amended pesticide law. Section 12 of the Act states that ". . . it shall be unlawful for any person in any State to distribute, sell, offer for sale, hold for sale, ship, deliver for shipment, or receive and (having so received) deliver or offer to deliver to any person . . . any registered pesticide the composition of which differs at the time of its distribution or sale from its composition as described in the statement required in connection with its registration under Section 3." This has been interpreted to mean that the formulations may contain neither less nor *more* than the declared quantity of any component within reasonable limits.

Section A, therefore, must include analytical methods for the active ingredients of the formulations. Samples of analytical standards of known purity and of a typical batch of the technical material must also be submitted. These analytical methods have been described for specific pesticides in previous volumes and will be found in subsequent chapters of this volume. In addition, many of the methods for formulations of registered pesticides are included in the *Official Methods of Analysis of the AOAC*.

The analytical techniques may be combinations of thin-layer chromatography, ion exchange separation, or gel filtration fractionation, together with gas and liquid chromatography or infrared and mass spectroscopy. Since the methods submitted are used for law enforcement, they must be accurate and reasonably specific. It is often necessary to include an alternate assay procedure or a semiquantitative identification technique such as thin-layer chromatography.

Methods also may be needed for highly toxic or persistent manufacturing impurities. Such methods are usually given for the technical material rather than for the formulation. The lack of methods for the latter may in some cases hamper enforcement of the law. However, it is recognized that the uniform requirement of methods for all manufacturing impurities in formulations could create complex analytical problems.

2. SECTION B

Section B contains all of the necessary information on how the product will be used. The labels and use directions are reviewed by EPA to ensure that when used as directed, the product will be efficaceous. In addition, the precautionary statements are evaluated for practicality and adequacy for preventing contaminative residues in the food supply or environment.

3. SECTION C

Section C contains the pertinent toxicity data. Mammalian studies are used to categorize the pesticide and determine appropriate precautions to protect the user. These normally include the determination of oral, dermal, and inhalation LD_{50} values. Other testing involves evaluations of oncogenic, mutagenic, and teratogenic properties, especially for food uses. For many uses, metabolism and subacute toxicity studies are appropriate. The degree of testing required increases with the possibility of the direct exposure to humans, livestock, and pets.

4. SECTIONS D–G

These sections, which normally are required only where tolerance proposals accompany the application, are discussed below.

5. SECTION H

Section H contains reports on the effectiveness and usefulness of the product. A full discussion of this section is outside the scope of this volume.

In enforcing the law, EPA routinely tests pesticide formulations to determine if they contain the declared amount of active ingredient. On occasion, samples also are analyzed to ensure that only the registered inert ingredients are present and in declared quantities. In addition, increasing emphasis is now being placed on biological testing to ensure that the pesticides being sold are as effective as the material used in the tests reported in Section H.

6. SECTIONS I AND J

Section I (hazards to fish and wildlife) and Section J (hazards to the environment) are closely related. The amount of data required on fish and wildlife obviously depends on the potential of the pesticide and/or its alteration products to spread and persist in the environment. Such data are provided in Section J. The reader is advised to review the following chapters in Volume V (1967) of this treatise in this connection: Chapter 4, on residue analysis of water resources; Chapter 5, on metabolism; and Chapter 6, on fish and wildlife. Proposed guidelines were issued on data requirements in these areas on June 25, 1975, which will guide the registrants in obtaining adequate data.

Section I is comprised primarily of toxicity data. Assurance is needed that a proposed use will not produce unreasonable adverse effects on wild mammals, birds, or aquatic environments, including the flora and fauna of water columns and bottom sediments. The acute and subacute studies provided under Section C above are usually adequate for wildlife. The type of additional data needed for Section I is based on considerations of the data discussed in Section C, the use pattern, the expected environmental contamination, and the species present in the habitats involved.

Normally, some data on fish and wildlife are required for any pesticide that is used outdoors or under conditions that could yield contamination of water or other environmental components. For such uses, the minimal requirements generally are:

1. Acute toxicity data (oral LD_{50}) on a waterfowl species (mallard duck) or an upland game bird species.

2. Determination of the dietary LC_{50} (parts per million in diet, five-day feeding protocol) for a waterfowl species and an upland game bird species. One of these must correspond to the species used in 1, above.

3. Acute toxicity data (96 hours, LC_{50}) on a cold- and a warmwater fish species (preferably the rainbow trout and bluegill).

In addition, the following types of data may be requested in some cases:

1. Reproduction studies on the bobwhite quail and the mallard duck, especially for persistent materials that tend to store in animal tissues or in cases where the use conditions would be expected to expose wild birds.

2. Subacute or chronic toxicity data on fish, particularly for pesticides tending to accumulate in fish tissue. Where residues could enter the food web, toxicity-residue studies might be needed on bottom feeders (such as catfish or carp), cold- and warmwater predators, mollusks, crustaceans, and insect larvae.

3. Where marine environments are involved, acute toxicity data (96 hours, LC_{50}) on shrimp, crabs, and oysters. Shell deposition data are another possibility.

Most of the above studies can be performed on a laboratory scale. Potential hazards relating to a proposed pattern of use may require special testing under actual or simulated field conditions. It is desirable to obtain sufficient data for statistical analysis. A description of adverse symptoms will serve to facilitate the review.

Section J is comprised primarily of environmental chemistry data. The basic requirements can be summarized in the two words "fate" and "movement." The official guidelines for environmental studies were presented in the PR Notice 70-15 (the fifteenth notice in the year 1970). The questions to be resolved along with some of the EPA comments are as follows:

1. What is the rate of dissipation of the pesticide in the soil?

2. What is the mechanism of degradation of the pesticide residues? Typical field studies include: (a) photodecomposition studies in or on soil and water, using sterile and nonsterile samples, (b) degradation and metabolism studies of the effect of microorganisms, and the effect of the residues on microorganisms, (c) degradation studies in water, (d) duration of the biological activity (for example, phytotoxicity).

3. Do the residues leach through the soil?

4. Are the residues moved from the site of application by runoff water?

5. Is the pesticide bound in soils; that is, are residues present that are not readily extractable?

6. What levels of the parent compound and principal metabolites will accumulate in fish, rabbit, and bird tissues and what dosage-related symptoms are exhibited during the laboratory test period?

Some of the considerations involved in the evaluation of Section J by EPA are the following:

1. Soil persistence studies are important in evaluating the potential for residues in follow-up crops and environmental contamination. Thus, they should reflect practical conditions most likely to yield persistent soil residues. Samples should be analyzed for the pesticide and its major degradation products.

2. In soil metabolism studies the pesticide should be radiolabeled in a portion of the molecule that would remain in the significant degradation products. The studies should be used to develop procedures for separating and identifying degradation products and metabolites.

3. When a pesticide use can yield residues in water, data usually are needed for irrigated crops and aquatic plants. Feeding studies should be carried out on fish, poultry, cattle, and other livestock fed water contaminated with residues. The studies on fish are often run in dynamic flow-through systems using radiolabeled pesticides.

Miscellaneous Registration Procedures

The above discussion dealt with the requirements for the registration of new pesticide chemicals. In addition, there are established procedures, and application forms, for such actions as registering new distributors, amending registrations by changing the composition of formulations, combining two registered pesticide chemicals, adding new uses, changing product names, etc. Naturally, most of these actions require less supporting data than applications for completely new products. During the present energy crisis, with the resultant shortages of petroleum solvents, EPA has expedited the handling of applications requesting formulation changes. Each registration is subject to review every five years under the amended law. EPA may request additional data and information to support renewed registration.

III. TOLERANCE SETTING PROCEDURES

Where a use of a pesticide involves the direct exposure of a food or feed crop, the establishment of appropriate tolerances is a prerequisite to registration. In uses that are classified as "nonfood" this requirement is waived. Such clearances are limited to cases where the patterns of use and the nature of the pesticide make it clear that there would be no residue in the crops. One example of this might be the application of a nonsystemic pesticide to fruit trees in a dormant state. In other cases, pesticide uses are granted an exemption from the requirement of a tolerance if it is shown that the proposed uses are safe even in the event of gross misapplication. Requests for such exemptions are processed much in the same manner as requests for tolerances. Normally, some residue data are required, but exemptions are based primarily on toxicity data.

Prior to 1965, it was possible to clear pesticide uses on a "no residue" basis. Usually, such clearances were granted by showing the absence of residues in the crops within the limits of sensitivity of the analytical methods used (usually 0.1 ppm). In some cases, where feed crops were involved, the "no residue" registrations were "reinforced" by establishing "zero tolerances," usually for residues in milk. The purpose of such "zero tolerances" was to facilitate government action if residues were found. In June 1965, a committee appointed by the National Academy of Sciences, National Research Council recommended that "The concepts of 'no residue' and 'zero tolerance' as employed in the registration and regulation of pesticides are scientifically and administratively untenable and should be abandoned." Subsequently, many of the old "no residue" registrations were cleared by the establishment of "negligible residue"* or regular tolerances. Others were abandoned and some are still under review or litigation. In any case, tolerances are now required for the majority of new uses involving food or feed crops.

There may be some confusion regarding the differences between the registration and tolerance setting processes. Where a tolerance is needed, all of the requirements for registration still apply before a pesticide can be used; but additional data and information must be submitted as well. Registration is regulated by the Federal Insecticide, Fungicide, and Rodenticide Act (FIFRA) as amended and tolerances are set under the Federal Food, Drug, and Cosmetic Act (see Chapter 1, Volume I of this treatise). There is some duplication between the infor-

* According to the *Pesticide Regulations* [Section 180.1 (i)], "The term 'negligible residue' means any amount of a pesticide chemical remaining in or on a raw agricultural commodity or group of raw agricultural commodities that would result in a daily intake regarded as toxicologically insignificant on the basis of scientific judgment of adequate safety data. Ordinarily this will add to the diet an amount which will be less than 1/2,000th of the amount which has been demonstrated to have no effect from feeding studies on the most sensitive animal species tested. Such toxicity studies shall usually include at least 90-day feeding studies on two species of mammals."

mation needed for item 11 of EPA Form 8570-1 (the application for a new registration discussed above) and the requirements for petitions as spelled out in the law regulating tolerances. In the course of time, the Food and Drug Administration and subsequently the Environmental Protection Agency have informally required increasing amounts of data and information prior to establishing tolerances.

In March 1968, the FDA published a monograph entitled "Guidelines for Chemistry and Residue Data Requirements of Pesticide Petitions," but in many respects this publication is out of date. An updated version of the monograph is in preparation and will be published in the forthcoming "Registration Guidelines," as amended.

Section 408 (d) (1) of the Federal Food, Drug, and Cosmetic Act in describing pesticide petitions states that "The petition shall contain data showing—

(A) the name, chemical identity, and composition of the pesticide chemical;

(B) the amount, frequency, and time of application of the pesticide chemical;

(C) full reports of investigations made with respect to the safety of the pesticide chemical;

(D) the results of tests on the amount of residue remaining, including a description of the analytical methods used;

(E) practicable methods for removing residue which exceeds any proposed tolerance;

(F) proposed tolerances for the pesticide chemical if tolerances are proposed; and

(G) reasonable grounds in support of the petition."

While these is no such thing as a typical petition, the usual requirements for a pesticide petition are indicated below in an outline of a petition for a fictitious carbamate pesticide, "Zag Insecticide," common name caranome, used to control the corn borer. The petition begins with a table of contents and contains index tabs for each section. There is a detailed index for Sections C and D. Pages are numbered with a dual system using the alphabetical designation of the section and arabic numerals assigned sequentially. After an introductory discussion, there follow summary tables of the residue and metabolism data before the presentation of Sections A–G.

1. SECTION A

Section A has the same information on manufacturing process, composition, chemical and physical properties, etc. that is required for Section A of item 11 of EPA Form 8570-1 (Fig. 1). In addition there would be information on the properties of the major metabolite 1-hydroxy-caranome. Purified analytical standards of the parent compound and the metabolite, 2 gm of each, would be submitted for use in trials of the residue method. If any of the inert ingredients were not cleared under Section 180.1001 of the pesticide regulations, a request for their clearance along with supporting data would accompany the petition.

2. SECTION B

Section B would be identical to Section B of item 11 of EPA Form 8570-1. Since residues in food are involved, we can anticipate that the label will be thoroughly scrutinized. The hypothetical use directions are:

> Zag insecticide, an emulsifiable concentrate, is to be applied to corn fields at rates of 1–2 lbs active ingredient per acre using air or ground equipment calibrated for low-volume spraying. The application rate will vary with the degree of infestation. An initial treatment is made 15 days after the borers appear, about 2 months after planting. A second application, if necessary, may be made 15–30 days later, but no application may be made within 7 days of harvest or the feeding of the corn forage to livestock.

Section B would include a presentation and discussion of the proposed use pattern in somewhat more detail than the above and sample labels.

3. SECTION C

The toxicity data submitted for Section C of item 11 of EPA Form 8570-1 would be included in the petition, but additional studies also would be given. The exact nature of these studies would have been determined, in part, after discussions with the reviewing toxicologists and chemists in EPA. (The appendix to the *Guidelines* will include information on the types of tests usually required.)

Section C includes oral LD_{50} values for the parent compound and metabolite determined on the rat and neurotoxicity tests for both compounds using the hen, because these compounds are both cholinesterase inhibitors. The results of 3-month feeding studies on the rat and dog are presented. In this case, based on the discussions with EPA personnel, only the parent compound was fed.

The above tests are usually appropriate for all tolerances. Since the tolerances involved here are "nonnegligible," Section C also includes the following studies (for the parent compound):

1. Lifetime feeding studies (with observations for tumors) on the rat and mouse.
2. A three-generation reproduction study on the rat.
3. A teratogenicity study on the rabbit.

4. Mutagenicity studies on rats and mice.

5. Metabolism studies on the rat, goat, and hen. (These are cross-referenced to Section D.)

Because of the high no-effect level indicated in the feeding studies above, a feeding study, acute or subacute, was not required to show safety to livestock but one was run for setting tolerances on meat and milk.

4. SECTION D

Section D, the section of most interest to analytical chemists is the key to the petition. The basic required information includes:

1. The nature of the residue in crops and any derivative foods that might contain residues as a result of the proposed use.
2. Validation of the method(s) used to obtain the residue data, one or more of which must be suitable for enforcement use.
3. The magnitude of food residues. The petition must indicate the maximum residues likely to be encountered for the purpose of establishing appropriate tolerance levels. It is also desirable to indicate the average levels so that the reviewing toxicologists can evaluate the additional margin of safety over and above that provided by the usual criteria.

Below are detailed discussions of the above Sections in Section D of the petition.

a. *Nature of Residues*

The nature of the residues in corn was determined using a combination of radiotracer and identification techniques, similar to those of the soil metabolism studies (Section J, item 11, EPA Form 8570-1) and the animal metabolism studies of Section C of the petition. Section D includes the studies of both types for the convenience of the reviewer. The same ^{14}C-ring labeled caranome was used in all of the metabolism studies. The corn studies reflect both greenhouse and field conditions simulating the proposed quantity and mode of application.

The results of the studies show a similar pattern for greenhouse and field conditions, except for the more rapid dissipation of residues in the field. The pesticide is absorbed by the corn plant. The major metabolite, 1-hydroxycaranome, reaches a peak at 3 days after application where the residue is 70% parent compound and 30% metabolite. Further degradation involves the splitting of the ring yielding fragments which are incorporated into the plant's metabolic pool. The 3:7 ratio of hydroxy derivative to parent compound remains fairly constant as long as any of the parent compound remains. These results are in line with the soil metabolism study where the hydroxy derivative was the only metabolite found.

The animal metabolism studies show a uniform pattern for the three species tested. 1-Hydroxycaranome is formed, and both parent compound and metabolite are excreted unchanged or in the form of glucuronides. Residues in tissues, milk, and eggs show the same 3:7 ratio found in corn plants.

b. *Validation of Methods*

Residue methods proposed for enforcement must employ adequate extraction and cleanup steps. The sensitivity preferably must be within an order of magnitude in relation to the proposed tolerance. The methods should require standard reagents and equipment; they should be quantitative, reproducible, and specific. Methods that do not meet all of these criteria nevertheless may be used to obtain supporting data.

The petition contains a complete step-by-step description of each method used. (Those for corn and milk will probably be tested in EPA or FDA laboratories.) The method for corn (ears and grain), corn forage, and fodder uses an acetonitrile extraction in the manner of Porter *et al.* (1969). A study on these substrates fortified with radiolabeled caranome and hydroxycaranome shows an extraction efficiency of 95–99%. The basic steps of the method include extraction of the residue, cleanup on a Florisil column, separation of the parent compound and the metabolite by differential elution, followed by determination by gas–liquid chromatography with electron capture detection.

Control values are given for each of the 11 residue studies, ranging from 0.02 to 0.1 ppm for the grain and ears and from 0.1 to 0.3 ppm for the forage or fodder. Recoveries on all substrates fortified with either compound or mixture thereof at 0.1–10 ppm range from 90 to 100%.

The procedure was tested on all other pesticides registered for use on corn that might be expected to interfere. The interfering compounds are removed in the cleanup or have gas–liquid chromatography retention times significantly different from those of caranome and 1-hydroxycaranome. In addition, the thin-layer chromatography procedure used in the metabolism studies, a variation of that used by Finocchiaro and Benson (1967), is available as a confirmatory procedure. This method can distinguish residues of the parent compound and metabolite from each other and those of all the carbamate pesticides tested.

Modified versions of the method for corn, using different extraction and cleanup steps, were applied to various tissues, eggs, and milk. Control values range from 0.03 to 0.1 ppm of either compound for tissues and eggs and from 0.005 to 0.01 ppm for milk. Recovery values at 0.005–5 ppm range overall from 87 to 98%.

c. *The Magnitude of Food Residues*

i. *For Corn and Corn Products*

The residue data for raw agricultural commodities should reflect the proposed use in formulation and application rate as well as mode and timing of applications. The sites should include the major growing areas for the crop and a sufficient number of studies and samples should be taken. A statistical treatment of the data may be used to express the precision of the analytical results.

Section D contains eleven field studies from nine states, including several in the midwestern cornbelt, California, Florida, and New York. There are data for field corn, sweet corn, and in one study, popcorn. Several studies reflect application rates of 1 and 2 lbs active per acre. Three studies reflect the proposed air applications. Most of the studies involved two applications at intervals of 15–30 days. Samples were taken of ears, grain, and forage at intervals of 0–30 days after the last application. All samples were placed in frozen storage and maintained in that state until analyzed. All samples were analyzed within 6 months of storage. A storage stability study shows no significant loss of residues for periods ranging up to 1 year.

In Section D, the residue data are presented in tabular form for most studies. Dissipation curves on semilog paper are included. The values for combined residues of parent compound and metabolite at the proposed 7-day preharvest interval range from 0.2 to 1 ppm (average 0.6 ppm) for corn grains and ears and from 1 to 5 ppm (average 3 ppm) for corn forage. The results of two additional studies on corn after processing are given.

In the first study, the canning process was simulated on a laboratory bench scale. The results indicate losses of 20–50% of parent compound and metabolite on heating. No attempt was made to obtain data on sweet corn cannery waste, a cattle feed, as this would serve as a replacement for the corn forage, which, if anything, would bear higher residues.

In the second study, the processing of corn to corn meal and oil was simulated on a small scale. The results show that the pesticide and metabolite are concentrated in the corn oil, but are destroyed in the alkaline wash and deodorization processes used in refining the oil.

ii. *For Derived Animal Products*

Section D also contains feeding studies on chickens, hogs, and cattle. In these studies, it is usual to medicate at 1, 3, and 10 times the maximum likely amount to be ingested. For chickens and hogs, which would be fed only the corn grain, the feeds would be fortified at 1, 3, and 10 ppm of a 7:3 mixture of caranome and 1-hydroxycaranome. For dairy cattle, which could be fed approximately 70% corn forage and 30% corn grain, the levels would be 4, 12, and 40 ppm. For beef cattle, at approximately 40% forage and 60% grain, the levels would be 2.5, 7.5, and 25 ppm. The calculations for the rations are on a dry basis for chickens and hogs and on an "as is" basis for cattle. Data on average intakes and the weights of the animals are obtained so that the feeding levels can be expressed in terms of milligrams per kilogram of body weight.

In these studies, three animals each were used as controls and at each feeding level. (For chickens there were ten in each group.) During a 10-day premedication period, samples of milk and eggs were taken to establish control levels. Medication was then initiated and continued for 28 days. At the end of the medication period, three chickens, one hog, and one beef animal representing the control animals and each feeding level were sacrificed for tissue analysis. The process was repeated 10 and 20 days later. In addition, samples of milk and eggs were taken periodically during the postmedication period.

The results for the milk analyses show no detectable residues at any time at the 1 × feeding level. There are residues however at the higher feeding levels, which plateau at 2 weeks at 0.02 ppm and 0.06 ppm, respectively, for the 3 × and 10 × levels. These levels drop to nondetectable somewhere between the tenth and twentieth day after medication is stopped.

The results for eggs and tissues are similar; i.e., no detectable residues (<0.1 ppm) at the 1 × level, but residues at the 3 and 10 × levels in eggs and in liver, muscle, and fat of chickens, hogs, and cattle. Here, too, the residues dropped to nondetectable levels in the feed-off period.

5. SECTION E

Section E of most petitions is of relatively little significance. Section E of the petition would state something along the line that residues would *not* exceed the tolerances, which are safe, and that in any case, residues on corn would be reduced by cooking before it is eaten.

6. SECTION F

There are several considerations involved here. Since the parent compound and metabolite are approximately of equivalent toxicity, there is no need to limit the amount of either component of the residues. In the case of the animal products, the use would fall into Category (a) (2) of Section 180.6 of the *Pesticide Regulations,* i.e., "that it is not possible to establish with certainty whether finite residues will be incurred, but there is a reasonable expectation of finite residues." Under these circumstances, it is appropriate to propose a tolerance at the sensitivity of the residue method if pharmacological considerations permit. Since the feeding studies show a uniform pattern, the results can be translated to other livestock. It is desirable to propose broad tolerances for meat to allow for possible future tolerances on feed items. In the case of milk, the low level would probably qualify as being a "negligible residue" tolerance. No food additive tolerances are necessary for corn meal or corn oil. Such tolerances are needed only if residues concentrate to a level exceeding that in the raw agricultural commodity. Based on these considerations, Section F would read as follows:

> The following tolerances are proposed for combined residues of the insecticide, caranome (chemical name), and its metabolite, 1-hydroxycaranome (chemical name), in or on raw agricultural commodities as follows:

5 ppm	in or on corn forage and fodder (including field corn, popcorn, and sweet corn)
1 ppm	in or on corn grain including field corn, popcorn and sweet corn (kernels plus cob with husk removed)
0.1 ppm	in eggs and meat, fat and meat by-products of cattle, goats, hogs, horses, poultry and sheep
0.01 ppm	(negligible residue) in milk

7. SECTION G

Section G is a catch-all for any arguments the petitioner may wish to advance in support of the proposed tolerance. It almost always includes a discussion of the safety of the proposed tolerances. Section G includes a discussion of the problem of inflation and the need to minimize the cost of food. It depicts the plight of the corn grower who is in danger of being wiped out by the ubiquitous corn borer, which only Zag insecticide can really control. This is followed by calculations of the maximum contribution of residues to the average daily diet that could result from the proposed use and a comparison of this level to the "no effect" levels found in the studies in Section C. Special attention is given to the proposed tolerance for milk, which could comprise the entire diet of infants.

The Processing of Petitions

There are two basic differences in the handling of applications for registrations and petitions for tolerances. The former are without charge or statutory deadlines. For petitions, on the other hand, there are fees which vary with the number of crops and tolerance levels, and by law the agency must complete action on the proposal within 90 days from the data of filing.

A new petition receives a preliminary review for completeness, at present on a clerical level, after which it is filed. The first step is a certification of usefulness followed by review by the chemists and toxicologists of the EPA. Simultaneously, the residue method is subjected to a trial in at least one laboratory. If deficiencies are discovered during the review process, the petitioner may be asked to amend the petition, which involves an additional fee and additional time.

As noted before, EPA is reviewing requirements for petitions and some of the items in the illustrated petition may be inappropriate in the future. For example, the concept of negligible residue tolerances will probably be modified or even eliminated. A problem area has been encountered when tolerances have been established, but subsequent registrations have been withheld pending the resolution of deficiencies in the other supporting data.

IV. RELATION BETWEEN FOOD AND DRUG ADMINISTRATION (FDA) AND EPA

From time to time, agreements on matters of mutual responsibility between FDA and EPA have been published (*Federal Register,* 1971, 1973). These agreements give EPA jurisdiction over the following:

1. Pesticide petitions for residues on raw agricultural commodities.
2. Food additive petitions for residues resulting from:
   (a) concentration of residues in processed foods manufactured from treated raw agricultural commodities;
   (b) the direct treatment of processed foods with pesticides;
   (c) the use of pesticide-impregnated food packaging materials.

FDA has jurisdiction over food additive petitions for the use of sanitizers on food-contact surfaces and the use of pesticides as preservatives in processed foods. In addition the agreements spell out which agency has primary jurisdiction in various cases where a chemical is both a pesticide and a drug. FDA has continued responsibility for agreements with states and foreign countries on surveillance and enforcement activities on pesticide residues in food. Although EPA establishes tolerances for food, it does not enforce these tolerances.

Tolerances for residues in meat and poultry are enforced by the USDA, Animal and Plant Health Inspection Service (APHIS). This is associated with the USDA enforcement of quality standards under the

Federal Meat Inspection Act as amended by the Wholesome Meat Act of 1967, and the Poultry Products Inspection Act as amended by the Wholesome Poultry Act of 1968. Where a pesticide petition proposes tolerances for meat or poultry, APHIS is usually asked for an opinion on the adequacy of the proposed residue method for enforcement use.

All other pesticide tolerances are enforced by FDA. EPA and FDA jointly publish the *Pesticide Analytical Manuals* (*PAM*) (Food and Drug Administration, 1973a,b). *PAM I,* which contains directions for sampling, multiresidue methods, and general information is primarily the responsibility of FDA. *PAM II,* which contains specific residue methods (usually based on those in petitions and published in this series) is primarily the responsibility of EPA. [In its enforcement activities, FDA usually uses multiresidue methods (see Chapters 5–7 in Volume VI of this treatise) to screen food samples. Where illegal residues are indicated, this is followed up by analyses using specific methods.] On occasion, FDA is asked to participate in method trials for new pesticides.

FDA also sets "action levels" for pesticide residues in raw agricultural commodities and fish which are not covered by formal tolerances. Action levels are permissible residues due to inadvertent exposure, e.g., pesticide residues in fish in rivers and lakes. Most of the pesticides involved in action levels are persistent chlorinated hydrocarbons, including aldrin and dieldrin, BHC, DDT, endrin, heptachlor, heptachlor epoxide, lindane, toxaphene, methoxychlor, chlordane, hexachlorobenzene, carbaryl, dioxathion, Diazinon, and diuron.

V. THE FEDERAL ENVIRONMENTAL PESTICIDE CONTROL ACT OF 1972 (FEPCA PUBLIC LAW 92-516)

A. General Information

The FEPCA is an amended version of the Federal Insecticide, Fungicide, and Rodenticide Act (FIFRA), which in one form or another dates back to 1910. The new law created new functions for the federal government in the regulation of pesticides and was a natural sequel to the establishment of EPA. The old FIFRA was administered by the USDA until 1970, when the responsibility passed to EPA. Under the provision of this act, a pesticide product could be registered if the manufacturer could show the efficaciousness of the product and its safety, including the absence of illegal residues in food and feed.

The new act institutes stronger, more extensive mechanisms to avoid potential harm to humans or the environment. Federal controls are extended to the *users* of pesticides and to the intrastate as well as the interstate marketing of pesticide products. All provisions of the act must be in effect within 4 years of enactment (October 1976). Some sections are effective immediately, while others have later deadlines, pending the establishment of regulations and the development of federal standards to guide the states in implementing the law.

B. Provisions of the Law

The major provisions of the law are:

1. The use of any registered pesticide in a manner inconsistent with its labeling is prohibited immediately.

2. Pesticides are to be classified for "general" or "restricted" use by October 1976. Restricted pesticides may be used only by, or under the supervision of, certified applicators or under such other conditions as the administrator of EPA may require.

3. The states will certify pesticide applicators for the use of restricted pesticides.

4. (a) Farmers and other private applicators may be fined and/or receive prison sentences upon criminal conviction for a knowing violation of the law. They are also liable for civil penalties. (b) A registrant, commercial applicator, or distributor who knowingly violates the law is liable to a civil or criminal fine and/or a prison sentence.

5. The administrator of EPA may issue a "stop sale, use, and removal" order when it appears that a pesticide violates the law or its registration has been suspended or finally cancelled.

6. Pesticide manufacturing establishments must be registered with EPA. Information on types and quantities of pesticides produced, distributed, and sold is required for registration and also must be submitted annually. EPA agents may inspect establishments and take samples.

7. The agency is authorized to assist the states in enforcing the law and help develop and administer applicator certification programs.

8. The agency may issue experimental use permits, conduct research on pesticides and alternatives, and monitor the use of pesticides and their presence in the environment.

Other provisions of the law include:

1. The agency is authorized to indemnify certain owners of registrations that are suspended and then cancelled.

2. Applicants for the registration of a pesticide product retain proprietary rights to the submitted supporting data. A system is authorized whereby a second applicant can use such data upon payment of reasonable compensation to the original applicant.

3. States may be authorized to issue limited registrations for pesticide uses intended for special local needs. All other uses will require federal registration.

C. Impact of the New Law

The amended FIFRA will have a profound impact on pesticide users and manufacturers. This discussion will focus on those areas that, directly or indirectly, would be expected to affect pesticide chemists. Since the enactment of the new law, EPA has issued a number of proposed and final regulations as well as interpretations. The publication most pertinent to pesticide chemistry is the proposed Section 3 regulations (*Federal Register,* 1974). The usual procedure is to allow 30 days for comments on a proposed regulation by interested parties followed by promulgation of the final regulations as revised.

Under the confidentiality of data clause, all applicants for registration have the right to specify which test data they consider to be confidential. On the other hand, applications must now be accompanied by a "Statement of Offer to Pay and Method of Support" statement. The supporting data can be submitted directly or by citing references. Alternately, EPA can be requested to evaluate the application on the basis of "established use patterns." Applications based on "established use patterns" are reviewed in the usual manner; but other parties will have a 60-day period during which they may present claims for compensation for the alleged use of confidential data. This confidentiality of data clause, which apparently was put into the act at the behest of industry, has already tended to discourage the publication of the results of pesticide research.

The prohibition against the use of a pesticide product in a manner inconsistent with its labeling presents several problems. There is a question about whether or not using lesser amounts than those recommended would violate the law. Offhand, it might seem that there should be no objection to using less pesticide. However, there is some risk that using quantities insufficient to totally eradicate the pests could result in the development of immunity on the part of the pest involved. Another problem area is the requirement of organizations such as university agricultural experiment stations to obtain experimental permits in order to conduct field research on pesticides that are already registered.

The classification of pesticides will be a large task, since each use of over 30,000 federally registered pesticide products and an additional approximately 15,000 products with state registrations will have to be considered individually. In the case of new products, applicants must now propose a "general" or "restricted" classification for each proposed use. Section 162.11(c) (3) of the proposed Section 3 Regulations cites five factors to be used in classification:

1. the complexity of the use directions,
2. severity of effects from minor deviations from the use directions,
3. "widespread and commonly recognized" use practices that may nullify the directions,
4. need for specialized application or protective equipment,
5. likelihood of adverse effects even where directions are complied with.

The proposed regulation also suggests that additional toxicity data such as "subacute, chronic, and delayed effects data including potential mutagenic, teratogenic, oncogenic, metabolic hazards . . ." may be required in some cases when determined by EPA. There is an inference that such data would also be required for tolerances. This raises a question regarding "negligible residue" tolerances, which presently require less supporting toxicity data than ordinary tolerances. The provisions of the law authorizing EPA to establish reentry standards and conduct monitoring studies and research on pesticides and alternatives are discussed below.

VI. MONITORING ACTIVITIES OF FEDERAL AGENCIES

The National Pesticides Monitoring Program arose from a report of the President's Science Advisory Committee entitled "Use of Pesticides" (see Bennett, 1967). The Committee proposed that various federal agencies "develop a continuing network to monitor residue levels in air, water, soil, man, wildlife, and fish." Most of the data on pesticide residues in the food supply and the environment is reported in the *Pesticides Monitoring Journal,* which is published quarterly by the Division of Pesticide Community Studies of EPA.

A. The Food Supply

The Department of Agriculture and FDA were monitoring the food supply long before concern developed about the environment. The Animal and Plant Health Inspection Service of USDA maintains resident inspectors at all meat packing plants and has been analyzing meat and poultry samples for pesticide residues for many years. APHIS has developed its own gas–liquid chromatography methods for meat and poultry analyses (see Chapter 5 of Volume VI of this treatise). Some of those which are included in the "Chemistry Laboratory Guidebook" of the Consumer and Marketing Service are substantially the same as those in the "Official Methods of Analysis of the AOAC."

Since 1962, FDA has conducted market-basket studies on pesticide residues in the food supply. The most recently reported data cover the year ending April 1971 (Manaske and Corneliussen, 1974). In these studies, typically, thirty composites each of twelve commodity groupings are analyzed for approximately thirty pesticides each year. The food samples are in ready-to-eat form and are chosen to reflect the daily diet of a 17-year-old male. In the previously reported studies, samples were taken and analyzed at five district laboratories of FDA. The results for 1971 (the latest published) reflect sampling in the various districts with all analyses performed at the Kansas City District. The methods used are those in *PAM I* and *II* (see Chapters 5–7 of Volume VI of this treatise).

The approach used in these studies tends to dilute high residues on individual commodities. Some pesticides, such as chlordane and toxaphene, could be missed at low levels. For other pesticides, recoveries are highly variable at the indicated residue levels. Nevertheless, the results are useful in appraising the national residue picture. Evaluations of the dietary intake of pesticides by FDA, based on these studies (Duggan and Corneliussen, 1972), show that levels in general are below the established acceptable daily intakes. (However, for some years, the calculated combined aldrin–dieldrin intake was close to the allowable daily intake.)

Another study (Duggan *et al.,* 1971) summarizes the analytical results on over 100,000 samples of raw agricultural commodities obtained by FDA in surveillance programs between 1963 and 1969. Included are USDA data for red meat from 1964 to 1969, and poultry from 1967 to 1969. Most of the samples were "objective samples," i.e., samples with no previous indications of excessive residues or misuse. More than half of the food or feed samples and over three quarters of the fish, meat and poultry samples contained residues of one or more pesticides, chiefly the chlorinated hydrocarbons. However, average levels were quite low (<0.11 ppm in animal fat, the highest group) and did not approach established tolerances, but in some cases residues were found that are not covered by tolerances.

B. The Environment

1. SOIL

The USDA initiated a soils monitoring program in 1964. Since 1971, EPA has assumed this responsibility. The results for 1970 (Crockett *et al.,* 1974) include data on pesticide applications, soil residues, and crop residues from 1500 cropland sites in 35 states. The data of this study and those for the preceding year will provide the baseline for evaluating future trends.

A pilot study was conducted in 1969 (Wiersma *et al.*, 1972) on soils in urban areas, and preliminary data show higher residues of chlorinated hydrocarbon pesticides in urban turf areas than are found in cropland.

2. WATER

The monitoring of water involves sample collection by the U.S. Geological Survey and the U.S. Army Environmental Health Command. Samples are analyzed at the EPA and U.S. Geological Survey. Data reported by the U.S. Public Health Service in the early 1960's involved absorbing residues from large volumes of water on activated carbon and analyses by methods that are now outdated. More recent studies for 1968–1971 show that residues of twelve pesticides (chiefly chlorinated hydrocarbons) were present at the fraction of a part per billion level in about one-third of the samples (Schulze *et al.*, 1973).

3. FISH AND SHELLFISH

The Bureau of Sport Fisheries and Wildlife of the USDI monitors residues in freshwater fish. The most recently published data (Henderson *et al.*, 1972) showed that almost all samples contained DDT and its metabolites, in a number of cases at levels in excess of the FDA action level of 5 ppm. Other common residue components are the other chlorinated hydrocarbon pesticides and PCBs.

A joint USDA–EPA study on commercially grown catfish in Arkansas and Mississippi (Crockett *et al.*, 1975) showed DDT in all samples, and dieldrin, endrin, and toxaphene in over 75% of the samples. Here too, there were some instances of residues in excess of FDA action levels.

The monitoring of residues in estuarine fish and shellfish involves sample collection by cooperating universities and state marine agencies with analyses by the EPA. Previous monitoring programs conducted by USDI (Butler, 1973) reported data for organochlorine residues in mollusks for the period 1965–1972. Here, too, DDT was commonly found, usually at under 1 ppm with values ranging up to 5 ppm.

4. BIRDLIFE

The Bureau of Sport Fisheries and Wildlife monitors residues in starlings (obtained by their field personnel), mallards and black ducks (supplied by hunters), and bald eagles (supplied by wildlife management personnel). (See Martin and Nickerson, 1972; Heath, 1969; Belise *et al.*, 1972.) The eagles that were found were dead or moribund, and in many cases pesticide poisoning by DDT or dieldrin residues was suspected. Residues of DDT ranged up to only 1.6 ppm, but DDE was as high as 93 ppm. Other types of birds also showed residues of DDT and dieldrin almost universally with the former (plus metabolites) ranging up to 23 ppm. There is some indication that residues are diminishing with time.

5. OTHER ENVIRONMENTAL SAMPLES

EPA is also considering monitoring pesticide residues in air, but there are problems in devising suitable protocols. There were previous attempts (1969–1971) to monitor pesticides in air, but methodology has not been standardized to give consistent results (Stanley *et al.*, 1971; Compton, 1973).

The Public Health Service initiated a program of determining the level of chlorinated hydrocarbon pesticides in human adipose tissue in 1967. This activity is now part of EPA's National Human Monitoring Program. The data for 1970 (Kutz *et al.*, 1974a) show that upward of 94% of human adipose tissue samples contain residues of all of the following: total DDT equivalent (ca. 77% in the form of DDE), β-BHC, dieldrin, and heptachlor epoxide. The maximum residues found (on a percent lipid basis) were 270, 27, 15, and 11 ppm, respectively. It is known that data for DDT may be distorted by the presence of PCBs. However, a recent study (Yobs, 1972) indicates that only about 30% of human adipose tissues contain PCBs, with 95% of the positive samples showing residue levels below 2 ppm. More limited data for oxychlordane, a metabolite of chlordane (Biros and Enos, 1973), show residues of 0.03–0.4 ppm to be present in about 80% of the samples. A new study (Kutz *et al.*, 1974b) reports the finding of mirex residues of up to 6 ppm in human adipose tissues collected in Georgia and Louisiana.

VII. OTHER EPA PESTICIDE-RELATED ACTIVITIES

EPA has several functions concerning pesticides derived from the amended pesticide law and other laws protecting the environment where pesticides are among the pollutants involved. The establishment of reentry standards for areas that have been recently treated with pesticides is being closely coordinated with the Occupational Safety and Health Administration of the U.S. Department of Labor. A brief summary of the problems so far encountered includes:

1. inadequate residue data for the foliage of treated crops;
2. meager subacute toxicity data for dermal and inhalation exposure to pesticides;

3. the need for different standards where the same crop is grown in different geographic areas or under different cultural conditions;

4. limitations on reentry times imposed by activities such as pruning or picking fruit, which cannot be postponed without reducing yields or crop quality—in some areas, high temperatures make currently available protective clothing impractical;

5. communicating the restrictions to migrant agricultural workers who may not understand English—in some cases the workers are also illiterate.

One area under study is the development of simple monitoring devices or systems to determine when it is safe for agricultural laborers to enter fields or orchards that have been treated with the organophosphate pesticides, which are replacing the safer (to the applicator) chlorinated hydrocarbon pesticides.

Another activity of the EPA is the substitute, or alternative, chemicals program. As the use of hazardous pesticides is phased out, the agency is attempting to develop lists of safer pesticides with equivalent functions. This program may lead to large-scale uses of pesticides whose uses at present are minor. Therefore, detailed reviews on all of the different facets of possible substitute chemicals are needed to ensure that their use would not cause presently unknown hazards to the food supply or to the environment. The social and economic costs and benefits also must be considered. EPA is interested in increasing patent protection so as to stimulate research and development on new pesticides by industry.

One important function of the EPA is setting quality criteria for contaminants, including pesticides in water under the Federal Water Pollution Control Act Amendment of 1972 (PL 92-500).

The enforcement of the Solid Waste Disposal Act as amended in 1973 (PL 93-14) involves pesticides. Jurisdiction under this act includes discarded materials from agricultural activities, such as pesticide containers. There is also a problem of disposing of surplus pesticides the uses of which have been cancelled. Many of these are highly persistent materials. Disposal by burial risks the transfer of these materials into water tables. The most effective means of disposal is high temperature incineration. However, this is quite expensive and in some cases may yield potentially toxic impurities in air.

The final activity that will be considered here is EPA's preparation of its own Environmental Impact Statements involving action on pesticides. EPA also reviews all of the Environmental Impact Statements issued in connection with large-scale pesticide programs by other federal agencies.

VIII. OUTLOOK FOR THE FUTURE

The trend toward eliminating the use of chlorinated hydrocarbon pesticides will probably continue. Most of the uses of aldrin, dieldrin, DDT, chlordane, and heptachlor have been or will be phased out.

There is an ever increasing use in herbicides. Only about half of the cropland in the United States is now being treated for weeds. The need for higher yields per acre, increased labor costs, and higher farm prices provide powerful incentives for using herbicides. More specific chemicals will be developed, like the new thiocarbamate sulfoxides (see Casida *et al.,* 1974), for which methods of analysis will be developed.

The fees and data requirements for the establishment of tolerances and granting of registrations have increased sharply in recent years and probably will continue to do so. The sharpest increase has been in the cost of developing the required toxicity data. While the cost of the environmental data are considerably lower at this time, the ultimate standards have not yet been developed. Possibly the approach of Metcalf *et al.* (1971) combined with "chemodynamics" (the study of chemicals in the environment) will provide the insight necessary to simplify the data requirements. Fair, effective regulation of pesticide uses depends ultimately on the development of reliable and specific methods of analysis. For example, there is a great need for better methods of analysis for bound pesticide residues in soil. Here, adequate extraction and identification procedures are the major problems. These methods are needed for setting standards for residue levels and persistence criteria. Another research need is the development of more sensitive methods of analysis for residues in water where sensitivities of the order of parts per trillion are sometimes needed for setting "safe" standards. Hopefully, some of the readers of this volume will play a role in the development of such methods.

REFERENCES

Association of Official Analytical Chemists. (1975). "Official Methods of Analysis," 12th ed. Washington, D.C.

Belise, A. A., Reichel, W. L., Locke, L. N., Lamont, T. G., Mulhern, B. M., Prouty, R. M., De Wolf, R. B., and Cromartie, E. (1972). *Pestic Monit. J.* **6,** 133.

Bennett, I. L. (1967). *Pestic, Monit, J.* **1,** (Foreword).

Biros, F. J., and Enos, H. F. (1973). *Bull. Environ. Contam. Toxicol.* **10,** 257.

Butler, P. A. (1973). *Pestic. Monit. J.* **6,** 238.

Casida, J. E., Gray, R. A., and Tilles, H. (1974). *Science* **184,** 573.

Compton, B. (1973). "Analysis of Pesticides in Air in Chemical Analysis of the Environment," pp. 133–152. Plenum, New York.

Crockett, A. B., Wiersma, G. B., Tai, H., Mitchell, W. G., Sand, P. S., and Carey, E. G. (1974). *Pestic. Monit. J.* **8,** 69.

Crockett, A. B., Wiersma, G. B., Tai, H., and Mitchell, W. G. (1975). *Pestic. Monit. J.* **8,** 235.

Duggan, R. E., and Corneliussen, P. E. (1972). *Pestic. Monit. J.* **5,** 331.

Duggan, R. E., Lipscomb, G. Q., Cox, E. L., Heatwole, R. E., and Kling, R. C. (1971). *Pestic. Monit. J.* **5,** 73.

Federal Register. (1971). "Agreement Between Department of Health, Education, and Welfare and Environmental Protection Agency, Notice Regarding Matters of Mutual Responsibility," 36 FR 24234. Environmental Protection Agency, Washington, D.C.

Federal Register. (1973). "Agreement Between Department of Health, Education, and Welfare and Environmental Protection Agency. Notice Regarding Matters of Mutual Responsibility," 38 FR 24233. Environmental Protection Agency, Washington, D.C.

Federal Register. (1974). "Pesticide Programs, Proposed Registration Reregistration and Classification Procedures," 39 FR 36973. Environmental Protection Agency, Washington, D.C.

Finocchiaro, J. M., and Benson, W. R. (1967). *J. Ass. Offic. Anal. Chem.* **50,** 888.

Food and Drug Administration. (1973a). "Pesticide Analytical Manual," Vol. I (2nd ed., 1968, with revisions dated 7/1/69, 7/1/70, 4/1/71, 1/1/72, 9/1/72 and 6/1/73). FDA, Washington, D.C.

Food and Drug Administration. (1973b). "Pesticide Analytical Manual," Vol. II (2nd ed., 1968, with revisions dated 7/15/68, 1/2/69, 7/1/69, 7/1/70 and 11/1/73). FDA, Washington, D.C.

Heath, R. G. (1969). *Pestic. Monit. J.* **3,** 115.

Henderson, C., Inglis, A., and Johnson, W. L. (1972). *Pestic. Monit. J.* **6,** 144.

Kutz, F. W., Yobs, A. R., Johnson, W. G., and Wiersma, G. B. (1974a). *Bull. Soc. Pharmacol. Environ. Pathol.* **2,** 4.

Kutz, F. W., Yobs, A. R., Johnson, W. G., and Wiersma, G. B. (1974b). *Environ. Entomol.* **3,** 882.

Manaske, D. D., and Corneliussen, P. E. (1974). *Pestic. Monit. J.* **8,** 110.

Martin, W. E., and Nickerson, P. R. (1972). *Pestic. Monit. J.* **6,** 33.

Metcalf, R. L., Sangha, G. K., and Kapoor, I. P. (1971). *Environ. Sci. Technol.* **5,** 709.

Porter, M. L., Gajan, R. J., and Burke, J. A. (1969). *J. Ass. Offic. Anal. Chem.* **52,** 177.

Schulze, J. A., Manigold, D. B., and Andrews, F. L. (1973). *Pestic. Monit. J.* **7,** 1973.

Stanley, C. W., Barney, J. E., Helton, M. R., and Yobs, A. R. (1971). *Environ. Sci. Technol.* **5,** 430.

Wiersma, G. B., Tai, H., and Sand, P. J. (1972). *Pestic. Monit. J.* **6,** 126.

Yobs, A. R. (1972). "Environmental Health Perspectives," Exp. Issue No. 1, p. 79. National Institute of Environmental Health Sciences, Research Triangle Park, N.C.

2

Analysis of Pheromones and other Compounds Controlling Insect Behavior

M. N. Inscoe and Morton Beroza

I. INTRODUCTION

A. Recent History and Scope of This Chapter

Scientists have been troubled by difficulties encountered in the use of insecticides, starting in the fifties with the finding of resistance in insects to insecticides to the subsequent finding of residues of insecticides and their metabolites in every facet of our environment including man. These scientists have responded by searching for suitable alternatives to control damaging insects, or at least ways to minimize the use of insecticides. Part of this search has focused on compounds affecting insect behavior; progress by chemists in this area has been heavily dependent on advances in the analytical methodology that this chapter will try to cover. The search has also heavily involved the entomologist as well as scientists of other disciplines, and their close cooperation has spawned new and more efficient approaches to solving the problems of dealing with insect pests safely and in an ecologically acceptable manner.

The use of insect attractants was among the earliest of the alternative approaches investigated. Synthetic attractants were and are widely used for detection. By getting caught in the attractant-baited traps, insects betray their presence in the vicinity and indicate where suppressive measures should be instituted or maintained. The successful role that siglure, an attractant for the Mediterranean fruit fly, played in the 1956–1957 eradication of this insect from a million acres found infested in Florida was a striking demonstration of the value of attractants in pest management (Steiner *et al.,* 1961). Siglure and subsequently the more potent lure, trimedlure, were found by testing a wide variety of compounds and then synthesizing compounds related to those found attractive to find a better attractant. Synthetic attractants found by this screening route are listed in Table I.

The use of attractants as a selective means of direct insect control has only recently been subjected to serious study. The potential of this approach may be gathered from a pilot study conducted in 1962

TABLE I

Compounds of Interest as Behavior-Controlling Chemicals

Common name or trivial name[a]	Chemical name and structure	Species affected	Reference
	a. Synthetic attractants		
Amlure	Propyl 1,4-benzodioxan-2-carboxylate 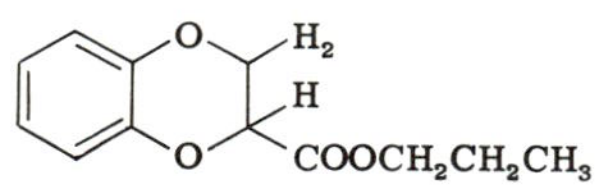	European chafer, *Amphimallon majalis* (Razoumowsky)	McGovern *et al.*, 1970a
	Butyl sorbate $CH_3CH{=}CHCH{=}CHCOOC_4H_9$	European chafer	Tashiro *et al.*, 1964
Cue-lure	4-(*p*-Hydroxyphenyl)-2-butanone acetate 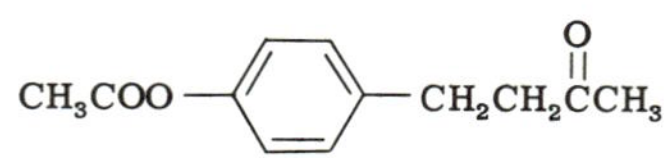	Melon fly, *Dacus cucurbitae* Coquillett	Beroza *et al.*, 1960
Ethyl dihydrochrysanthemumate, chrislure	Ethyl 3-isobutyl-2,2-dimethylcyclopropanecarboxylate 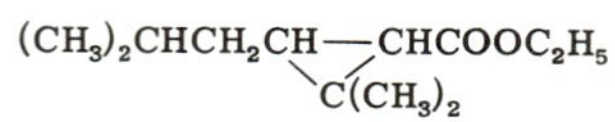	Coconut rhinoceros beetle, *Oryctes rhinoceros* (L.)	Barber *et al.*, 1971

Eugenol[b]	4-Allyl-2-methoxyphenol 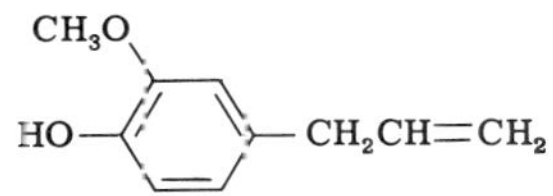	Japanese beetle, *Popillia japonica* Newman	McGovern *et al.*, 1970b,c
	Heptyl butyrate $C_3H_7COOC_7H_{15}$	Yellow jacket wasps, *Vespula* spp.	Davis *et al.*, 1969
	2,4-Hexadienyl butyrate $C_3H_7COOCH_2CH{=}CHCH{=}CHCH_3$	Yellow jacket wasps	Davis *et al.*, 1967
	Methyl cyclohexanepropionate[b] $C_6H_{11}CH_2CH_2COOCH_3$	Japanese beetle	McGovern *et al.*, 1970b
Methyl eugenol	4-Allyl-1,2-dimethoxybenzene 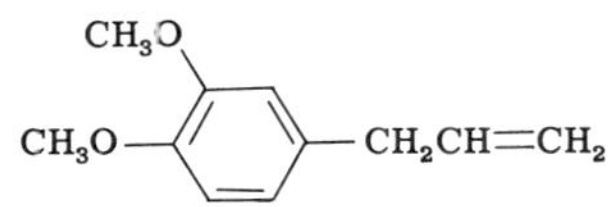	Oriental fruit fly, *Dacus dorsalis* Hendel	Steiner, 1952
	Octyl butyrate $C_3H_7COOC_8H_{17}$	Yellow jacket wasps	Davis *et al.*, 1972
	Phenethyl propionate[b] $C_6H_5CH_2CH_2OC(=O)C_2H_5$	Japanese beetle	McGovern *et al.*, 1970c

(*Continued*)

TABLE I (*Continued*)

Common name or trivial name[a]	Chemical name and structure	Species affected	Reference
Ethyl chrysanthemumate, rhinolure	Ethyl 2,2-dimethyl-3-(2-methylpropenyl)cyclopropanecarboxylate 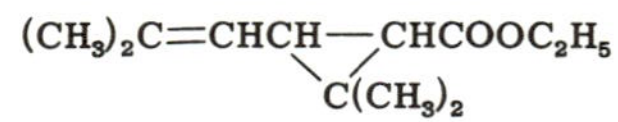	Coconut rhinoceros beetle	Maddison *et al.*, 1973
Siglure	*sec*-Butyl 6-methyl-3-cyclohexenecarboxylate 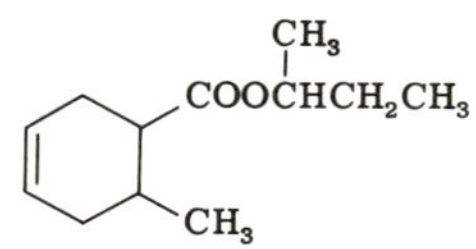	Mediterranean fruit fly, *Ceratitis capitata* (Wiedemann)	Gertler *et al.*, 1958
Trimedlure	*tert*-Butyl 4(or 5)-chloro-2-methylcyclohexanecarboxylate 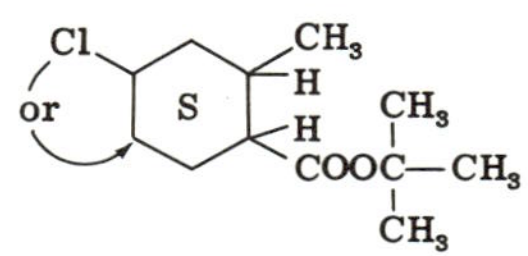	Mediterranean fruit fly	Beroza *et al.*, 1961
	b. Sex attractant pheromones,[c] aggregation pheromones,[c] and compounds with similar action		
	4-Methyl-3-heptanol 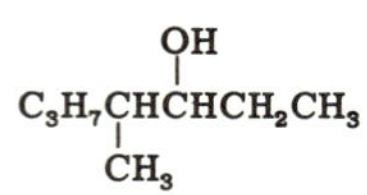	Smaller European elm bark beetle, *Scolytus multistriatus* (Marsham)[d,e,f]	Pearce *et al.*, 1975

Sulcatol	6-Methyl-5-hepten-2-ol $(CH_3)_2C{=}CHCH_2CH_2CH(OH)CH_3$	*Gnathotrichus sulcatus*[e,g]	Byrne *et al.*, 1974
Ipsenol	(−)-2-Methyl-6-methylene-7-octen-4-ol $CH_2{=}CHC({=}CH_2)CH_2CH(OH)CH_2CH(CH_3)_2$	California fivespined ips, *Ips paraconfusus* (Lanier) (formerly *I. confusus* LeConte) *I. confusus* (LeConte)[e,f,g] *I. grandicollis* Eichhoff	Silverstein *et al.*, 1966 Young *et al.*, 1973b Vité and Renwick, 1971
Ipsdienol	(+)-2-Methyl-6-methylene-2,7-octadien-4-ol $CH_2{=}CHC({=}CH_2)CH_2CH(OH)CH{=}C(CH_3)_2$	California fivespined ips[e,f,g] *I. confusus*[e,f,g] *I. calligraphus* (Germar)[d,e,f,g] *I. sexdentatus* (Börner)[e,g]	Silverstein *et al.*, 1966 Young *et al.*, 1973b Renwick and Vité, 1972 Vité *et al.*, 1974
	(*E*)-6-Nonen-1-ol $C_2H_5CH\overset{E}{=}CH(CH_2)_5OH$	Mediterranean fruit fly[f,g]	Jacobson *et al.*, 1973
	(*Z*)-7-Dodecen-1-ol $CH_3(CH_2)_3CH\overset{Z}{=}CH(CH_2)_6OH$	*Harpipteryx xylostella auct.* (L.)[h,i] *Dicentria semirufescens* (Walker)[h] *Raphia frater* Grote[h] *Exartema* sp.[h]	Roelofs and Comeau, 1971a Roelofs and Comeau, 1971a Roelofs and Comeau, 1971a Weatherston *et al.*, 1974
	(*Z*)-8-Dodecen-1-ol $CH_3(CH_2)_2CH\overset{Z}{=}CH(CH_2)_7OH$	*Epiblema scudderiana* (Clemens)[h] *Celypha striana* (Schiffermüller)[h]	Roelofs and Comeau, 1971a Arn *et al.*, 1974
	(*E*)-8-Dodecen-1-ol $CH_3(CH_2)_2CH\overset{E}{=}CH(CH_2)_7OH$	*Hedia chionosema* (Zeller)[h,i]	Roelofs and Comeau, 1971a

(*Continued*)

TABLE I (*Continued*)

Common name or trivial name[a]	Chemical name and structure	Species affected	Reference
	(*E*)-9-Dodecen-1-ol $CH_3CH_2CH\overset{E}{=}CH(CH_2)_8OH$	*Dichrorampha* sp.[h,i]	Roelofs and Comeau, 1971a
Codlelure, Codlemone®	(*E*,*E*)-8,10-Dodecadien-1-ol $CH_3CH\overset{E}{=}CHCH\overset{E}{=}CH(CH_2)_7OH$	Codling moth, *Laspeyresia pomonella* (L.)	Roelofs *et al.*, 1971b; Beroza *et al.*, 1974
		Hedia nubiferana (Haworth)[h]	Arn *et al.*, 1974
	(*Z*)-11-Tetradecen-1-ol $CH_3CH_2CH\overset{Z}{=}CH(CH_2)_{10}OH$	*Choristoneura fractivittana* (Clemens)[h]	Roelofs and Comeau, 1971a
		Clepsis melaleucana (Walker)[h,i]	Roelofs and Comeau, 1971a
		Nedra ramosula (Guenée)[h]	Roelofs and Comeau, 1970
		Sparganothis niveana (Walsingham)[h]	Roelofs and Comeau, 1970, 1971a
		Zygaena transalpina Esper[h]	Benz and von Salis, 1973
	(*E*)-11-Tetradecen-1-ol $CH_3CH_2CH\overset{E}{=}CH(CH_2)_{10}OH$	Tufted apple budmoth, *Platynota idaeusalis* (Walker)	Hill *et al.*, 1974
	(*Z*)-(−)-14-Methyl-8-hexadecen-1-ol $C_2H_5\underset{\substack{\vert\\ CH_3}}{C}H(CH_2)_4CH\overset{Z}{=}CH(CH_2)_7OH$	*Trogoderma inclusum* LeConte	Rodin *et al.*, 1969
Bombykol	(*E*,*Z*)-10,12-Hexadecadien-1-ol $CH_3(CH_2)_2CH\overset{Z}{=}CHCH\overset{E}{=}CH(CH_2)_9OH$	Silkworm, *Bombyx mori* (L.)	Butenandt *et al.*, 1959

Grandisol, grandlure I	*cis*-2-Isopropenyl-1-methylcyclobutaneethanol 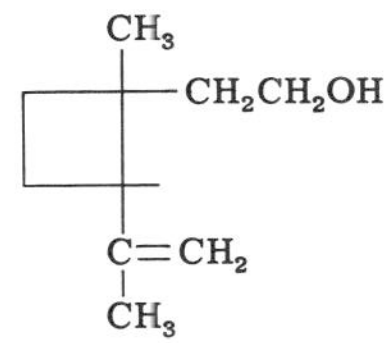	Boll weevil, *Anthonomus grandis* Boheman[f,g]	Tumlinson *et al.*, 1969
Grandlure II	(*Z*)-3,3-Dimethyl-$\Delta^{1,\beta}$-cyclohexaneethanol 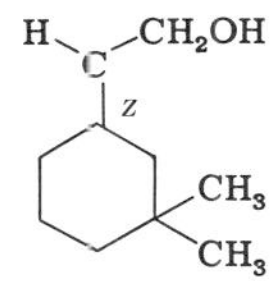	Boll weevil[f,g]	Tumlinson *et al.*, 1969
cis-(+)-Verbenol	*cis*-(+)-2-Pinen-4-ol 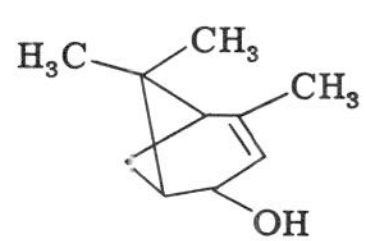	California fivespined ips[e,f,g] *I. confusus*[e,f,g] *I. calligraphus*[d,e,f,g]	Silverstein *et al.*, 1966 Young *et al.*, 1973b Renwick and Vité, 1972
trans-Verbenol, pondelure[d]	*trans*-2-Pinen-4-ol 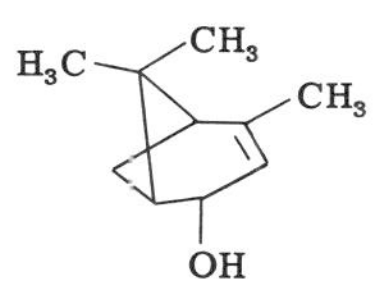	Mountain pine beetle, *Dendroctonus pondorosae* Hopkins	Pitman and Vité, 1969

(*Continued*)

TABLE I (*Continued*)

Common name or trivial name[a]	Chemical name and structure	Species affected	Reference
	(*Z*)-7-Decen-1-ol acetate $CH_3CH_2CH\overset{Z}{=}CH(CH_2)_6OAc$	*Battaristis* sp.[h]	Roelofs and Comeau, 1971a
	(*Z*)-5-Dodecen-1-ol acetate $CH_3(CH_2)_5CH\overset{Z}{=}CH(CH_2)_4OAc$	Bean leafskeletonizer, *Autoplusia egena* (Guenée)[h]	Kaae *et al.*, 1973b
		Chionodes fuscomaculella (Chambers)[h]	Roelofs and Comeau, 1971a
		Scrobipalpa atriplicella (F.R.)[h]	
Looplure	(*Z*)-7-Dodecen-1-ol acetate $CH_3(CH_2)_3CH\overset{Z}{=}CH(CH_2)_6OAc$	Cabbage looper, *Trichoplusia ni* (Hübner)	Berger, 1966
		Soybean looper, *Pseudoplusia includens* (Walker)	Tumlinson *et al.*, 1972b
		Trichoplusia oxygramma (Geyer)[h]	Mitchell *et al.*, 1972
		Alfalfa looper, *Autographa californica* (Speyer)[h]	Kaae *et al.*, 1973a
		Anagrapha falcifera (Kirby)[h]	Roelofs and Comeau, 1970
		Autographa ampla (Walker)[h]	Roelofs and Comeau, 1971a
		Autographa biloba (Stephens)[h]	Roelofs and Comeau, 1971a
		Autographa precationis (Guenée)[h]	Roelofs and Comeau, 1971a
		Epinotia zandana (Kearfott)[h]	Roelofs and Comeau, 1971a
		Chrysaspidia contexta (Grote)[h]	Roelofs and Comeau, 1970
		Harpipteryx xylostella auct. (L.)[h,i]	Roelofs and Comeau, 1971a
		Phlyctaenia terrealis (Treitschke)	Roelofs and Comeau, 1971a
		Plusia aereoides Grote[h]	Roelofs and Comeau, 1970
		Pterophorus tenuidactylus Fitch[h]	Roelofs and Comeau, 1971a
	(*E*)-7-Dodecen-1-ol acetate $CH_3(CH_2)_3CH\overset{E}{=}CH(CH_2)_6OAc$	False codling moth, *Argyroploce leucotreta* (Meyrick)[*Cryptophlebia leucotreta* Meyrick]	Read *et al.*, 1968, 1974

Orfralure[j], Orframone®	(*Z*)-8-Dodecen-1-ol acetate $CH_3(CH_2)_2CH\overset{Z}{=}CH(CH_2)_7OAc$	Oriental fruit moth, *Grapholitha molesta* (Busck)[k]	Roelofs *et al.*, 1969; Beroza *et al.*, 1973b,c
		Aphania infida (Hein.)[h]	Roelofs and Comeau, 1971a
		Epiblema desertana (Zeller)[h]	Roelofs and Comeau, 1971a
Funemone®		*Grapholitha funebrana* (Treitschke)[h]	Granges and Baggiolini, 1971
		Lesser appleworm, *Grapholitha prunivora* (Walsh)[h,k]	Roelofs *et al.*, 1969; Roelofs and Cardé, 1974
		Pseudexentera maracana (Kearfott)[h]	Roelofs and Comeau, 1971a
	(*E*)-8-Dodecen-1-ol acetate $CH_3(CH_2)_2CH\overset{E}{=}CH(CH_2)_7OAc$	*Ecdytolopha insiticiana* (Zeller)[h]	Roelofs and Comeau, 1971a
		Grapholitha packardi (Zeller)[h]	Roelofs *et al.*, 1969
		Hedia chionosema (Zeller)[h,i]	Roelofs and Comeau, 1971a
Grapemone®	(*Z*)-9-Dodecen-1-ol acetate $CH_3CH_2CH\overset{Z}{=}CH(CH_2)_8OAc$	Grape berry moth, *Paralobesia viteana* (Clemens)	Roelofs *et al.*, 1971c
		Episimus argutanus (Clemens)[h]	Roelofs and Comeau, 1971a
		Fall armyworm, *Spodoptera frugiperda* (J. E. Smith)	Sekul and Sparks, 1976
	(*E*)-9-Dodecen-1-ol acetate $CH_3CH_2CH\overset{E}{=}CH(CH_2)_8OAc$	European pine shoot moth, *Rhyacionia buoliana* (Schiffermüller)	Smith *et al.*, 1974
		Dichrorampha sp.[h,i]	Roelofs and Comeau, 1971a
		Loxostege chortalis (Grote)[h]	Roelofs and Comeau, 1971a
		Caradrina morpheus (Hufnagel)[h]	Arn *et al.*, 1974
	(*Z*)-10-Dodecen-1-ol acetate $CH_3CH\overset{Z}{=}CH(CH_2)_9OAc$	*Argyrotaenia quadrifasciana* (Fernald)[h]	Comeau and Roelofs, 1973
	(*E*)-10-Dodecen-1-ol acetate $CH_3CH\overset{E}{=}CH(CH_2)_9OAc$	*Argyroploce aurofasciana* (Haworth)[h]	Arn *et al.*, 1974
	(*E*,*Z*)-7,9-Dodecadien-1-ol acetate $CH_3CH_2CH\overset{Z}{=}CHCH\overset{E}{=}CH(CH_2)_6OAc$	*Lobesia botrana* (Schiffermüller)	Roelofs *et al.*, 1973; Buser and Arn, 1975

(*Continued*)

TABLE I (*Continued*)

Common name or trivial name[a]	Chemical name and structure	Species affected	Reference
	(*E*)-9,11-Dodecadien-1-ol acetate $CH_2{=}CHCH\overset{E}{=}CH(CH_2)_8OAc$	Red bollworm, *Diparopsis castanea* Hampson	Nesbitt *et al.*, 1973a,b
	(*Z*)-11-Tridecen-1-ol acetate $CH_3CH\overset{Z}{=}CH(CH_2)_{10}OAc$	*Argyrotaenia velutinana* (Walker)[h]	Roelofs and Comeau, 1970
	(*Z*)-7-Tetradecen-1-ol acetate $CH_3(CH_2)_5CH\overset{Z}{=}CH(CH_2)_6OAc$	*Amathes c-nigrum* (L.), large[h] *Lacinipolia lorea* (Guenée)[h]	Roelofs and Comeau, 1971a Roelofs and Comeau, 1971a
	(*E*)-7-Tetradecen-1-ol acetate $CH_3(CH_2)_5CH\overset{E}{=}CH(CH_2)_6OAc$	*Amathes c-nigrum* (L.), small[h] Larch bud moth, *Zeiraphera diniana* (Guenée)[h]	Roelofs and Comeau, 1971a Roelofs *et al.*, 1971a
	(*Z*)-8-Tetradecen-1-ol acetate $CH_3(CH_2)_4CH\overset{Z}{=}CH(CH_2)_7OAc$	*Spilonota ocellana* Fabricius[h]	Arn *et al.*, 1974
Prodenialure A	(*Z*)-9-Tetradecen-1-ol acetate $CH_3(CH_2)_3CH\overset{Z}{=}CH(CH_2)_8OAc$	Southern armyworm, *Spodoptera eridania* (Cramer)[f] Smaller tea tortrix, *Adoxyphes fasciata* Walsingham[l] Summerfruit tortrix, *Adoxyphes orana* (Fischer von Röslerstamm)[l] *Amphipoea interoceanica* (Smith)[h] *Bryotropha similis* (Stainton)[h] *Cucullia intermedia* Speyer[h]	Jacobson *et al.*, 1970 Tamaki *et al.*, 1971a Meijer *et al.*, 1972; Tamaki *et al.*, 1971b Roelofs and Comeau, 1971a Roelofs and Comeau, 1969 Roelofs and Comeau, 1971a

		Leucania phragmitidicola Guenée[h]	Roelofs and Comeau, 1971a
		Nemapogon apicisignatellus (Dietz)[h]	Roelofs and Comeau, 1971a
		Pyreferra citromba Franclemont[h]	Roelofs and Comeau, 1971a
	(*E*)-9-Tetradecen-1-ol acetate	*Bryotropha* sp.[h]	Roelofs and Comeau, 1969
		Bryotropha terrella (Hübner)	Arn *et al.*, 1974
	$CH_3(CH_2)_3CH\overset{E}{=}CH(CH_2)_8OAc$	*Loxostege neobliteralis* Capps[h]	Roelofs and Comeau, 1971a
		Polia grandis (Boisduval)[h]	Roelofs and Comeau, 1971a
	(*Z*)-10-Tetradecen-1-ol acetate	*Exartema* sp.[h]	Comeau and Roelofs, 1973
		Oak leafroller, *Archips semiferanus* (Walker)	Hendry *et al.*, 1974
	$CH_3(CH_2)_2CH\overset{Z}{=}CH(CH_2)_9OAc$		
		Apotomis corticana (Hübner)[h]	Arn *et al.*, 1974
		Endothenia carbonana (Dougl.)[h]	Arn *et al.*, 1974
Riblure, Redlamone®	(*Z*)-11-Tetradecen-1-ol acetate	Redbanded leafroller, *Argyrotaenia velutinana* (Walker)[k]	Roelofs and Arn, 1968; Klun *et al.*, 1973
	$CH_3CH_2CH\overset{Z}{=}CH(CH_2)_{10}OAc$	European corn borer, *Ostrinia nubilalis* (Hübner)[k]	Klun and Brindley, 1970; Klun *et al.*, 1973
		Obliquebanded leafroller, *Choristoneura rosaceana* (Harris)	Roelofs and Tette, 1970
		Fruittree leafroller, *Archips argyrospilus* (Walker)[k]	Roelofs *et al.*, 1974b
		Smartweed borer, *Ostrinia obumbratalis* (Lederer)[k]	Klun and Robinson, 1972
		Smaller tea tortrix[l]	Tamaki *et al.*, 1971a
		Summerfruit tortrix[l]	Meijer *et al.*, 1972; Tamaki *et al.*, 1971b
		Archips podana (Scopoli)[k]	Persoons *et al.*, 1974
		Argyrotaenia pulchellana (Haworth)[h]	Maini, 1973
		Zebra caterpillar, *Ceramica picta* (Harris)[h]	Roelofs and Comeau, 1971a
		Clepsis melaleucana (Walker)[h,i]	Roelofs and Comeau, 1971a
		Thyris maculata (Harris)[h]	Roelofs and Comeau, 1971a
		Zygaena transalpina Esper[h]	Benz and von Salis, 1973
		Yponomeuta padellus-malinellus (Gl.)[h]	Arn *et al.*, 1974

(*Continued*)

TABLE I (*Continued*)

Common name or trivial name[a]	Chemical name and structure	Species affected	Reference
	(*E*)-11-Tetradecen-1-ol acetate	Fruittree leafroller[k]	Roelofs *et al.*, 1974b
		Smartweed borer[k]	Klun and Robinson, 1972
	$CH_3CH_2CH\overset{E}{=}CH(CH_2)_{10}OAc$	*Archips podana*[k]	Persoons *et al.*, 1974
		European corn borer (New York strain)[h]	Roelofs and Comeau, 1971a
		Choristoneura viridis Freeman[h]	Sanders *et al.*, 1974
		Dichomeris ligulella (Hübner)[h]	Roelofs and Comeau, 1971a
		Phalonia sp.[h]	Roelofs and Comeau, 1971a
		Pyrausta ochosalis Dyar[h]	Roelofs and Comeau, 1971a
		Sparganothis albicaudana Busck[h]	Roelofs and Comeau, 1971a
		Sparganothis sulfureana (Clemens)[h]	Roelofs and Comeau, 1971a
		Larch bud moth, *Zeiraphera diniana* (Guenée)[h]	Roelofs *et al.*, 1971a
		Croesia holmiana (L.)	Arn *et al.*, 1974
	(*Z,E*)-9,11-Tetradecadien-1-ol acetate	*Spodoptera littoralis* (Boisduval)[l]	Nesbitt *et al.*, 1973a; Tamaki and Yushima, 1974
Litlure[m]	$CH_3CH_2CH\overset{E}{=}CHCH\overset{Z}{=}CH(CH_2)_8OAc$	*Spodoptera litura* (Fabricius)[l]	Tamaki *et al.*, 1973
Prodenialure B	(*Z,E*)-9,12-Tetradecadien-1-ol acetate	Southern armyworm[f]	Jacobson *et al.*, 1970
	$CH_3CH\overset{E}{=}CHCH_2CH\overset{Z}{=}CH(CH_2)_8OAc$	Indian meal moth, *Plodia interpunctella* (Hübner)	Brady *et al.*, 1971; Kuwahara *et al.*, 1971b
		Almond moth, *Cadra cautella* (Walker)	Brady *et al.*, 1971; Kuwahara *et al.*, 1971b
		Mediterranean flour moth, *Anagasta kuehniella* (Zeller)	Kuwahara *et al.*, 1971a; Dahm *et al.*, 1971b
		Spodoptera littoralis[l]	Tamaki *et al.*, 1973; Tamaki and Yushima, 1974

Litlure[m]		*Spodoptera litura*[l]	Tamaki *et al.*, 1973
		Spodoptera dolichos (Fabricius)[h]	Mitchell and Tumlinson, 1973
Hexalure	(Z)-7-Hexadecen-1-ol acetate $CH_3(CH_2)_7CH\overset{Z}{=}CH(CH_2)_6OAc$	Pink bollworm, *Pectinophora gossypiella* (Saunders)[h]	Green *et al.*, 1969
		Euxoa tessellata (Harris)[h]	Roelofs and Comeau, 1971a
	(Z)-11-Hexadecen-1-ol acetate $CH_3(CH_2)_3CH\overset{Z}{=}CH(CH_2)_{10}OAc$	*Amphipoea velata* (Walker)[h]	Roelofs and Comeau, 1971a
		Morrisonia confusa (Hübner)[h]	Roelofs and Comeau, 1971a
		Orthodes crenulata (Butler)[h]	Roelofs and Comeau, 1971a
		Pseudorthodes vecors (Guenée)[h]	Roelofs and Comeau, 1971a
		Scotogramma trifolii (Rott.)[h]	Roelofs and Comeau, 1971a
Gossyplure[n]	(Z,Z)-7,11-Hexadecadien-1-ol acetate $CH_3(CH_2)_3CH\overset{Z}{=}CH(CH_2)_2CH\overset{Z}{=}CH(CH_2)_6OAc$	Pink bollworm[k]	Hummel *et al.*, 1973; Bierl *et al.*, 1974
Gossyplure[n]	(Z,E)-7,11-Hexadecadien-1-ol acetate	Pink bollworm[k]	Hummel *et al.*, 1973; Bierl *et al.*, 1974
Angoulure	$CH_3(CH_2)_3CH\overset{E}{=}CH(CH_2)_2CH\overset{Z}{=}CH(CH_2)_6OAc$	Angoumois grain moth, *Sitotroga cereallela* (Olivier)	Vick *et al.*, 1974
	(Z,Z)-3,13-Octadecadien-1-ol acetate $CH_3(CH_2)_3CH\overset{Z}{=}CH(CH_2)_8CH\overset{Z}{=}CHCH_2CH_2OAc$	Peachtree borer, *Sanninoidea exitiosa* (Say)	Tumlinson *et al.*, 1974b
	(E,Z)-3,13-Octadecadien-1-ol acetate $CH_3(CH_2)_3CH\overset{Z}{=}CH(CH_2)_8CH\overset{E}{=}CHCH_2CH_2OAc$	Lesser peachtree borer, *Synanthedon pictipes* (Grote and Robinson)	Tumlinson *et al.*, 1974b

(*Continued*)

TABLE I (*Continued*)

Common name or trivial name[a]	Chemical name and structure	Species affected	Reference
	(*Z*)-5-Decenyl 3-methylbutyrate $CH_3(CH_2)_3CH\overset{Z}{=}CH(CH_2)_4O\overset{O}{\overset{\|}{C}}CH_2CH(CH_3)_2$	Pine emperor moth, *Nudaurelia cytherea* (Fabricius)	Henderson *et al.*, 1973
	(−)-Methyl (*E*)-2,4,5-tetradecatrienoate $CH_3(CH_2)_7CH{=}C{=}CHCH\overset{E}{=}CHCOOCH_3$	Dried bean beetle, *Acanthoscelides obtectus* (Say)[g]	Horler, 1970; Halstead, 1973
	(−)-Methyl (*Z*)-14-methyl-8-hexadecenoate $CH_3CH_2\underset{CH_3}{\underset{\|}{C}}H(CH_2)_4CH\overset{Z}{=}CH(CH_2)_6COOCH_3$	*Trogoderma inclusum*	Rodin *et al.*, 1969
	Methyl (*E*)-6-nonenoate $CH_3CH_2CH\overset{E}{=}CH(CH_2)_4COOCH_3$	Mediterranean fruit fly[f,g]	Jacobson *et al.*, 1973
	Valeric acid $CH_3(CH_2)_3COOH$	Sugarbeet wireworm, *Limonius californicus* (Mannerheim)	Jacobson *et al.*, 1968
	(*Z*)-3-Decenoic acid $CH_3(CH_2)_5CH\overset{Z}{=}CHCH_2COOH$	Furniture carpet beetle, *Anthrenus flavipes* LeConte	Fukui *et al.*, 1974

Megatomoic acid	(*E,Z*)-3,5-Tetradecadienoic acid	Black carpet beetle, *Attagenus megatoma* (Fabricius)	Silverstein *et al.*, 1967
	$CH_3(CH_2)_7CH\overset{Z}{=}CHCH\overset{E}{=}CHCH_2COOH$		
Queen substance	(*E*)-9-Oxo-2-decenoic acid	Honey bee, *Apis mellifera* L.	Gary, 1962
	$CH_3C(=O)(CH_2)_5CH\overset{E}{=}CHCOOH$	*Apis cerana* Fabricius	Sannasi and Rajulu, 1971
		Apis dorsata Fabricius	Sannasi and Rajulu, 1971
		Apis florea Fabricius	Sannasi and Rajulu, 1971
	Nonanal	Greater wax moth, *Galleria mellonella* (L.)[f,g]	Leyrer and Monroe, 1973
	$CH_3(CH_2)_7CHO$		
	Undecanal	Greater wax moth[f,g]	Röller *et al.*, 1968; Leyrer and Monroe, 1973
	$CH_3(CH_2)_9CHO$	Lesser wax moth, *Achroia grisella* (Fabricius)[f,g]	Dahm *et al.*, 1971a
	(*Z*)-9-Tetradecenal	Tobacco budworm, *Heliothis virescens* (Fabricius)[f]	Roelofs *et al.*, 1974a; Tumlinson *et al.*, 1975
	$CH_3(CH_2)_3CH\overset{Z}{=}CH(CH_2)_7CHO$		
	(*Z*)-11-Tetradecenal	Orange tortrix, *Argyrotaenia citrana* (Fernald)	Hill *et al.*, 1975
	$CH_3CH_2CH\overset{Z}{=}CH(CH_2)_9CHO$		
Soolure	(*E*)-11-Tetradecenal	Eastern spruce budworm, *Choristoneura fumiferana* (Clemens)	Weatherston *et al.*, 1971
	$CH_3CH_2CH\overset{E}{=}CH(CH_2)_9CHO$	*Choristoneura biennis* Freeman[h]	Sanders *et al.*, 1974
		Choristoneura occidentalis Freeman[h]	Weatherston *et al.*, 1971
		Acleris emargana (Fabricius)[h]	Weatherston *et al.*, 1974
	(*Z*)-11-Hexadecenal	Tobacco budworm[f]	Roelofs *et al.*, 1974a; Tumlinson *et al.*, 1975
	$CH_3(CH_2)_3CH\overset{Z}{=}CH(CH_2)_9CHO$		

(*Continued*)

TABLE I (*Continued*)

Common name or trivial name[a]	Chemical name and structure	Species affected	Reference
	(*Z*)-11-Octadecenal $CH_3(CH_2)_5CH\overset{Z}{=}CH(CH_2)_9CHO$	Lesser wax moth[f,g]	Dahm *et al.*, 1971a
Grandlure III	(*Z*)-3,3-Dimethyl-$\Delta^{1,\alpha}$-cyclohexane-acetaldehyde 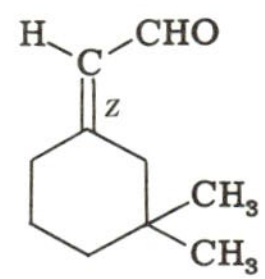	Boll weevil[f,g]	Tumlinson *et al.*, 1969
Grandlure IV	(*E*)-3,3-Dimethyl-$\Delta^{1,\alpha}$-cyclohexane-acetaldehyde 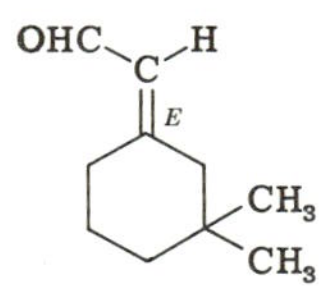	Boll weevil[f,g]	Tumlinson *et al.*, 1969
	(*Z*)-6-Heneicosen-11-one $CH_3(CH_2)_9\overset{O}{\overset{\parallel}{C}}(CH_2)_3CH\overset{Z}{=}CH(CH_2)_4CH_3$	Douglas-fir tussock moth, *Orgyia pseudotsugata* (McDunnough)	Smith *et al.*, 1975

Muscalure	(*Z*)-9-Tricosene	House fly, *Musca domestica* L.	Carlson *et al.*, 1971
	$CH_3(CH_2)_{12}CH\overset{Z}{=}CH(CH_2)_7CH_3$		
	2-Methylheptadecane	*Holomelina* (spp.)	Roelofs and Cardé, 1971
	$CH_3(CH_2)_{14}CH(CH_3)_2$		
Disparlure	*cis*-7,8-Epoxy-2-methyloctadecane	Gypsy moth, *Porthetria dispar* (L.)	Bierl *et al.*, 1970
	$CH_3(CH_2)_9HC\underset{O}{—}CH(CH_2)_4CH(CH_3)_2$	Nun moth, *Lymantria monacha* (L.)	Schönherr, 1972
		Lymantria obfuscata Walker[h]	Beroza *et al.*, 1973a
		Lymantria fumida Butler[h]	Beroza *et al.*, 1973d
Brevicomin	*exo*-7-Ethyl-5-methyl-6,8-dioxabicyclo[3.2.1]octane	Western pine beetle, *Dendroctonus brevicomis* LeConte[d,e]	Silverstein *et al.*, 1968
	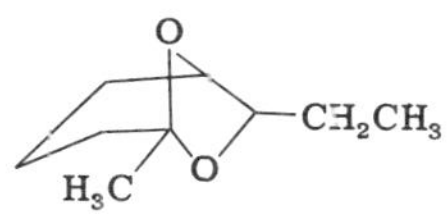	*Temnochila chlorodia* (Mannerheim)[h]	Pitman and Vité, 1971
Frontalin	1,5-Dimethyl-6,8-dioxabicyclo-[3.2.1]octane	Southern pine beetle, *Dendroctonus frontalis* (Zimmerman)[d,e]	Kinzer *et al.*, 1969
Douglure[d]	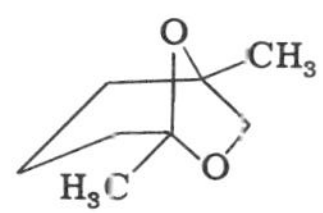	Douglas fir beetle, *Dendroctonus pseudotsugae* (Hopkins)[d,e]	Pitman and Vité, 1970
		Western pine beetle[d,e,g]	Vité and Pitman, 1970
		Thanasimus undatulus Wolcott[h]	Pitman and Vité, 1970
		Thanasimus dubius (Fabricius)[e,h]	Vité and Williamson, 1970
		Medetera bistriata Parent[d,e,h]	Williamson, 1971

(*Continued*)

TABLE I (*Continued*)

Common name or trivial name[a]	Chemical name and structure	Species affected	Reference
Multistriatin	5-Ethyl-2,4-dimethyl-6,8-dioxabicyclo-[3.2.1]octane 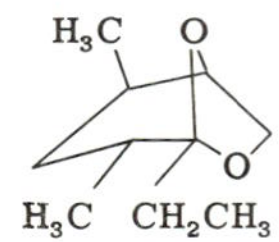	Smaller European elm bark beetle, *Scolytus multistriatus* (Marsham)[d,e,f]	Pearce *et al.*, 1975
	Phenol C_6H_5OH	Common grass grub beetle, *Costelytra zealandica* (White)	Henzell and Lowe, 1970
	2,6-Dichlorophenol $C_6H_3Cl_2OH$	Lone star tick, *Amblyomma americanum* (L.)	Berger, 1972
	c. Sex attractant inhibitors		
	(*Z*)-7-Dodecen-1-ol $CH_3(CH_2)_3CH\overset{Z}{=}CH(CH_2)_6OH$	Cabbage looper[h] Soybean looper[h]	Tumlinson *et al.*, 1972a McLaughlin *et al.*, 1974
	(*Z*)-8-Dodecen-1-ol $CH_3(CH_2)_2CH\overset{Z}{=}CH(CH_2)_7OH$	Oriental fruit moth[h]	Beroza *et al.*, 1973b
	(*Z*)-11-Tetradecen-1-ol $CH_3CH_2CH\overset{Z}{=}CH(CH_2)_{10}OH$	Redbanded leafroller[h] Tufted apple budmoth[h]	Roelofs and Comeau, 1971b Hill *et al.*, 1974

(*E*)-11-Tetradecen-1-ol $CH_3CH_2CH\overset{E}{=}CH(CH_2)_{10}OH$	Eastern spruce budworm	Sanders *et al.*, 1972
(*Z*,*E*)-9,12-Tetradecadien-1-ol $CH_3CH\overset{E}{=}CHCH_2CH\overset{Z}{=}CH(CH_2)_8OH$	Almond moth[h]	Sower *et al.*, 1974
(*Z*)-7-Hexadecen-1-ol $CH_3(CH_2)_7CH\overset{Z}{=}CH(CH_2)_6OH$	Pink bollworm (hexalure)[h]	Neumark *et al.*, 1973
Dodecanol acetate $CH_3(CH_2)_{11}OAc$	European pine shoot moth Pink bollworm (hexalure)[h]	Smith *et al.*, 1974 Staten *et al.*, 1973; Neumark *et al.*, 1974
(*Z*)-7-Dodecen-1-ol acetate $CH_3(CH_2)_3CH\overset{Z}{=}CH(CH_2)_6OAc$	Fall armyworm[h]	Mitchell *et al.*, 1974
(*Z*)-8-Dodecen-1-ol acetate $CH_3(CH_2)_2CH\overset{Z}{=}CH(CH_2)_7OAc$	Codling moth[h]	Arn *et al.*, 1974
(*E*)-8-Dodecen-1-ol acetate $CH_3(CH_2)_2CH\overset{E}{=}CH(CH_2)_7OAc$	Oriental fruit moth[h]	Roelofs and Comeau, 1970
(*Z*)-9-Dodecen-1-ol acetate $CH_3CH_2CH\overset{Z}{=}CH(CH_2)_8OAc$	European pine shoot moth[h]	Smith *et al.*, 1974; Lange and Hoffmann, 1972

(*Continued*)

TABLE I (*Continued*)

Common name or trivial name[a]	Chemical name and structure	Species affected	Reference
	(*Z*)-10-Dodecen-1-ol acetate $CH_3CH\overset{Z}{=}CH(CH_2)_9OAc$	Codling moth[h]	Arn *et al.*, 1974
	(*E*)-10-Dodecen-1-ol acetate $CH_3CH\overset{E}{=}CH(CH_2)_9OAc$	Codling moth[h]	Arn *et al.*, 1974
	(*E,E*)-8,10-Dodecadien-1-ol acetate $CH_3CH\overset{E}{=}CHCH\overset{E}{=}CH(CH_2)_7OAc$	Codling moth[h]	Hathaway *et al.*, 1974
	Tetradecanol acetate $CH_3(CH_2)_{13}OAc$	Pink bollworm (hexalure) (gossyplure)	Beroza *et al.*, 1971a; Bierl *et al.*, 1974
	(*E*)-9-Tetradecen-1-ol acetate $CH_3(CH_2)_3CH\overset{E}{=}CH(CH_2)_8OAc$	Summerfruit tortrix[h]	Minks and Voerman, 1973
	(*Z*)-11-Tetradecen-1-ol acetate $CH_3CH_2CH\overset{Z}{=}CH(CH_2)_{10}OAc$	Larch bud moth[h]	Benz and von Salis, 1973
	(*E*)-11-Tetradecen-1-ol acetate $CH_3CH_2CH\overset{E}{=}CH(CH_2)_{10}OAc$	Summerfruit tortrix[h] Eastern spruce budworm[h] European corn borer[h] Redbanded leafroller[h] Obliquebanded leafroller	Minks and Voerman, 1973 Sanders *et al.*, 1972 Klun and Robinson, 1971 Roelofs and Comeau, 1971b Roelofs and Tette, 1970

11-Tetradecyn-1-ol acetate	European corn borer[h]	Klun and Robinson, 1971
$CH_3CH_2C{\equiv}C(CH_2)_{10}OAc$		
(*Z*,*Z*)-9,12-Tetradecadien-1-ol acetate	Indian meal moth[h]	Vick and Sower, 1973
$CH_3CH\overset{Z}{=}CHCH_2CH\overset{Z}{=}CH(CH_2)_8OAc$		
(*Z*,*E*)-9,12-Tetradecadien-1-ol acetate	Fall armyworm[h]	Mitchell *et al.*, 1974
$CH_3CH\overset{E}{=}CHCH_2CH\overset{Z}{=}CH(CH_2)_8OAc$		
(*E*,*Z*)-7,11-Hexadecadien-1-ol acetate	Pink bollworm (gossyplure)[h]	Bierl *et al.*, 1974
$CH_3(CH_2)_3CH\overset{Z}{=}CHCH_2CH_2CH\overset{E}{=}CH(CH_2)_6OAc$		
(*E*,*E*)-7,11-Hexadecadien-1-ol acetate	Pink bollworm (gossyplure)[h]	Bierl *et al.*, 1974
$CH_3(CH_2)_3CH\overset{E}{=}CHCH_2CH_2CH\overset{E}{=}CH(CH_2)_6OAc$		
(*Z*,*Z*)-3,13-Octadecadien-1-ol acetate	Lesser peachtree borer[h]	Tumlinson *et al.*, 1974b
$CH_3(CH_2)_3CH\overset{Z}{=}CH(CH_2)_8CH\overset{Z}{=}CHCH_2CH_2OAc$		
(*E*)-11-Tetradecenal	Orange tortrix[h]	Hill *et al.*, 1975
$CH_3CH_2CH\overset{E}{=}CH(CH_2)_9CHO$		

(*Continued*)

TABLE I (*Continued*)

Common name or trivial name[a]	Chemical name and structure	Species affected	Reference
	(*Z*)-2-Methyl-7-octadecene $CH_3(CH_2)_9CH\overset{Z}{=}CH(CH_2)_4CH(CH_3)_2$	Gypsy moth	Cardé *et al.*, 1973
	d. Trail pheromones and related compounds		
	(*Z*)-4-Phenyl-3-buten-1-ol $C_6H_5CH\overset{Z}{=}CHCH_2CH_2OH$	Various termites[h]	Matsumura *et al.*, 1972
	(*Z,Z*)-3,6-Dodecadien-1-ol $CH_3(CH_2)_4CH\overset{Z}{=}CHCH_2CH\overset{Z}{=}CHCH_2CH_2OH$	Various termites[h]	Matsumura *et al.*, 1972
	(*Z,Z,E*)-3,6,8-Dodecatrien-1-ol $CH_3(CH_2)_2CH\overset{E}{=}CHCH\overset{Z}{=}CHCH_2CH\overset{Z}{=}CHCH_2CH_2OH$	*Reticulitermes virginicus* (Banks) *Reticulitermes santonensis* Feytaud[h] *Reticulitermes flavipes* (Kollar) and other termites[h]	Matsumura *et al.*, 1968 Ritter and Coenen-Saraber, 1969 Matsumura *et al.*, 1972
Farnesol	(*E,E*)-3,7,11-Trimethyl-2,6,10-dodecatrien-1-ol $(CH_3)_2C{=}CH(CH_2)_2C(CH_3)\overset{E}{=}CH(CH_2)_2C(CH_3)\overset{E}{=}CHCH_2OH$	*Zootermopsis nevadensis* (Hagen)[h]	Karlson *et al.*, 1968

Attalure	Methyl 4-methylpyrrole-2-carboxylate 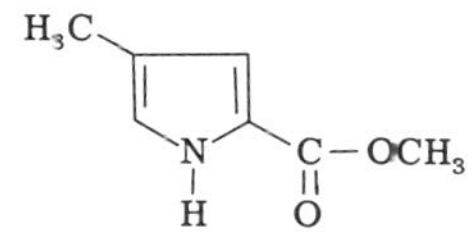	Texas leafcutting ant, *Atta texana* (Buckley) *Atta cephalotes* (L.)	Tumlinson *et al.*, 1972c Riley *et al.*, 1974a
	Methyl 4-chloropyrrole-2-carboxylate 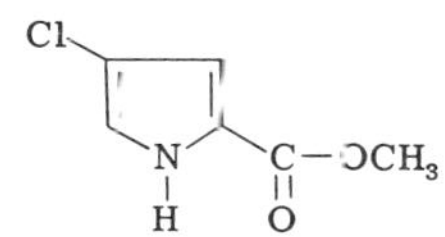	*Atta texana*[h]	Sonnet and Moser, 1973
	Methyl 4-bromopyrrole-2-carboxylate 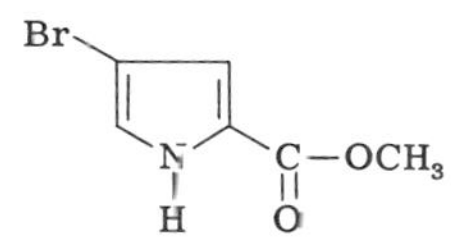	*Atta texana*[h]	Sonnet and Moser, 1973
	Hexanoic acid $CH_3(CH_2)_4COOH$	*Zootermopsis nevadensis* (Hagen)	Karlson *et al.*, 1968
	Dihydroagarofuran $C_{15}H_{26}O$	*Reticulitermes santonensis*[h]	Ritter *et al.*, 1973b
	Tricycloekasantalal $C_{12}H_{18}O$	*Reticulitermes santonensis*[h]	Ritter *et al.*, 1973b

(Continued)

TABLE I (*Continued*)

Common name or trivial name[a]	Chemical name and structure	Species affected	Reference
Neocembrene-A, formerly nasutene	12-Isopropenyl-1,5,9-trimethyl cyclotetradeca-1,5,9-triene 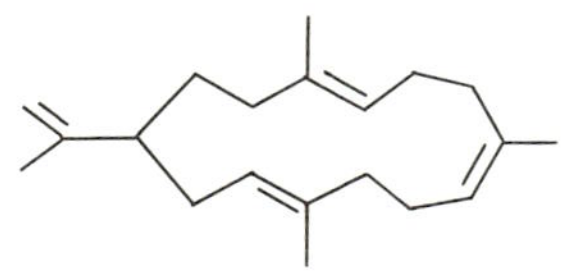	*Nasutitermes exitiosus* (Hill)	Moore, 1966; Birch *et al.*, 1972
	5-Methyl-3-butyloctahydroindolizine 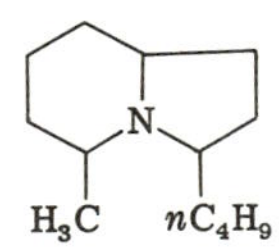	Pharaoh's ant, *Monomorium pharaonis* (L.)	Ritter *et al.*, 1973a
	e. Alarm pheromones		
	3-Octanol $CH_3(CH_2)_4CH(OH)CH_2CH_3$	*Myrmica incompleta* Provancher *Crematogaster peringueyi* Emery	Crewe and Blum, 1970 Crewe *et al.*, 1970
	Isopentyl acetate $(CH_3)_2CHCH_2CH_2OAc$	Honey bee	Boch *et al.*, 1962

	Decanol acetate $CH_3(CH_2)_9OAc$	*Formica subintegra* Emery	Regnier and Wilson, 1971
	Dodecanol acetate $CH_3(CH_2)_{11}OAc$	*Formica subintegra*	Regnier and Wilson, 1971
	Tetradecanol acetate $CH_3(CH_2)_{13}OAc$	*Formica subintegra*	Regnier and Wilson, 1971
	(*E*)-2-Hexenal $CH_3(CH_2)_2CH\overset{E}{=}CHCHO$	*Crematogaster africana* Mayr Bed bug, *Cimex lectularius* L.	Bevan *et al.*, 1961 Levinson *et al.*, 1974
	(*E*)-2-Octenal $CH_3(CH_2)_4CH\overset{E}{=}CHCHO$	Bed bug	Levinson *et al.*, 1974
Citronellal	3,7-Dimethyl-6-octenal $(CH_3)_2C{=}CHCH_2CH_2CH(CH_3)CH_2CHO$	*Acanthomyops claviger* (Roger)	Ghent, 1961
Citral	3,7-Dimethyl-2,6-octadienal $(CH_3)_2C{=}CHCH_2CH_2C(CH_3){=}CHCHO$	*Lestrimelitta limao* (F. Smith) *Acanthomyops claviger*	Blum, 1970 Ghent, 1961

(*Continued*)

TABLE I (*Continued*)

Common name or trivial name[a]	Chemical name and structure	Species affected	Reference
	4-Methyl-2-hexanone 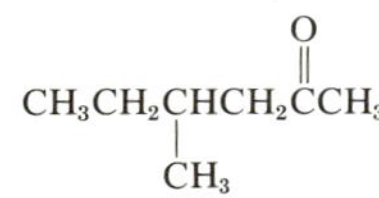	*Dolichoderus clarki* Wheeler	Cavill and Hinterberger, 1962
	2-Heptanone 	*Iridomyrmex pruinosus* (Roger)	Blum *et al.*, 1963
	4-Methyl-3-heptanone 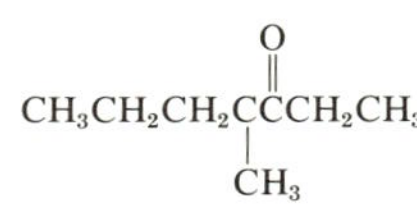	*Pogonomyrmex barbatus* (F. Smith) *Atta texana* *Atta cephalotes*	McGurk *et al.*, 1966 Moser *et al.*, 1968; Riley *et al.*, 1974b Riley *et al.*, 1974b
	2-Methyl-4-heptanone 	*Tapinoma nigerrimum* (Nylander)	Trave and Pavan, 1956
	6-Methyl-5-hepten-2-one 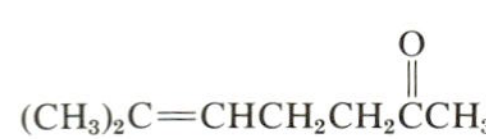	*Dolichoderus scabridus* Roger	Cavill and Hinterberger, 1960

	3-Octanone	*Crematogaster peringueyi* Emery	Crewe *et al.*, 1969
		Myrmica incompleta	Crewe and Blum, 1970
	$CH_3(CH_2)_4\overset{O}{\overset{\|}{C}}CH_2CH_3$		
Manicone	4,6-Dimethyl-4-octen-3-one	*Manica mutica* (Emery)	Fales *et al.*, 1972a
		Manica bradleyi (Wheeler)	Fales *et al.*, 1972a
	$CH_3CH_2\underset{CH_3}{\underset{\|}{C}}HCH{=}\underset{CH_3}{\underset{\|}{C}}{-}\overset{O}{\overset{\|}{C}}CH_2CH_3$		
	2-Nonanone	*Trigona postica* Latr.	Blum, 1970
	$CH_3(CH_2)_6\overset{O}{\overset{\|}{C}}CH_3$		
	2-Tridecanone	*Acanthomyops claviger*	Regnier and Wilson, 1968
	$CH_3(CH_2)_{10}\overset{O}{\overset{\|}{C}}CH_3$		
	2-Methylcyclopentanone	*Azteca* nr. *velox* Forel	Wheeler *et al.*, 1975
		Azteca nr. *nigriventris* Forel	Wheeler *et al.*, 1975
	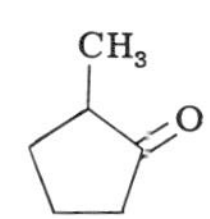		

(*Continued*)

TABLE I (*Continued*)

Common name or trivial name[a]	Chemical name and structure	Species affected	Reference
	Methyl *cis*-2-methylcyclopentyl ketone	*Azteca* nr. *velox*	Wheeler *et al.*, 1975
		Azteca nr. *nigriventris*	Wheeler *et al.*, 1975
	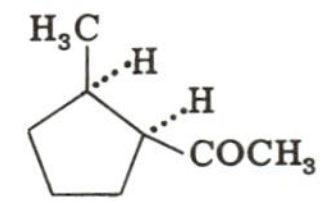		
	Methyl 5-methyl-1-cyclopenten-1-yl ketone	*Azteca* nr. *nigriventris*	Wheeler *et al.*, 1975
	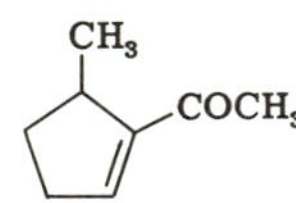		
	Undecane	*Lasius niger* (L.)	Bergström and Löfqvist, 1970
	$CH_3(CH_2)_9CH_3$	*Lasius alienus* (Foerster)	Regnier and Wilson, 1969
(*E*)-*β*-Farnesene	(*E*)-7,11-Dimethyl-3-methylene-1,6,10-dodecatriene	Rose aphid, *Macrosiphum rosae* (L.)	Bowers *et al.*, 1972
		Pea aphid, *Acyrthosiphon pisum* (Harris)	Bowers *et al.*, 1972
	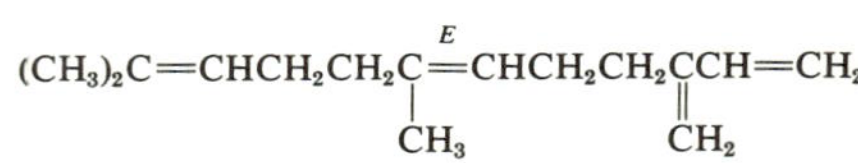	Greenbug aphid, *Schizaphis graminum* (Rodani)	Bowers *et al.*, 1972
		Cotton aphid, *Aphis gossypii* Glover	Bowers *et al.*, 1972
		Green peach aphid, *Myzus persicae* (Sulzer)	Edwards *et al.*, 1973
		Five other aphid species[h]	Bowers *et al.*, 1972
Terpinolene	*p*-Mentha-1,4(8)-diene	*Amitermes herbertensis* Mjöberg	Moore, 1968
	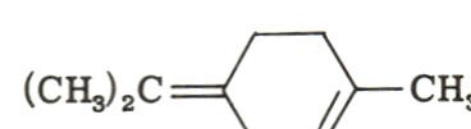		

Limonene	*p*-Mentha-1,8-diene 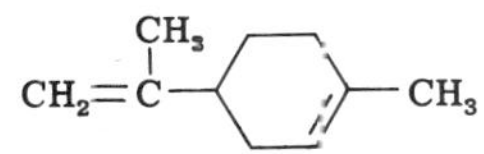	*Drepanotermes rubriceps* (Froggatt)	Moore, 1968
	f. Other related compounds		
(*E*)-2,3-Dihydro-farnesol, terrestrol	(*E*)-3,7,11-Trimethyl-6,10-dodecadien-1-ol $(CH_3)_2C{=}CHCH_2CH_2\underset{\underset{CH_3}{\vert}}{C}\overset{E}{=}CHCH_2CH_2\underset{\underset{CH_3}{\vert}}{C}HCH_2CH_2OH$	Territorial marking secretion, *Bombus terrestris* L. Territorial marking secretion, *Bombus jonellus* K.	Bergström *et al.*, 1968 Bergström and Svensson, 1973
	Ethyl dodecanoate $CH_3(CH_2)_{10}COOC_2H_5$	Territorial marking secretion, *Bombus lucorum* L.	Bergström *et al.*, 1973
	Ethyl (*Z*)-9-tetradecenoate $CH_3(CH_2)_3CH\overset{Z}{=}CH(CH_2)_7COOC_2H_5$	Territorial marking secretion, *Bombus lucorum*	Bergström *et al.*, 1973
	3-Methyl-2-cyclohexen-1-one 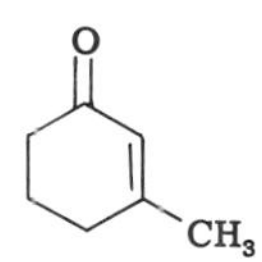	Douglas fir beetle, "antiaggregative pheromone"	Rudinsky *et al.*, 1973
	Undecane $CH_3(CH_2)_9CH_3$	*Blaberus cranifer*, aggregation pheromone	Brossut *et al.*, 1974

(*Continued*)

TABLE I (*Continued*)

Common name or trivial name[a]	Chemical name and structure	Species affected	Reference
	Tetradecane $CH_3(CH_2)_{12}CH_3$	*Blaberus cranifer*, aggregation pheromone	Brossut *et al.*, 1974
Locustol	3-Ethyl-6-methoxyphenol 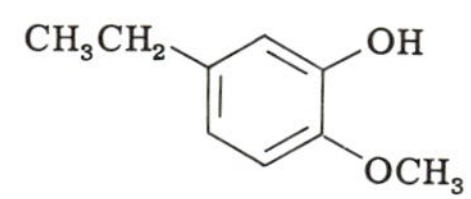	*Locusta migratoria migratorioides*, gregarization pheromone	Nolte *et al.*, 1973
	(*E*)-2-Hexenal $CH_3CH_2CH_2CH\overset{E}{=}CHCHO$	Polyphemus moth, *Antheraea polyphemus* (Cramer), mating stimulant from oak leaves	Riddiford, 1967
(*Z,E*)-α-Farnesene	(*Z,E*)-3,7,11-Trimethyl-1,3,6,10-dodecatetraene $(CH_3)_2C{=}CHCH_2CH_2\underset{CH_3}{C}\overset{E}{=}CHCH_2CH\overset{Z}{=}\underset{CH_3}{C}CH{=}CH_2$	Attractant for codling moth larvae and codling moth oviposition stimulant from apple skins	Sutherland, and Hutchins, 1972, 1973; Wearing and Hutchins, 1973
(*E,E*)-α-Farnesene	(*E,E*)-3,7,11-Trimethyl-1,3,6,10-dodecatetraene $(CH_3)_2C{=}CHCH_2CH_2\underset{CH_3}{C}\overset{E}{=}CHCH_2CH\overset{E}{=}\underset{CH_3}{C}CH{=}CH_2$	Attractant for codling moth larvae and codling moth oviposition stimulant from apple skins	Sutherland and Hutchins, 1972, 1973; Wearing and Hutchins, 1973

Triacetin $CH_2OAc-CHOAc-CH_2OAc$	Synthetic oviposition stimulant for *Heliothis zea* (Boddie)	Jones *et al.*, 1970
Tricosane $CH_3(CH_2)_{21}CH_3$	Compound from *Heliothis zea* scales stimulating parasitization of *H. zea* eggs by *Trichogramma evanescens* Westwood	Jones *et al.*, 1973
13-Methylhentriacontane $CH_3(CH_2)_{11}CH(CH_3)(CH_2)_{17}CH_3$	Compound from *Heliothis zea* triggering host-seeking response by *Microplitis croceipes* (Cresson)	Jones *et al.*, 1971

[a] Some coined names and commercial names are included. Mention of a proprietary product is for information only and does not constitute endorsement by the USDA.

[b] The Japanese beetle attractants are used in ratios of 3:7 phenethyl propionate:eugenol or 9:1 methyl cyclohexanepropionate:eugenol.

[c] Produced by females and attractive to males unless indicated otherwise.

[d] Synergized by host volatiles.

[e] Acts as aggregating pheromone.

[f] More than one compound required for activity or maximum attractancy.

[g] Produced by males.

[h] Found to be effective in field tests or bioassays (also EAG tests for some); not shown to be present in the insect.

[i] Both alcohol and acetate needed for attractancy.

[j] Orfralure contains about 6% *E* isomer.

[k] Certain proportion of opposite geometric isomer required for optimal attractancy.

[l] Regioisomer also required for activity.

[m] Litlure contains the 9,11 and 9,12 isomers in the ratio 10:1.

[n] Gossyplure contains the two compounds in about equal amounts.

(Steiner *et al.*, 1965). In it, the oriental fruit fly was eradicated from a small Pacific island by periodically distributing small porous boards containing methyl eugenol, a powerful attractant for the insect, plus an insecticide. While responding insects were eliminated, other biota were unaffected.

Success with the synthetic chemicals generated interest in the possibility that the sex attractants manufactured by insects themselves to promote their survival might be identified, synthesized, and then used to combat injurious species. The identification in 1959 of the sex attractant pheromone of the silkworm moth (Butenandt *et al.*, 1959), albeit the chemical has no known value in insect control, showed that such identifications could be made. (A pheromone is a chemical emitted by an insect that influences the behavior of another member of its species.) However, it was not until the latter half of the 1960s that microchemical techniques and instrumentation, along with improvements in biological procedures, had advanced enough that identifications of pheromones potentially useful against important species were made; at least fifty such chemicals are now known (see Table I).

Concurrent with the development of sex pheromone technology, other behavior-controlling chemicals were investigated and are now at various stages of development. Some of these are inhibitors of sex pheromones, trail-following pheromones, alarm pheromones, repellents, kairomones (a substance given off by a member of one species that is of benefit to members of another species, e.g., the scent by which a parasitoid locates its host), oviposition attractants, and feeding deterrents. Many such compounds are also included in Table I.

B. Scope of Analytical Problems

Analytical problems encountered in this field can be divided into three major categories. The first relates to isolation and identification of an active natural chemical (or chemicals) that comprises a pheromone or any other active agent. Most attention will be devoted to this category. The second relates to trace analysis, e.g., the determination of the active material(s) in a natural or synthetic product, its residues, degradation products, emission rates. The third category deals with the analysis of synthetic materials, either neat or in formulations; such tests are essential to avoid the presence of impurities or degradation products that can affect performance.

This survey presupposes knowledge of general analytical practices, including chromatography, spectroscopy, and instrumental analysis. Accordingly, details of accepted practices and interpretations will not be given; rather, applications and specialized techniques reported to be used in the field of behavior-controlling chemicals will be discussed to alert the analyst to available technology; references should be consulted for complete details. Coverage will be illustrative rather than exhaustive. No special effort has been made to cite the originators of methodology. Instead, recent references have usually been cited mainly to highlight current methodology and trends. With a variety of methods and techniques cited, the analyst can select the procedures he is best equipped to follow or those he thinks best to solve his problem.

II. ISOLATION AND IDENTIFICATION

A. Background

Much of the difficulty encountered in pheromone identifications arises from the extremely low concentrations at which these materials occur. Thus, to obtain 6 mg of bombykol, Butenandt *et al.* (1959) processed the abdominal tips of 500,000 silkworm moths; Beroza *et al.* (1974) estimated that one female codling moth may contain about 5 ng of its sex attractant; and Moore (1966) concluded that the concentration of trail pheromone in an Australian termite is about one part in four million. Obviously, then, a major problem will be the separation of a minute amount of unknown active material from the mass of contaminants. Even if an adequate supply of insects is available, the small amount of material isolated will require careful manipulation and application of microtechniques to obtain the information needed to ascertain the structure.

In the attempt to identify a physiologically active substance by classic methodology, the first objective is to isolate the active ingredient(s) in pure form. One danger of using insufficiently pure material is that the chemical and spectral probes that are applied to provide the data needed for identification will be confused by signals originating from the extraneous material. It may also happen that a substance thought to be bioactive is isolated and identified, only to be subsequently found inactive, with the observed activity resulting from an impurity. The decision as to whether a compound is pure enough may be a difficult one, but from a practical standpoint (not always correct), purity has been defined as the inability to fractionate by the most rigorous attempted fractionation techniques. For example, pheromones emerging as a single peak on polar and nonpolar columns in capillary gas chromatography are apt to be pure. The many advances made in the sixties in instrumentation and technology—such as high-resolution gas and liquid chromatography, thin-layer and gel chromatography, and ancillary techniques and devices—have greatly facilitated the separation of minute amounts of chemical in a purity suitable for identification.

Reviews dealing specifically with the identification of compounds controlling insect behavior are given in Jacobson (1972), Shorey *et al.* (1968), Beroza (1972), Silverstein (1970), and Bedoukian (1970). Helpful references covering the various techniques include those by Heftmann (1974), Stahl (1969), Kirchner (1967), Kirkland (1971), Karger *et al.* (1973), Silverstein *et al.* (1974), Budzikiewicz *et al.* (1967), and Bentley and Kirby (1972).

B. Bioassay and Biological Considerations

Since the bioassay is generally devised by the entomologist, it will be discussed only briefly to provide an understanding and appreciation of its function. Suffice it to state that progress in identification of a behavior-controlling chemical depends heavily on the availability of a suitable and reliable method of bioassay.

By distinguishing between active and inactive fractions, the bioassay guides the analyst in his efforts to isolate and purify the active components of his source material. Bioassays of behavior chemicals are generally based on the stereotyped response of a given insect species to a specific chemical, e.g., a pheromone. To educe the appropriate response, the bioassay must be built around the natural habits and idiosyncracies of each species. Parameters to be considered in setting up a sex pheromone bioassay, particularly in the laboratory, are age, maturity, sex, and vitality of the insect, whether it has been mated, temperature, relative humidity, timing of test as related to circadian rhythm or light–dark cycle (to establish duration of responsiveness as well as time of peak responsiveness), light intensity, speed of air movement, odor concentration (too great a concentration may repel), possible habituation, air contamination, and means of presenting the chemical to the insect. To cite a simple example, many insects that respond to their sex attractant at night are indifferent to the same stimulus during the day.

If insects can be reared and made to respond to lures in the laboratory, a laboratory bioassay has certain advantages: conditions of the bioassay, such as light, temperature, and humidity, may be regulated; nocturnal insects can be made to respond during the day by manipulating their light–dark cycle; tests can often be rapid and can be well replicated; bioassays can be conducted throughout the year instead of only when insects are available in the field; insects can be reared under controlled conditions so their response will be more uniform. But laboratory testing can also have some disadvantages or pitfalls: laboratory-reared insects readily succumb to disease; their responses may not parallel those of wild insects in the field; more important, the criteria chosen as indicators of activity may not be appropriate. Thus, in several studies aimed at isolating an attractant, laboratory bioassays based on a wing-flutter or excitation response, which is part of precopulatory behavior, apparently provided improper guidance in the purification process because the compound isolated did not attract. For sex pheromones to be useful, they must attract in the field. Indeed, attraction in the field is the ultimate test of efficacy. All attractants found in laboratory testing must prove themselves in the field in competition with natural odors, colors, light conditions, and other stimuli. While direct field testing of fractions can be employed if insects cannot be reared and laboratory bioassay is therefore not feasible, adequate control of such tests often is difficult because of changes in the weather, season, insect density, predation, etc. Usually a standard attractant, if available, is included in such field bioassays to minimize the many variables encountered.

Even with a good attractant at hand, many details usually have to be worked out to secure good attraction to a trap in the field. Design of the trap can be crucial; e.g., some insects like the Mexican fruit fly, *Anastrepha ludens* (Loew), enter a trap only from the bottom, while other species prefer to enter from the side. The method of entrapment may be an adhesive in the trap, a volatile insecticide, or a mechanical baffle. The bait dispenser can be anything from a piece of filter paper to a cylinder of corrugated cardboard, rubber bands, or adsorbent cotton. Wick size and geometry affect volatilization rate and lure capacity. Other parameters affecting trap performance are height above ground, amount of lure emitted, wind direction and speed, and distance between traps. However, even the most efficient trap will not function well if it is placed in an area of restricted air movement, such as in a hollow or surrounded by dense foliage.

Bioassay results, from both laboratory or field tests, can be seriously compromised by a pheromone's complexity. The multicomponent nature of many, if not most, pheromones has become evident in recent years. For example, as many as ten species have been reported to respond to the same chemical (Roelofs and Comeau, 1971a), and at least several of the species coexist in the same vicinity. Therefore, other chemicals undoubtedly are present in the pheromones of those species to modify behavioral response. In effect, each species utilizes its own unique blend of chemicals, which helps maintain the reproductive isolation of the species. The behavior-modifying, or secondary chemicals, as they are sometimes called, can be synergists (improve attraction without being attractive themselves) or inhibitors (mask, diminish, or cancel attraction), or they can be an obligatory constituent of the pheromone; i.e., the ingredients of the pheromone are individually inactive, and only the combination is active. With some multicomponent combinations, only certain ratios of the components will attract, and, of course, it is desirable to utilize the ratio giving optimum attraction. In one subtle combination of this kind, which may be quite common, a small amount of stereoisomer opposite to that of the primary pheromonal ingredient is required for optimum attraction; what is more, the percentage of the isomer has to be within narrow limits or attraction does not occur (Beroza *et al.*, 1973c; Klun *et al.*, 1973). In another instance, the situation is reversed; the presence of even 1% of an opposite geometrical isomer seriously interferes with the attraction of the pheromone (Tumlinson *et al.*, 1974b).

Multicomponent pheromones, particularly those with ingredients that are individually inactive, pose difficulties in utilizing the behavioral-type bioassay to choose the appropriate fractions in the isolation process. The multiplicity of components becomes evident when an active extract becomes inactive on fractionation but regains activity upon recombination of fractions; this gives rise to the frequent caveat against discarding any fraction before checking out the possibility that it may contain some component necessary for optimal activity.

Bioassays may also be conducted by an electrophysiological procedure which depends on the response of an insect antenna (intact or excised) to an odor. In this procedure, devised and developed by Schneider (1957, 1969), electrodes are connected to the base and sensory hairs of an insect antenna (its organ of smell), and the reaction of antennal receptor cells, suitably amplified, are registered on an oscilloscope; exposure of the antenna to a chemical such as a sex pheromone produces a characteristic pattern called an electroantennogram (EAG), the shape and intensity of which depends upon the stimulus. Thus, when chromatographic fractions from an attractive extract (e.g., from abdominal tips of female Lepidoptera) are tested with an appropriate antenna, the greatest response is given by the fractions containing the pheromone; these are then selected for further purification. The high specificity of the antennal response may be judged from the fact that closely related isomers of a pheromone produce responses much weaker than the pheromone. The method has the advantages of generally requiring far fewer insects than the behavioral bioassay and not involving all the specialized criteria to educe the proper behavioral response for each species. The EAG procedure does have the disadvantage of giving responses to chemicals other than the target ones, so some judgment in interpreting EAG results is necessary. Furthermore, the results of EAG tests must be verified eventually with a suitable behavioral test, preferably in the field. When applicable, this technique is one of the most efficient means of selecting active fractions for isolation and purification.

Another important aspect of the EAG bioassay is that individual components of a multicomponent pheromone elicit an electrophysiological response; in contrast, a behavioral bioassay may require a combination of components for activity. Aside from helping find primary pheromonal compounds that are individually inactive, the EAG can be potentially invaluable in helping elucidate the role of secondary pheromonal components; e.g., synergists and inhibitors, which are generally EAG active, although seldom as active as the primary pheromonal components required for attraction.

While the foregoing discussion has centered on sex attractant pheromones, similar considerations must be kept in mind in devising a bioassay to guide the isolation of other behavior-controlling chemicals, whether the action under investigation be repellency, stimulation or attraction for oviposition, attraction of a parasite to its host, induction of trail following, feeding deterrence or stimulation, or alarm.

C. Source Material and Preliminary Separation

1. PHEROMONE SOURCES

Before an appropriate separation procedure can be devised, experimentation is needed to select the best source material. With an attractant pheromone produced by one of the sexes, active extracts have sometimes been obtained without separating the sexes (e.g., Kuwahara *et al.,* 1971a,b); however, more often the sexes are separated to eliminate extraneous material that may complicate cleanup of the extract or interfere with observation of activity in the bioassay. Further exclusion of inactive material from pheromone-containing tissues can be accomplished by selective dissection; e.g., by clipping terminal abdominal segments ("tips") of Lepidoptera or excising hind guts of beetles. Unfortunately, general rules are not available in regard to the amount of labor that can profitably be expended in physical separation as a means of cleanup, and the procedure adopted often depends upon the options at hand and the ingenuity of the investigator, with the amount of information available being important. Perhaps one general rule that can be applied is that many pheromones are labile and their vulnerability to degradation requires that extra or unnecessary steps in the isolation should be avoided. Other than insects themselves, sources of sex pheromones have been insect feces (Tumlinson *et al.,* 1968, 1969; Carlson *et al.*, 1971), frass and boring dust produced by bark beetles (Silverstein *et al.*, 1968; Peacock *et al.*, 1973; Young *et al.*, 1973b; Byrne *et al.*, 1974), and filter paper used to line the containers in which insects are housed (see Section II,C,5).

The stage of insect development or status must be considered for optimum yield. With many insects, pheromone production is greatest 3 or 4 days after emergence from pupation (eclosion), although Steinbrecht (1964) reported that the pheromone content of the female silkworm glands was highest at eclosion. Diurnal variations in the rate of pheromone production have been noted, and it is sometimes advantageous to process insects at a specific time in the light–dark cycle. For example, extracts of female beet armyworm tips cut near the end of the dark cycle were more active than those cut about 3 hours after the start of the light cycle (Brady and Ganyard, 1972). Since pheromone production or content terminates or declines after mating, virgin insects are commonly used.

Similar considerations are involved in obtaining suitable sources for other types of active compounds. Trail pheromones have been obtained from ant parts (Riley *et al.*, 1974a), whole ants (Ritter *et al.*, 1973a) or termites, as well as from fungus-infected wood (Matsumura *et al.*, 1968). Compounds attracting parasites to their host (kairomones) have been isolated from insect feces (Jones *et al.*, 1971) and moth scales (Jones *et al.*, 1973). Alarm pheromones and defensive compounds have been isolated from whole insects, insect parts, or specific glands. In some cases, the gland contents themselves are obtained by provocation of the insects, by gentle pressure ("milking"), or by immersing aphids in hexane to cause them to release their cornicle droplets (Wientjens *et al.*, 1973). The secretions can then be collected in capillary tubing or on filter paper for further study (Benn *et al.*, 1973; Blum *et al.*, 1973; Brand *et al.*, 1973; Duffield *et al.*, 1974).

Like the trail compounds from fungus-infected wood, active compounds can be of noninsect origin; some may even be essential to the action of a pheromone. A host factor, without which polyphemus moths, *Antheraea polyphemus* (Cramer), will not mate, has been isolated from oak leaves (Riddiford, 1967). A larval attractant and oviposition stimulant is found in apple skin (Sutherland and Hutchins, 1972; Wearing and Hutchins, 1973). Methyl eugenol, used as a synthetic attractant for male oriental fruit flies, is found in blossoms of the golden shower, *Cassia fistula* L. (Kawano *et al.*, 1968). Mosquito repellent lipids (Skinner *et al.*, 1970) and a mosquito attractant (Acree *et al.*, 1968) have been isolated from solvent washes of human skin. A compound inducing oviposition by the corn earworm was found to be a component of felt-tipped marking pens (Jones *et al.*, 1970).

2. EXTRACTION

Extraction has been the most generally used procedure for the preliminary separation of active material. Depending on the substrate, the method employed may range from a brief solvent rinse of the whole insect to exhaustive Soxhlet extraction of a macerated substrate. (Soxhlet extraction is avoided if the active principle is affected by heat.) Tissues have been broken up by grinding with a mortar and pestle (sometimes with sand or sodium sulfate added) or in a tissue grinder. Freezing in liquid nitrogen has been used to make it possible to grind female abdominal tips to a fine powder (Henderson *et al.*, 1973). Very often, the substrate is homogenized with solvent in a laboratory blender such as the Waring Blendor™. Ultrasonic cell disruption has also been used (Roelofs *et al.*, 1971c).

The choice of solvent may be critical to the success of the study. Most of the sex attractant pheromones tend to be nonpolar and are soluble in the common organic solvents. However, variable results have been reported when a series of solvents has been tested, and the solvent of choice usually is found by trial. Gaston *et al.* (1966) found each of six solvents (ether, methylene chloride, carbon disulfide, methyl formate, benzene, and hexane) to be equally effective in extracting pheromone from cabbage looper tips. Of seven solvents tested by Roelofs and Feng (1967) for the extraction of the sex pheromone of the redbanded leafroller, all gave active extracts but the ether and methylene chloride extracts were more potent at low concentrations than were those with acetone, benzene, chloroform, methanol, and 95% ethanol. Similarly, methylene chloride extracts of the tips of female pine moths, *Dendrolimus spectabilis* Butler, were the most active in laboratory and field trapping tests, with acetone extracts somewhat less active and benzene, ether, hexane, and ethyl alcohol extracts showing much less activity (Sakurai *et al.*, 1973). Methanol was found to be more effective than methylene chloride, pentane, or distilled water in extracting an attractant pheromone from females of a mushroom-infesting gnat, *Lycoriella mali* (Fitch) (Girard *et al.*, 1974).

The solvent most commonly used for these extractions has been methylene chloride; it can be purified adequately, it extracts the desired materials efficiently, it is sufficiently low-boiling to volatilize without excessive heating, and in the large volumes often required, it does not constitute a fire hazard. For work with acid-sensitive materials, it may be advisable to shake the methylene chloride with potassium carbonate (e.g., 15 gm/liter) to remove traces of acid (Tumlinson *et al.,* 1972b).

In general, the least efficient solvent that will still adequately extract the compounds of interest is preferred because the extracts then contain less extraneous material.

3. LOW-TEMPERATURE PRECIPITATION OF IMPURITIES FROM EXTRACT

Inactive lipids have been removed from acetone solutions of insect pheromones by cooling, usually between −20°C (Solomon *et al.,* 1972) and −70° (Klun, 1968); several days may be required for complete precipitation. The precipitate is removed by decantation, filtration, or centrifugation and washed with cold solvent. This series of steps is usually repeated several times.

Dahm and Röller (1970) devised a refined purification technique for juvenile hormone (JH) which made it possible to isolate JH from a single insect abdomen; precipitation from diethyl ether–methanol (1 : 1) at −78°C was one step in this procedure. Metzler *et al.* (1971), in studying the recovery of JH by adding radiolabeled JH to the crude extracts, found that the major loss of JH in the procedure occurred in this step, where 30% of the JH coprecipitated. Changes in the solvent ratio could decrease this loss, but subsequent purification then became more difficult. This low-temperature precipitation may therefore be inefficient as a separation process.

Birch *et al.* (1972) used saponification to remove inactive lipids from extracts of a termite trail pheromone. After repeated washing with water (with care to avoid undue agitation, which could cause intractable emulsions) the concentrated petroleum ether extracts were chilled at 4°C for several hours. The flocculent soaps which precipitated were removed by centrifugation.

4. DISTILLATION

For substances not adversely affected by elevated temperatures, distillation may be an effective means of separating and concentrating active material. Steam distillation of aqueous suspensions or homogenates, followed by extraction of the aqueous distillates with a suitable solvent, has been used to obtain sex pheromones from insects (Hedin *et al.,* 1974; Mody *et al.,* 1973), an insect attractant from flowers (Kawano *et al.,* 1968), and a trail substance from termites or fungus-infected wood (Ritter and Coenen-Saraber, 1969), to list a few examples. Henderson *et al.* (1973) found steam distillation yielded additional active sex pheromone from the pulp remaining after methylene chloride extraction of tips of a female moth.

Direct distillation is seldom used. Riddiford (1967) collected the polyphemus host factor by vacuum distillation of freshly ground oak leaves. Morgan and Wadhams (1972), in studies on constituents of the Dufour's glands of *Myrmica rubra* L., placed freshly killed ants in a short-path vacuum distillation apparatus; volatile compounds were trapped on activated charcoal for subsequent elution and analysis. Bergström and co-workers (Bergström *et al.*, 1968, 1973; Bergström and Löfqvist, 1970) have separated volatile materials from various glands of ants and bumblebees by vacuum distillation either in a round-bottom flask connected to a glass spiral immersed in an alcohol–Dry Ice mixture (Andersson *et al.*, 1967) or in a small sealed tube heated at one end and cooled at the other (Bergström and Löfqvist, 1968; Ställberg-Stenhagen and Stenhagen, 1968).

The precolumn separation of volatile materials from a solid sample before gas chromatography may also be regarded as a distillation process. Bergström (1973) has described a precolumn tube that has been used with a variety of materials, such as abdominal tips of ants (Bergström and Löfqvist, 1970), heads of bees (Bergström *et al.*, 1973), or butterfly wings (Bergström and Lundgren, 1973). Carrier gas passing through the heated precolumn tube transports the volatile materials to the cooled anterior portion of the gas chromatography column. After the transfer is complete, the precolumn tube is replaced and gas chromatography conducted as usual. Morgan and Wadhams (1972) have used a solid injection technique for gas chromatography of volatile material from individual Dufour's glands of ants. In another application, a sex pheromone inhibitor was identified in a volatile material similarly obtained from two abdominal tips of female eastern spruce budworms (Weatherston and Maclean, 1974).

5. RECOVERY OF VOLATILES FROM AIR

Because most of the compounds of interest are volatile, collection of active material from air that has been passed through containers in which insects are housed is a frequent practice and is sometimes more successful than the direct extraction of insects or other materials. Experience has shown that when the method is applicable, it has the advantage of yielding material requiring far less cleanup. Furthermore, insects can often be "milked" of their emissions over long periods and thereby provide more attractant than can be obtained from extraction of tissues. Thus, Yamamoto (1963) collected sex attractant from female cockroaches by directing air passed through 10-gallon cans containing about 1000 insects into a condensing flask immersed in an alcohol–Dry Ice bath; a number of groups (Jacobson *et al.*, 1973; Jones *et al.*, 1965;

Leyrer and Monroe, 1973) have used modifications of this system. Complete trapping of the air and its volatiles at about −193°C followed by controlled evaporation yielded essentially quantitative recovery of components, according to Browne *et al.* (1974). These condensates, as well as those from conventional cold-trapping, contain large amounts of water that sometimes clog the trap. The condensed compounds are generally salted out with excess sodium chloride, extracted with solvent, dried, and appropriately concentrated.

Removal of volatiles from an air stream by adsorption often has the advantage of allowing water vapor to pass. Columns or tubes containing gas chromatography packings (Silverstein and Rodin, 1966) were reported superior to cold traps in tests with model compounds; in trapping nonanal, incorporation of an antioxidant in the adsorbent reduced loss by oxidation. Collection in tubes containing Porapak Q™ preconditioned by heat treatment and solvent extraction has proven highly effective (Byrne *et al.,* 1975); the volatiles are recovered by solvent extraction. Similar trapping arrangements have been used for bark beetle pheromones (Rudinsky *et al.,* 1973, 1974; Libbey *et al.,* 1974).

Röller *et al.* (1968) collected a sex attractant of the greater wax moth by drawing air sequentially over male moths and a battery of glass plates coated with purified fat containing an antioxidant. The attractant was isolated from the fat and identified as undecanal. Later, Leyrer and Monroe (1973) showed that nonanal, another component of the pheromone, was missed, probably because of the inefficiency of the collection system.

Although aeration and removal of volatiles from the air stream is the more usual procedure, volatile substances can also be obtained by adsorption from air in a static system; transfer of the insects' secretions by direct contact with the adsorbent may also be involved in such systems. Weatherston *et al.* (1971) obtained the sex attractant of the eastern spruce budworm from ether rinses of the glass containers that had held virgin females. Filter paper liners of insect containers are often a source of pheromone (e.g., Bell *et al.,* 1974; Brady, 1973; Burkholder and Dicke, 1966; Jones *et al.,* 1965). To study factors affecting the rate of emission of sex pheromones by female Indian meal moths, Nordlund and Brady (1974) moved the moths into clean jars with fresh filter paper every 12 hours; rate of emission was then estimated by bioassay of the filter paper extracts. In another variation of this technique, parts of insects or flowers were placed in contact with a gas chromatography packing coated with a silicone high-vacuum grease (Bergström, 1973). After several days the packing was placed in the precolumn tube (Section II,C,4) for analysis by gas chromatography–mass spectroscopy. In this way, isoprenoid compounds active as hymenopteran attractants and excitants could be identified from orchids when amounts obtained by direct analyses of the flowers were inadequate for mass spectral determination.

6. GENERATION OF ADDITIONAL MATERIAL FROM PRECURSORS

In the course of isolation and in preliminary probes to elucidate the chemical structure of pheromones (see Section II,D), the functional groups present in target molecules may often be determined. Knowledge of the presence of a functional group can then be utilized to attempt to generate additional active material from an inactive precursor or related material present in the source material. Thus, when the sex pheromone of the gypsy moth was shown to be an epoxide but there was insufficient material available for characterization, a compound with a double bond in place of the epoxide was postulated as a likely precursor. A search for the double-bonded compound in the inactive monoolefin fraction did indeed turn up a compound that became highly active upon epoxidation; this fraction not only yielded sufficient precursor for identification, it was considerably simpler to identify the olefin than the epoxide. When the structure of the precursor became known, the identity of the pheromone was clear (Bierl *et al.,* 1970). Similarly, active acetates could be obtained from the inactive alcohol fractions of pink bollworm extracts (Bierl *et al.,* 1974). Butler *et al.* (1972) obtained attractive material, apparently identical to the attractant emitted by female *Ceramica picta* (Harris), by saponification of an inactive extract of the males, followed by acetylation. In light of several other observations of inactive fractions that become active upon reaction (e.g., Kuwahara *et al.,* 1971b; Solomon *et al.,* 1972; Yushima *et al.,* 1973), activation of inactive fractions by reaction should be considered as a possible source of active material in other applications, especially if the starting material is expensive or unavailable in sufficient quantity.

Examples of precursors or related compounds that one might expect for behavior-controlling chemicals based on functional groups shown to be present are given in Table II.

7. CHOICE OF METHOD

Determination of the method of choice for the preliminary separation of a desired active material has generally been empirical; a method satisfactory in one case may yield no active material in another. Some examples from sex pheromone studies will illustrate this point.

In early work on the boll weevil (Tumlinson *et al.,* 1968), aeration of males with cold trapping of volatiles yielded no active material, and

TABLE II
FUNCTIONAL GROUPS OF BEHAVIOR-CONTROLLING CHEMICALS AND POSSIBLE PRECURSORS

Functional group	Possible precursors
Epoxide	Olefin
Ester	Alcohol, phenol, carboxylic acid
Alcohol	Ester, carboxylic acid, aldehyde, ketone
Carboxylic acid	Ester, alcohol, salt
Aldehyde	Alcohol, carboxylic acid, acetal
Ketone	Alcohol, ketal
Amine	Amide, salt
Methyl ether	Phenol, alcohol
Methyl ester	Carboxylic acid

material obtained by adsorbent trapping was only slightly active; best results were obtained by steam distillation of concentrated methylene chloride extracts of weevils or feces. Likewise, materials from air that had been passed over female moths of the carpenterworm trapped on activated charcoal or in a cold trap were considerably less active than extracts of abdominal tips or of filter paper on which the insects had rested (Solomon *et al.,* 1972). On the other hand, in studies on the sex pheromone of the Mediterranean fruit fly (Jacobson *et al.,* 1973), solvent extraction of males or of filter paper to which the males had been exposed gave inactive or weakly active extracts; steam distillation of aqueous homogenates of males yielded highly active material, but better results were obtained by aeration and cold trapping.

Extraction of the abdominal tips of female moths is often considered the best starting point for lepidopteran sex pheromones. However, Brady *et al.* (1971) found that extracts of filter paper that had been in contact with female Indian meal moths for 3 days contained considerably more sex pheromone than extracts of abdominal tips of the same number of females. Extracts of female abdominal tips of the eastern spruce budworm not only failed to elicit any response from male moths, they reduced the response to active extracts from jars which had housed the females (Sanders and Lucuik, 1972); the diminished response was caused by a pheromone inhibitor subsequently found in the abdominal tips (Weatherston and Maclean, 1974).

A method need not be ruled out even if inactive material is obtained, because purification may remove inhibiting substances. Thus, tips from female false codling moths yielded active extracts, while an extract of mixed populations of male and female moths showed little pheromone activity; yet chromatography of the latter extract on alumina gave a highly active fraction; for preparative purposes, it was possible to work with extracts of mixed populations and avoid the difficulties of separating pupae under the microscope and clipping tips (Read *et al.*, 1974).

The many variables encountered when working with bioactive materials make it essential for the chemist to work closely with the biologist at all stages of the isolation.

D. Preliminary Probes

With the aid of a reliable, sensitive bioassay, simple microchemical tests are often carried out early in an investigation with small portions of the active extract or fractions, and important information on the functional groups in the active molecules may be obtained. The general idea is that the chemical test destroys or generates active material by a reaction with one of the functional groups in the target molecule, and the bioassay on the product indicates the resulting loss or gain in activity. (An important proviso is that the chemical test must be allowed to go to completion in order to secure a definitely positive or negative bioassay.) For example, if a crude pheromonal extract is treated with alkali, and the bioassay shows a loss of activity, a tentative conclusion may be that the pheromone contains an ester. Should activity then be restored by acetylation of the neutral fraction of the saponified product (presumably the alcohol moiety of the ester), the pheromone is most likely an acetate. In another test, ozone is passed into a solution of the extract in hexane or carbon disulfide; loss of activity indicates the presence of unsaturation in the pheromone or in one of its essential components.

Table III lists some of the simple chemical tests that have been used to shed light on the presence or absence of a variety of functional groups. Undoubtedly other chemical tests will occur to the analyst depending on the functional groups he believes are present.

Because the microchemical tests are performed on impure materials, conclusions drawn from these tests are best considered tentative and should be confirmed by other tests. Chemical reactions can also be conducted on thin-layer chromatographic (TLC) plates or during gas chromatography (GC) followed by bioassay. With thin-layer chromatography, the sample spotted on a plate may be superimposed with a chemical reagent (or exposed to vapors or light) that reacts with a given functional group; after an appropriate reaction period, the plate is developed. The adsorbent scraped from the R_f zone to which the pheromone normally migrates is then eluted, and the eluted material is bioassayed. Loss of activity indicates the reaction occurred and the target functional group is present. This procedure was used in deciphering the structure

TABLE III
TYPICAL FUNCTIONAL GROUP TESTS USING A SENSITIVE BIOASSAY TO DETERMINE ACTIVITY OF THE PRODUCT

Functional group present	Chemical test	Effect on activity
Double bond	A. Hydrogenation, ozonolysis, or oxidation with $KMnO_4$, perbenzoic acid	Lost
	B. (1) Bromination	Lost
	(2) Treatment with Zn	Regained
Ester	A. Saponification	Lost
	B. Treatment with $LiAlH_4$	Lost
Acetate	Acetylation of neutral fraction of saponified product (A, above)	Regained
Methyl ester	Treatment of acid fraction of saponified product (A, above) with diazomethane	Regained
Carboxylic acid	A. (1) Extraction with alkali	Lost
	(2) Acidification of alkaline extract	Regained
	B. (1) Treatment with diazomethane	Lost
	(2) Hydrolysis of methylated product	Regained
	C. (1) Treatment with $LiAlH_4$ or $NaBH_4$	Lost
	(2) Oxidation with CrO_3 and H_2SO_4	Regained
	D. ZnO subtraction loop (GC)[a]	Lost
Alcohol	A. Acetylation	Lost
	B. Boric acid subtraction loop (GC)[a]	Lost
Carbonyl	A. Treatment with $LiAlH_4$, semicarbazide HCl, 2,4-dinitrophenylhydrazine HCl, or hydroxylamine HCl	Lost
	B. Benzidine subtraction loop (GC)[a]	Lost
Aldehyde	*o*-Dianisidine subtraction loop (GC)[a]	Lost
Acetal	Treatment with mineral acid	Lost
Epoxide	A. Treatment with dilute acid (H_3PO_4), HOAc, or $LiAlH_4$	Lost
	B. Phosphoric acid subtraction loop (GC)[a]	Lost
	C. Treatment with HIO_4	Lost
Amine	(1) Extraction with acid	Lost
	(2) Neutralization of acid layer	Regained

[a] See Section II,F,3b.

of the gypsy moth sex pheromone (Bierl *et al.,* 1970). Several portions of the active extract on a TLC plate were treated *in situ* with various reagents, e.g., neutralized semicarbazide hydrochloride solution to react with a carbonyl group and phosphoric acid, which reacts with epoxides. Only the reaction for epoxide was positive; i.e., activity was destroyed by the reaction. This result was confirmed by a chemical reaction conducted within a gas chromatograph in what is called a subtractive loop (Bierl *et al.,* 1969). The loop, to be described later (Section II,F,3b), is a short tube containing a reactive chemical that retains compounds with certain functional groups and allows other compounds to pass. The active fraction was injected into a gas chromatograph with each of the various subtraction loops, and the effluent was allowed to impinge upon 30 male gypsy moths held on a rack by their wings. With no loop, as little as 1 pg of attractant caused about half of the moths to exhibit characteristic copulatory excitment at the time the sex attractant emerged. Only the loop that retained epoxides gave a positive reaction; i.e., with all other loops, the moths became excited by the GC effluent. A summary of chemical reactions used in combination with TLC for functional group identification is given in Table IV. For a more extensive listing of

TABLE IV
TYPICAL REACTIONS ON TLC PLATES FOR FUNCTIONAL GROUP IDENTIFICATION

Reagent	Reaction	References
Pd or Pt on plate, H_2	Hydrogenation (e.g., unsaturated fatty acids)	Elgamal and Fayez, 1967
$LiAlH_4$	Reduction of carbonyls and esters to alcohols	Miller and Kirchner, 1953
Br_2 (as vapor, in mobile phase, or on spot)	Bromination (e.g., unsaturated compounds, phenols)	Cargill, 1962; Copius Peereboom and Beekes, 1962
CrO_3 in acetic acid	Oxidation of alcohols to carbonyl compounds	Rusiecki and Henneberg, 1963
Peracetic acid	Unsaturated acids oxidized and separated from saturated ones	Mangold, 1959; Malins and Mangold, 1960
H_2SO_4	Dehydration of alcohols (not primary) to olefins	Mathis, 1965; Miller and Kirchner, 1953
HCl (as vapor or in solution)	Hydrolysis of esters	Baggiolini and Dewald, 1967; Curtius and Müller, 1968
KOH	Saponification of esters	Mathis, 1965; Miller and Kirchner, 1953
Acetyl chloride or acetic anhydride–pyridine	Acetylation of alcohols	Holloway and Challen, 1966
3,5-Dinitrobenzoyl chloride	Alcohol derivatization	Minyard *et al.*, 1967
BF_3-MeOH or CH_2N_2 in ether	Methylation of acids	Holloway and Challen, 1966; Riess, 1965
2,4-Dinitrophenylhydrazine HCl, semicarbazide HCl, etc.	Carbonyl derivatization	Miller and Kirchner, 1953; Tumlinson *et al.*, 1967
Phosphoric acid	Hydrolysis of epoxides	Bierl *et al.*, 1971

reactions used in conjunction with chromatographic procedures, see Beroza (1970a).

Chromatographic mobility (migration of a compound in a chromatographic system) may also serve to detect or rule out the presence of certain functional groups. Thus, the polarity of compounds with various functional groups differs, and the migration rates of these compounds when chromatographed on polar phases differ concomitantly; e.g., with silica gel TLC, the lower the R_f, the greater the polarity; or in GC with a polar phase (other things being equal), the greater the polarity of a functional group and the more functional groups present, the greater the retention time. (See Section II,F,1a for a discussion of Kovats' retention indices in regard to GC mobility of compounds.) Usually, the mobility of model compounds and of the unknown active material are compared (e.g., by trapping and bioassaying fractions of the unknown from a GC column) to seek evidence for the presence of one or more functional groups. It helps to know the properties of the various chromatographic media, so the proper medium will be chosen; e.g., on silica gel TLC, the number of double bonds present will not affect the R_f value appreciably, while retention time in GC will be changed on a polar but not on a nonpolar phase. Thus, the presence of double bonds in a compound will cause no appreciable interference in silica gel TLC, while other functional groups will register their effects on the R_f value.

Once the functional groups are known, GC can provide additionally an estimation of the size of the molecule (see Section II, F, 1a), i.e. whether the compound is in the C_{12}, C_{14}, or C_{16} range.

E. Isolation and Purification

For the removal of contaminants and the resulting enrichment of the target component or components, as well as for the final purification, a variety of processes are available. The present tendency is to utilize the high-resolution processes—GC, high-pressure liquid chromatography (LC)—once the bulk of the extraneous material has been eliminated. When applicable, the high-resolution processes can sometimes provide pure or almost pure material after only three or four separation steps.

Information on the nature of the target compounds derived from preliminary probes (Section II,D.) as well as from any preliminary separation is, of course, utilized in devising separation schemes. With valuable material, separation trials are generally carried through with small amounts of material to be certain that the active ingredients do not lose their potency in the purification process. In the workup of larger quantities, all fractions, including inactive ones, are retained. As already noted, the inactivity of a fraction need not mean absence of active material, since an inhibitor may be present or more than one component may be needed for activity. Further, it is becoming increasingly apparent that pheromonal emissions are complex, and it is important to be able to check out combinations of fractions to search for synergists, inhibitors, or secondary chemicals that can be used to enhance or inhibit activity of the major component(s).

1. EXTRACTION

Partitioning a crude extract into acid, neutral, and basic fractions is common in purification processes. For example, extraction with dilute NaOH may be an effective way of removing inactive acidic material (Rodin *et al.,* 1969) or of separating out active substances bearing phenolic or carboxylic functional groups (e.g., Butler *et al.,* 1961; Berger, 1972; Chow *et al.,* 1972; Silverstein *et al.,* 1967; Skinner *et al.,* 1970). In the latter case, the active material is recovered by careful acidification of the aqueous extract and extraction with a lipophilic solvent. In the recovery of a phenol, the solution may be neutralized with gaseous CO_2 before extraction with the organic phase.

Partitioning between two immiscible solvents, by taking advantage of differences in polarity between an active substance and some of its contaminants, may also serve as a means of concentration or of transferring material from one solvent to another. The latter transfer is often accomplished by addition of water to a solution of the active material in a water-miscible solvent such as alcohol or acetone, followed by extraction with a lipophilic solvent. The technique of countercurrent distribution, a systematic procedure for repeated extractions between immiscible phases, has not been used for the concentration of the behavior-controlling chemicals.

In the recovery of compounds from an aqueous phase, especially those with appreciable water solubility, or in recoveries from mixtures containing much water (e.g., a steam distillate or a condensate of airborne volatiles), salting out by the addition of excess NaCl, followed by solvent extraction, is usually effective (e.g., Browne *et al.,* 1974). Extracts are usually dried (e.g., with anhydrous sodium sulfate) before concentration of the solvent. A continuous extraction device for the extraction of a steam distillate with recirculated ether was used in obtaining the pheromone of *Nudaurelia cytherea cytherea* (Henderson *et al.,* 1973).

2. DISTILLATION AND SUBLIMATION

Because thermal degradation must be avoided, vacuum distillation of extracts in a short-path distillation apparatus is a common cleanup procedure; it has been used in the purification of termite and ant trail pheromones (Hummel and Karlson, 1968; Riley *et al.,* 1974a), of coleopteran sex pheromones and aggregation pheromones (Borden *et al.,* 1968; Rodin *et al.,* 1969; Silverstein *et al.,* 1967, 1968; Young *et al.,* 1973b), and of a mosquito repellent from the skin of humans (Skinner *et al.,* 1970).

Similarly, vacuum sublimation has been used to isolate the pheromone of a grass grub beetle (Henzell and Lowe, 1970). Meinwald and Meinwald (1966) found that preferential vacuum sublimation of extracts of male "hair pencils" enabled them to avoid column chromatography in isolating butterfly courtship pheromones. Block and McChesney (1974), in attempting to isolate the aggregation pheromone of a cockroach, found the active material to be too volatile for sublimation at 60°C.

The most common use of distillation is in the removal of excess solvent from extracts before proceeding to subsequent steps. Although this concentration can often be accomplished by conventional distillation or evaporation, the active materials may be heat-sensitive, and it is then necessary to concentrate the extract at low temperatures by vacuum distillation in a rotary evaporator. A fractionating column is sometimes used to avoid loss of active material along with the solvent in a distillation. When the succeeding step is liquid chromatography, it is sometimes unnecessary to remove the solvent; e.g., an acetate in hexane may be chromatographed directly on a silica gel column because the acetate accumulates at the head of the column.

Steam distillation, already mentioned as a means of preliminary separation, has also been used to purify extracts of sex pheromones (Henderson *et al.,* 1973; Read *et al.,* 1974; Tumlinson *et al.,* 1968, 1969) and termite trail pheromones (Hummel and Karlson, 1968; Smythe *et al.,* 1967). However, Matsumura *et al.* (1969) found that steam distillation of a termite trail substance gave erratic results, possibly because of incomplete distillation or thermal degradation.

Sweep codistillation (Storherr and Watts, 1965; Storherr *et al.,* 1967) is a cleanup method in which extracts and solvent are injected into a heated tube packed with silanized glass wool; vaporized components are swept by nitrogen carrier gas and solvent through the heated tube and condensed in a cooled collection tube. This technique has been used to collect sex pheromones (Jacobson *et al.,* 1970; Vick *et al.,* 1974). With extracts of the carpenterworm sex attractant, sweep codistillation effluents were found active in a laboratory bioassay but not in the field; column chromatography, which produced field-active material, was therefore used (Solomon *et al.,* 1972).

Matsumura *et al.* (1968, 1969) used vacuum codistillation with mineral oil to purify a termite trail pheromone. Another technique that has been used with insect pheromones is "falling film distillation" (Gaston *et al.*, 1966).

3. CHROMATOGRAPHIC METHODS

Various combinations of liquid chromatography (LC) (including high-pressure LC), thin-layer chromatography (TLC), and gas chromatography (GC) are almost invariably used to purify pheromones and similar substances.

a. *Liquid Chromatography on Columns*

Probably the most commonly used cleanup procedure and very often the first step in the purification of a crude extract is liquid chromatography on adsorbent-filled columns. Columns are generally of glass, and depending on the amount of extract, they may contain 5–1000 gm or more of adsorbent. (A 1% loading of sample on adsorbent is common.) Silica gel is the adsorbent most often used in these applications, with Florisil® (a coprecipitated mixture of silica gel and magnesia) and alumina also being employed frequently. Adsorbents may or may not be activated by heating. Modifications of the procedure of Hirsch and Ahrens (1958), i.e., elution with solvents of increasing polarity, separate crude mixtures on the basis of compound polarity; thus, the so-called class separations are readily made with solvent systems such as increasing amounts of ether in hexane; and saturated hydrocarbons, olefins, ethers, esters, ketones and aldehydes, alcohols, and acids emerge in sequence from the column. Accordingly, as with other chromatographic systems (Section II,D), the point of elution (elution volume) of active material in an adsorbent–solvent system, particularly in a system calibrated with standards or model compounds, can shed light on the functional group or groups in the active molecule(s), or at a minimum, rule out the presence of certain functional groups. Thus, the active pheromone from an extract of pink bollworm female moths chromatographed on silica gel with hexane containing increasing amounts of ether (Bierl *et al.*, 1974) emerged in the 5% and the succeeding 15% ether–hexane fractions. The pheromone was subsequently shown to consist of two alkadienyl acetates, the structure of which was consistent with elution in the 5 and 15% ether–hexane fractions. Elution of the pheromone in these fractions eliminated as structural possibilities hydrocarbons (which are far less polar and emerge with hexane) and alcohols and acids, which are too polar, since the alcohol analogs of the acetates emerged in the 30% and the succeeding 50% ether–hexane fractions.

The use of columns of silica gel or Florisil impregnated with silver nitrate (usually ca. 25%, e.g., Adsorbosil CABN® or Hi-Flosil-Ag®), called argentation chromatography, has proven highly effective in the separation of olefinic compounds based on number of double bonds present, as well as in the separation of *E* and *Z* isomers,* which normally do not separate in conventional liquid–solid chromatography. Even though the need to separate *E,Z* isomers does not rise too often in purification of natural pheromones, the separation is generally required for cleanup of synthetic preparations of pheromones, especially if the concentration of the geometric isomers in a synthetic pheromone must be critically adjusted for optimum performance. In the separation of *E,Z* alkenes, 3% ether in hexane is used as the eluent (hexane only elutes the paraffins); the *E* isomer is eluted first, and with a sufficiently long column, the *E,Z* separation can be complete. With polyolefins and compounds of greater polarity, increasing concentrations of ether in hexane are used to effect separation. Columns developed with ether–hexane cannot be reused. However, in a high-pressure LC method described by Tumlinson *et al.* (1974b), complete separation of alkenyl acetates was readily obtained on columns (0.62 cm o.d. × 25 cm) of silver nitrate on silica gel (1–11 μm particles) with benzene as solvent and using index of refraction for detection. These columns, which can be reused, also provided sufficient resolution to separate alkadienyl acetates into their *E,E* and *Z,Z* isomers with an intervening separate peak for the unresolved *E,Z* and *Z,E* isomers.

Houx *et al.* (1974) have recommended a cation-exchange resin treated with silver nitrate for separating the geometric isomers of some alkenyl acetates with insect sex attractant activity. The major advantages of this procedure are that the column has a high capacity (up to 600 mg per run) and can be reused.

Anion exchange chromatography was used for the purification of a carboxylic acid ingredient of the carpet beetle sex attractant (Silverstein *et al.*, 1967).

Partition (sometimes called liquid–liquid) chromatography, which utilizes liquid phases on microparticulate supports in place of the adsorbents used in liquid–solid chromatography, has not found much use in pheromone isolation or analysis. However, the separation of the boll weevil pheromone ingredients (Tumlinson *et al.*, 1969) by Carbowax 20M (a solid) on silica gel may be regarded as partition chromatography.

* For unambiguous specification of isomerism around a double bond, the system described by Blackwood *et al.* (1968) is now used by *Chem. Abstracts* and elsewhere. In many olefinic compounds *Z* and *E* correspond respectively to *cis* and *trans* in earlier terminology.

In gel chromatography (also referred to as gel filtration, gel permeation chromatography, or exclusion chromatography) mixtures are separated on the basis of molecular size and shape. Molecules that fit into the pores of the chromatographic packings (organic polymers or siliceous materials with controlled pore sizes) pass through a column more slowly than larger molecules, which are partially or completely excluded from the pores. Although Sephadex, one of the first of the polymeric gel packings, is often thought of as being used primarily with aqueous solutions, applications with organic solvents are also possible; Horler (1970), Meijer *et al.* (1972), and Persoons *et al.* (1974) have used Sephadex for isolation of sex attractants. Tumlinson and Heath (1976), by removing higher molecular weight impurities on a Poragel 60A® column, which can be used with organic solvents, were able to obtain from a crude moth extract a fraction pure enough for direct GC analysis and separation.

Although conventional liquid chromatographic procedures have wide application and proven effectiveness, the technique has been generally slow and cumbersome. With the introduction of high-pressure liquid chromatography and the improved gradient elution devices, specially prepared packings (particularly those with 5- and 10-μm particles), specific detectors, and automated instrumentation, the scope of the process has been greatly widened (see Kirkland, 1971). Because of its high resolution and speed as well as the advantages of low-temperature operation and versatility, this technique is gaining considerable favor and wide applicability in the area of insect pheromone chemistry, and reports of applications of high-resolution liquid chromatography in pheromone work are beginning to appear (e.g., Meijer *et al.,* 1972). Tumlinson and Heath (1976) have described three applications in pheromone purification, including the argentation chromatography mentioned earlier (Tumlinson *et al.,* 1974b; Heath *et al.,* 1975).

b. *Thin-Layer Chromatography*

Because of the small amounts of material required and ease of recovery of fractions of a mixture, TLC has established itself as a rapid, versatile, and reliable procedure for following the progress of steps in the separation of material from the crude to the pure state. Efficiency of a chromatographic separation or content of a fraction for bioassay can be quickly checked and adjustments in subsequent stages made. Preparative TLC is also a frequently used procedure. Numerous chromogenic and other spray reagents, vapors (Br_2, I_2), and light (UV) are available to aid in detection of specific structures in a molecule. Reactions can also be performed on TLC plates to demonstrate the presence or absence of specific functional groups, as has been described in Section II,D. Stahl (1969) and Kirchner (1967), among others, provide data on chromogenic reagents, solvent systems, and general methodology on the use of this versatile technique.

c. *Gas Chromatography*

Like liquid chromatography, gas chromatography is almost invariably used at some stage(s) in bioassay or in the purification scheme, and the use of polar and nonpolar packings (preferably in glass columns) in purification is standard practice, e.g., Young *et al.* (1973b) list six different columns used in the isolation of different compounds of a bark beetle aggregation pheromone; up to four of these were used in sequence for each component. Because polar columns bleed far more than nonpolar ones, it is customary to have the final cleanup of a compound done with a nonpolar column. A precolumn containing GC support alone may be used to prevent high-boiling materials from contaminating the column (Silverstein *et al.,* 1968; Tumlinson *et al.,* 1972b; Riley *et al.,* 1974b). Detection of compound has generally been with a flame-ionization detector.

In some cases, the insects themselves have been used as detectors, indicating by their response when an active material was present (e.g., Bayer and Anders, 1959; Berger, 1966, 1972; Bierl *et al.,* 1970; Hummel *et al.,* 1973). Detection by means of electroantennograms (EAG) has also been employed either directly (Beevor *et al.,* 1973; Moorhouse *et al.,* 1969) or by collecting 1-minute fractions and then assaying these fractions (Roelofs *et al.,* 1971a).

Several methods are available for collection of the effluent. A cooled capillary tube is often used (e.g., Meijer *et al.,* 1972; Wientjens *et al.,* 1973; Roelofs *et al.,* 1971b). A thermal gradient device (Brownlee and Silverstein, 1968) to cool the capillary has had wide application (e.g., Tumlinson *et al.,* 1972c; Young *et al.,* 1973a,b; Riley *et al.,* 1974a). Trapping of effluents on gas chromatographic packing (Hornstein and Crowe, 1962; Beroza, 1964; Bierl *et al.,* 1968) can also be used. Houghton (1974) has described a device in which a thin-walled stainless steel tube packed with glass wool or coated support can be alternately cooled for collection or heated for reinjection or transfer of the sample. Cronin and Gilbert (1973) used short lengths of porous layer open tubular (PLOT) capillaries to collect GC effluents for subsequent hydrogenation or ozonolysis. Duffield *et al.* (1974) adsorbed effluent components of millipede defensive secretions directly onto powdered graphite for mass spectral analysis. Compounds have also been trapped by passing the effluent through a small amount of cooled solvent (Bierl *et al.,* 1972; Henderson *et al.,* 1973; Read *et al.,* 1974).

4. REACTIONS FOR PURIFICATION

It is sometimes possible to separate a substance from impurities by reactions that either simplify the process of removing contaminants or change the target material to a form in which it is more easily concentrated. Thus, saponification has been frequently useful for removing fats, as was done with extracts of termite trail pheromones (Ritter and Coenen-Saraber, 1969; Birch *et al.*, 1972) and the gypsy moth attractant (Bierl *et al.*, 1970). On the other hand, many lepidopteran sex pheromones are acetates of long-chain unsaturated alcohols; after it has been established that the substance being dealt with is indeed such a compound, the purification scheme can sometimes be facilitated by saponification and chromatography of the neutral fraction, followed by reacetylation (e.g., Bierl *et al.*, 1974; Kuwahara *et al.*, 1971a,b; Read *et al.*, 1974; Roelofs *et al.*, 1969, 1971c; Tamaki *et al.*, 1971a,b, 1973; Tamaki and Yushima, 1974).

Other examples of reactions used to simplify purification steps are methylation of fatty acids with diazomethane (Huwyler *et al.*, 1973; Skinner *et al.*, 1970) and the use of Girard's reagent to separate ketones (Butler *et al.*, 1961; Ritter and Coenen-Saraber, 1969).

F. Identification and Confirmation

Many of the procedures for identification and confirmation have already been discussed. For example, a successful scheme devised for purifying an active compound sheds light on its chemical structure. Behavior on extraction or partitioning between solvents as well as chromatographic behavior can provide an indication of polarity and some idea of molecular weight, and the results of preliminary probes on small portions of crude extract or of partially purified fractions give indications of the functional groups present, as well as suggest what cleanup procedures can be used without loss of activity. With the pure material in hand, more detailed study of its properties is undertaken.

The minute amounts of material available often require the use of specialized microtechniques, and considerable ingenuity may be needed to build up firm evidence for conclusive identification. There are many ways to identify a compound, some more efficient than others and some requiring more or less sample than others. The present survey presents methods actually utilized in the pheromone field.

1. CHROMATOGRAPHIC METHODS

a. *Gas Chromatography*

By comparing the retention times (t_R) of a compound obtained on a number of different GC columns with those of model compounds, the analyst can often obtain considerable information on the molecular size and functional groups in a compound.

The widely used Kovats' retention index system (Kovats, 1965; Ettre, 1964) refines this procedure. It is based on the adjusted retention time (t_R', which is t_R minus the dead volume) of a compound relative to the t_R' of *n*-alkane standards expressed as the carbon number of the standards $\times$ 100; e.g., hexane = 600, nonane = 900. Since in GC the carbon number of *n*-alkanes is proportional to the log of their t_R', a plot of carbon number of *n*-alkanes versus log t_R' gives a straight line, and with such a plot the retention index of any compound may be read off the graph from the t_R' found experimentally with any given column operated isothermally under a fixed set of conditions. As part of this system, it has been shown that homologous series (e.g., alkyl acetates, alcohols, and aldehydes and methyl esters) also give straight-line plots (except for the lowest members). In effect, this demonstrates that the retention index is made up of incremental contributions from the alkyl chain and the functional group(s), and that the sum of the increments equals the retention index of a compound. Although this explanation is a gross oversimplification of the system (e.g., position of functional group in a molecule and interaction of groups affect the retention index), the system has been invaluable in deducing chemical structure in the pheromone field when indices are determined on polar and nonpolar columns along with standards containing functional groups suspected to be present. The system has been especially effective with the pheromones of the Lepidoptera because so many of these are long straight-chain compounds (often unsaturated) with single primary functional groups. Thus, in the chromatography of such a pheromone on polar and nonpolar columns, it is possible from the retention indexes to make reliable estimates of the length of the carbon chain and the number of double bonds and generally to verify the type of functional group(s) present. As indicated, model compounds, which are generally of the type under investigation, are utilized to derive the data upon which identification is based. A variety of similar systems are based on the comparison of relative retention times with those of standards associated with different fields of endeavor [e.g., "theoretical nonane" (Smith, 1960), equivalent chain length of straight-chain fatty acid methyl esters (Miwa *et al.*, (1960), steroid number of VandenHeuvel and Horning (1962)].

When standards are available, comparison of t_R values or cochromatography on several columns is often used to provide evidence for identity. Young *et al.* (1973a) caution against putting too much reliance in identification by chromatographic behavior; two terpene alcohols (linalool and 2-methyl-6-methylene-7-octen-4-ol) had identical t_R values on the first three of five GC columns tried and would not have been distinguished by cochromatography on those columns. Comparison of t_R values with those of standards is often useful in establishing the configuration of unsaturated compounds whose general structure has been elucidated. Already mentioned for their value in proving purity, capillary columns, with their great resolving power, are excellent for establishing identity of compounds by t_R especially when the analyst uses both polar and nonpolar liquid phases. Although the conventional capillary injection system requires a substantial amount of sample for analysis because of a high split ratio, with the appropriate setup and conditions, it is possible to inject an entire sample into a capillary column and therefore use far less sample.

Identification of natural products is often facilitated by coupling chemical reactions with GC (see Section II,F,3b).

b. *Other Chromatographic Procedures*

With other chromatographic procedures (TLC and LC), cochromatography or comparison chromatography with compounds of known structure can similarly be used in establishing identity of compounds or for deducing information on the structure of a pure substance.

A major application of TLC in identification is the use of argentation TLC to establish geometric configuration around double bonds by comparison of the R_f values of the unknown with those of *E* and *Z* isomers of closely related compounds (e.g., Bierl *et al.,* 1970; Carlson *et al.,* 1971; Roelofs *et al.,* 1971b). Prasad *et al.* (1974) report that impregnation of plates with silver perchlorate instead of silver nitrate gave improved separations of some terpene olefins.

Procedures already described which deserve special mention here also are reaction-plus-chromatography combinations (see Section II,F,3) and the use of high-pressure liquid chromatography with its attendant advantages of operation at nonelevated temperatures, ease of collection of fractions, recycle capabilities for purification, the high resolving power of the recent 5–10 μm packings, and the new types of packings becoming available.

2. SPECTROMETRIC METHODS

Spectrometric methods are the procedures of choice in identification provided that sufficient pure compound is available. Ancillary techniques and equipment are now available to secure valid spectra on samples in the nanogram and microgram range. Since most of the behavior-controlling chemicals do not possess conjugated double bonds, ultraviolet spectrometry is not as valuable as mass spectrometry, nuclear magnetic resonance spectrometry, and infrared spectrometry. With spectrometric methods, it is particularly important that the material be as pure as possible, since residual contaminants (from incomplete purification) or introduced ones (e.g., from solvent impurities) can give rise to incorrect interpretations of spectra. Silverstein *et al.* (1974) have presented a very helpful survey of the application of spectrometric methods in identification.

a. *Mass Spectrometry*

Perhaps the most valuable technique in identifying or confirming identity of small amounts of compounds is mass spectrometry (MS), particularly when combined with GC and more particularly with a computer attached to the instrument.

Mass spectra obtained with samples of 10 ng or less are now common with modern mass spectrometers. Mass spectra usually give the molecular weight of the compound, and with the relative abundances and masses of the different fragments, the presence of functional groups and branching in molecules can often be deduced. For example, Bierl and Beroza (1974) have shown that fragmentation patterns can be used for location of epoxide position in long-chain epoxides. With a high-resolution (usually double-focusing) instrument, the elemental composition of the molecule (from the parent mass) and of the various fragments may be determined, greatly facilitating structural identification. Also helpful is chemical ionization mass spectrometry (CIMS), which has become popular recently (e.g., Fales *et al.,* 1972b; Tumlinson *et al.,* 1974a,b; Vick *et al.,* 1974); CIMS gives molecular weights of compounds and information on functional groups not provided by the usual electron-impact mass spectrometer. When very little information is available on a compound, the exact molecular weight can be most helpful for structure elucidation.

Tentative identifications of compounds by comparison of major peaks in their mass spectra with those of standard compounds are facilitated by compilations of spectra such as those edited by Stenhagen *et al.* (1969, 1974). However, discretion in making comparisons is needed. In addition to the expected variations in spectra obtained from different instruments, Young *et al.* (1973a) observed that spectra of certain thermally unstable compounds showed marked variations on the same instrument at different times. Also, different compounds may give identical spectra, e.g., those with different double-bond position or configuration.

When suitable standards or adequate compilations of spectra are available, it has sometimes been possible to characterize extensive series of insect constituents by gas chromatographic retention times and mass spectra (e.g., Bergström and Löfqvist, 1970; Brand *et al.,* 1972, 1973; Cavill and Houghton, 1974; Crewe and Blum, 1970).

Mass spectra of reaction products are also used to aid in identifications. Hydrogenation of double bonds (e.g., Bergström *et al.,* 1968; Bierl *et al.,* 1970; Carlson *et al.,* 1971; Horler, 1970; Moore, 1966) and hydrogenolysis for determination of the carbon skeleton (e.g., Silverstein *et al.,* 1968) are probably the most common reactions used.

Excellent GC–MS combinations are now available commercially, and they are undoubtedly the most powerful tool today for identifying pheromones. The GC provides the means of introducing very small samples (in a solvent), and also separation of mixtures of compounds, allowing spectra of the individual constituents to be recorded. Usually high-temperature, nonpolar GC packings are used because they bleed far less than polar packings at elevated temperatures and thereby avoid contamination of the MS and interference with spectral recording. With a capillary GC column, the entire effluent is usually fed into the MS, and the spectral scans are made every few seconds. With a packed GC column, an interface is normally provided to remove as much of the carrier gas as possible.

The spectra in GC–MS are generated so rapidly that a computer is necessary to store and analyze the data. Computers attached to mass spectrometers can also increase the sensitivity of mass spectral scans by directing the scan time on the less intense peaks to be increased.

In some instances, it is possible to secure a full spectrum of a minute amount of compound even though the compound is not separated in pure form with GC prior to entering the MS. This was shown recently (Beroza *et al.,* 1974) to demonstrate that a highly active synthetic sex attractant for the codling moth, (*E,E*)-8,10-dodecadien-1-ol, was present in an extract of female abdominal moth tips. The compound had been identified by EAG tests with the aid of model compounds (Roelofs and Comeau, 1971b) (see Section II,F,4). The extract, partially purified by a single chromatographic step on silica gel, was injected into the GC–MS, and a computerized plot of the parent mass (m/e 182) amplitude versus spectrum number (spectrum recorded every 2 seconds) showed a peak at the retention time of the compound. The computer was then directed to obtain a differential spectrum by subtracting the spectrum at the base of the peak from the spectrum at the apex of the peak. In effect, this subtraction cancels out spectral contributions of the background materials because the two spectra are generally only seconds apart, an interval short enough for the background to remain relatively unchanged; yet during this same interval, the pheromonal component rises from almost zero concentration to its maximum, thereby providing a maximum differential spectrum. The fact that the differential spectrum was in good agreement with the spectrum of synthetic (*E,E*)-8,10-dodecadien-1-ol was good evidence that the natural and synthetic compounds were identical, i.e., that the compound was present in the extract.

Regnier (1972) discussed applications of MS in determining structures of "semiochemicals"; many of the compounds he considered are ones affecting insect behavior. Mass fragmentography, a technique that should prove valuable in investigations of the behavior chemicals, was used with GC and MS (Buser and Arn, 1975) to demonstrate the presence of (*E,Z*)-7,9-dodecadien-1-ol acetate in female *Lobesia botrana* moths. (See also Section III, Trace Analysis.)

b. *Nuclear Magnetic Resonance Spectrometry*

Although in principle NMR studies can be made with any atomic nuclei that have a nonzero spin, in practice, because of sample limitations, current applications in pheromone identification involve proton magnetic resonance (PMR) almost exclusively. The PMR spectrum of a compound, which results from the magnetic interactions of protons with neighboring groups, can give a large amount of structural information [e.g., demonstration of 2 different methyl groups, and 3 CH groups adjacent to N in a trail pheromone (Ritter *et al.,* 1973a), or two partial structures, $(CH_3)_2C{=}CHCH_2{-}$ and $-CHOHCH_3$, and assignment of the absolute configuration of sulcatol (Byrne *et al.,* 1974)].

Until recently, sample size requirements for reliable NMR spectra have been a minimum of 100 μg. Special microtubes (which allow placement of the compound dissolved in a minimum of solvent in the most sensitive part of the probe) and a CAT (computer of average transients, which scans the solution repeatedly and averages results, thereby eliminating noise) are necessary to secure good spectra at the 50–100 μg level. However, in the past few years, amounts needed have been progressively lowered. For example, good NMR spectra (with just slight interference from solvent impurities at 1.04 ppm) have been reported on 2 μg of sample with a Bruker HX-90 spectrometer, a Nicolet 1080 Fourier transform data system, and a high sensitivity proton probe insert (Tumlinson and Heath, 1976). Spectra were recorded on samples in 75 μl of solvent in a special microtube manufactured by the Wilmad Glass Company. Solvents were commercially available spectroquality carbon disulfide or carbon tetrachloride, purified by passage through columns of silica gel and basic alumina (both activity grade 1 Woelm) followed by distillation from an all-glass apparatus to remove silica and alumina particles. As has been the case in the past, appropriate equipment and solvents are likely to be made available commercially to allow the sample size needed for NMR spectra to be substantially decreased from present levels.

Carbon-13-NMR, a powerful tool for determining chemical structure, now requires samples in the milligram range. If sample requirements can be reduced to the low microgram level, this technique will be invaluable in pheromone identification.

c. *Infrared Spectrometry*

The usefulness of infrared (IR) spectra in detection as well as in confirmation of functional groups in a molecule is well established, and this technique has been invaluable in pheromonal identifications. Of the many diagnostic IR absorption bands that have found application [e.g., those for C=O, (C=C), (C=O)ester, (OH), (NH), (CH_3), (C=CH), (CHO), and (CH_2OH)], the band at 970 cm^{-1} for the *E* configuration of a double bond is perhaps most frequently referred to. Micro cells, micro KBr disks, and beam condensers allow usable spectra to be obtained on as little as 1 μg of compound. With the aid of Fourier transform equipment, the acquisition of spectra on submicrogram amounts has been reported (King, 1973).

Raman spectrometry, although almost unexplored in pheromonal research, does appear to be a potentially valuable tool, and its use would be worth considering in future work.

d. *Other Spectrometric Procedures*

Ultraviolet spectrometry in isolation and identification of pheromone-type chemicals has, as noted, been used on those infrequent occasions when a UV-absorbing chromophore was found in the target molecule. This use will persist. With UV detection well established in high-pressure LC, the detection, isolation, and identification of UV-absorbing compounds should be facilitated.

Potential use exists for X-ray crystallography for elucidating structure of pheromones if techniques can be developed for the production of microcrystals and appropriate scanning at low temperatures. Measurements of optical rotation, optical rotatory dispersion (ORD), and circular dichroism (CD) come into play with optically active compounds. Recent evidence indicates that enantiomers can be differentiated by odor (e.g., Russell and Hills, 1971; Friedman and Miller, 1971), and that insects may be able to distinguish between enantiomers of pheromones. Thus, Iwaki *et al.* (1974) have recently reported that male gypsy moths are much more responsive to synthetic (+)-disparlure than to the (−)-enantiomer. ORD and CD may, therefore, become increasingly useful in pheromone studies.

3. REACTIONS IN IDENTIFICATION

a. *Structural Information from Reaction Products*

Many of the reactions used as preliminary probes (Section II,D) have also been used with purified materials to acquire additional structural information, and identification of the reaction products often makes it possible to deduce the structure of the starting substance. Catalytic hydrogenation is probably the most commonly used of the reaction procedures. For example, after the cabbage looper sex attractant was shown to be an acetate of an unsaturated alcohol, the alcohol was hydrogenated, oxidized with potassium permanganate, and methylated. Identification of the product as methyl dodecanoate indicated that the starting alcohol was a dodecenol (Berger, 1966). Other reactions (e.g., reduction with $LiAlH_4$, oxidation with CrO_3 or MnO_2, carbonyl derivatizations) have been used; in all cases, sample size often limits the usefulness of a reaction.

Location of double-bond position is frequently required in structure elucidation. Until recently, spectral analyses have not been effective with determinations at the level of 1 μg. Ozonolysis followed by GC analyses of the products, however, has proved effective in locating double bonds at this low level in a wide variety of compounds, e.g., sex attractants (e.g., Brady *et al.,* 1971; Smith *et al.,* 1974); or defensive secretions of beetles (Brand *et al.,* 1973). The ozonizer may be built from simple materials available in most laboratories (Beroza and Bierl, 1966); a much improved version has also been described (Beroza and Bierl, 1969). All of the reagents required for ozonolysis may be purchased from Supelco Corp. (Bellefonte, Pa.). The procedure is simple and rapid (Beroza and Bierl, 1967). In a typical analysis, ozone is passed into a solution of the compound in carbon disulfide, cooled to −70°C. After addition of triphenylphosphine to cleave the ozonide, the solution is warmed to room temperature and a portion is injected into the gas chromatograph. The double-bond position is then determined from the t_R values of the aldehyde and ketone fragments produced:

$$\mathrm{RCH{=}CHR' \longrightarrow RCHO + R'CHO}$$

$$\mathrm{R'RC{=}CHR'' \longrightarrow R\overset{\overset{\displaystyle O}{\|}}{C}R' + R''CHO}$$

With the CS_2 solvent, peaks with t_R values equal to or greater than valeraldehyde are visible on the chromatogram. Earlier emerging peaks, which are hidden by the CS_2, may be observed by repeating the analysis with pentyl acetate (emerges after valeraldehyde) as solvent. Conjugated structures react more slowly than nonconjugated ones, and a few unsaturated structures (triple bonds and α,β-unsaturated nitriles) do resist ozonolysis. Failure to react may be utilized in the identification of such nonreacting structures.

An ozonolysis procedure described by Moore and Brown (1971) has been used in the identification of a termite trail pheromone (Birch *et al.*, 1972).

Double-bond positions have also been located by oxidation with potassium permanganate and potassium periodate; the methyl esters of the acids thus obtained are then identified by GC retention times (e.g., Horler, 1970; Meinwald and Meinwald, 1966; Read *et al.*, 1974). This method requires larger amounts of sample.

For location of position of epoxide groups, epoxides may be cleaved to aldehyde and ketone fragments just as unsaturates are cleaved in ozonolysis (Bierl *et al.*, 1971). The compound in a halogenated solvent is reacted with dry, powdered HIO_4 for 5 minutes before GC analysis of the fragments. Reaction of a compound with phosphoric acid on a silica gel TLC plate can be used to determine the geometric configuration of many epoxides, in addition to demonstrating the presence of the epoxide group (Bierl *et al.*, 1971).

Epoxides are readily obtained from olefins by oxidation with *m*-chloroperbenzoic acid and the reaction is stereospecific (e.g., *cis*-2,3-epoxydecane is obtained from (*Z*)-2-decene). Accordingly, methods for determining epoxide position and configuration can be used to determine the position and configuration of double bonds. In addition to the cleavage method just cited, epoxide positions can be determined by electron-impact MS (Bierl and Beroza, 1974) or chemical-ionization MS (Tumlinson *et al.*, 1974a) (Section II,F,2a). Olefin configurations have been determined from GC retention times of the corresponding epoxides (e.g., McDonough and George, 1970; Nesbitt *et al.*, 1973b; Smith *et al.*, 1974). The configurations of the two double bonds in synthetic (*Z,E*)-7,11-hexadecadien-1-ol acetate, the pheromone of the Angoumois grain moth, were established by stopping the epoxidation at the monoepoxide stage and establishing configuration of the two monoepoxides by IR, NMR, and chemical-ionization MS (Vick *et al.*, 1974; Tumlinson and Heath, 1976).

b. *Reaction Chromatography*

The use of chromatography in combination with chemical reactions can generate much useful information. For a review of this general topic in GC, see Beroza and Inscoe (1969) and GC methods for specific functional groups discussed by these authors in a compilation on functional group analysis (Siggia, 1972).

Changes in chromatographic behavior caused by a reaction (appearance or disappearance of a peak in a chromatogram or increase or decrease in bioactivity of a fraction at a given t_R) can often help determine the nature of functional groups with a minute amount of compound, and analysis of the product(s) can also give information on the compound itself. The use of an on-line GC reactor (pre- or post-column), referred to as reaction gas chromatography, simplifies the manipulations required. With one such reaction, hydrogenation is accomplished on a short precolumn containing a hydrogenation catalyst in a GC with hydrogen carrier gas (Beroza and Sarmiento, 1966).

Since many of the compounds of interest are unsaturated, such hydrogenation is one of the more frequently used reaction GC techniques that simplify identifications (e.g., Carlson *et al.,* 1971; Meijer *et al.,* 1972; Tumlinson *et al.,* 1971, 1974b). For example, it was used in combination with mass spectral analysis to hydrogenate 1 μg of olefin because the methyl group position could be determined by MS on the saturated compound but not on the olefin (Bierl *et al.,* 1970). In an example of hydrogenation–GC–MS (Issenberg *et al.,* 1969), a switching valve was placed at the end of the GC column to direct the effluent either through a hydrogenation catalyst or through a bypass before entering the mass spectrometer. Comparison of the parent masses of the hydrogenated and unhydrogenated compounds in submicrogram amounts discloses the number of double bonds (assuming no hydrogenolysis). In addition, this procedure facilitates identification of *E* and *Z* isomers, because they are reduced to the same compounds; it also helps distinguish unsaturated compounds from those with rings or branches.

The GC hydrogenation procedure (carried out at 150–250°C with 1% palladium on a GC support as catalyst) is in most instances instantaneous and quantitative, so that t_R values are reliable even though the reaction occurs on-line. The procedure has been shown to be applicable to a wide variety of unsaturated compounds (e.g., alcohols, amides, amines, ketones, esters, ethers, nitriles). With some compounds (halides, aldehydes, mercaptans), some hydrogenolysis (cleavage of functional group from a molecule) occurs, and while quantitative estimation is destroyed, structural determinations are not affected because the hydrogenolysis products emerge far ahead of the hydrogenated products.

Cronin and Gilbert (1973) have described an off-line system in which a compound in a GC effluent is trapped in a capillary coated with palladium catalyst; hydrogenation is then accomplished by crushing the sealed capillary in the inlet of a gas chromatograph.

Treatment of a compound with a hydrogenation catalyst under more severe conditions (200–350°C and longer (up to 22 cm) catalyst lengths)

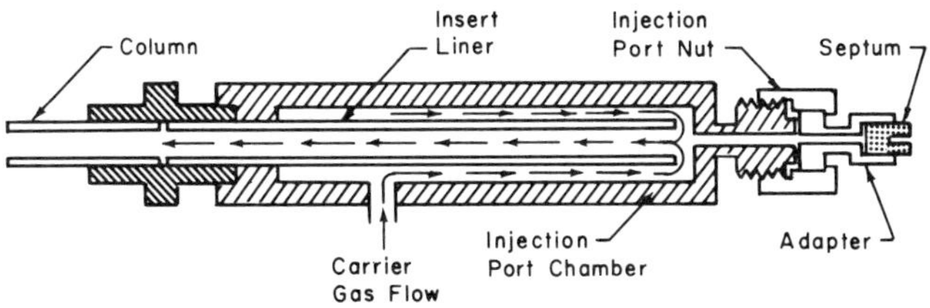

FIG. 1. Cross section of injection port device for holding catalyst in carbon-skeleton chromatography (Beroza, 1970b).

in a gas chromatograph with hydrogen as the carrier gas results in hydrogenolysis as well as hydrogenation (Beroza, 1962; Beroza and Acree, 1964; Beroza and Sarmiento, 1963, 1964). This process is called carbon-skeleton chromatography. An injection-port device for holding the catalyst just ahead of the GC column is shown in Fig. 1. In passing the hot catalyst (precolumn containing 1% palladium on a GC support at 300°C), the injected compound is stripped of its functional groups, and the resulting hydrocarbon products pass into the GC column and are carried to the detector. Products of the reaction, which are the parent hydrocarbon (the carbon skeleton) and/or the next lower homolog, are identified by their retention times. A summary of the reaction products is given in Fig. 2. Typically, the parent hydrocarbon is obtained with halides, sulfides, and compounds with secondary oxygen or nitrogen functions. For example, 2-octanol gives octane, 2-methyl-2-pentanol

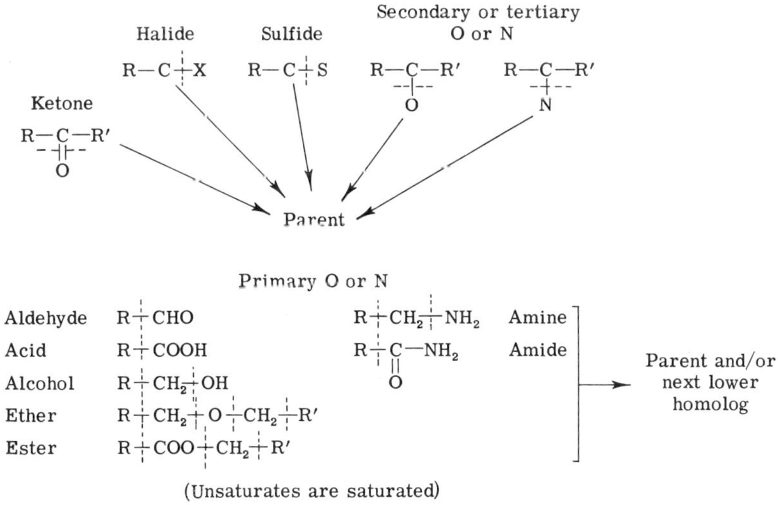

FIG. 2. Reactions observed in carbon-skeleton chromatography (Beroza and Inscoe, 1969).

gives 2-methylpentane. Only when an oxygen or nitrogen function is primary (i.e., on the end carbon of a chain, as in aldehydes, primary alcohols, esters, ethers, amines, amides, carboxylic acids) is the next lower homolog obtained, although parent hydrocarbon may be produced concurrently depending on type of compound and conditions of reaction. In the carbon-skeleton reaction, inorganic products (e.g., HBr, H_2S, H_2O, CO, CO_2) do not register on the flame-ionization detector. Needless to state, carbon-skeleton chromatography can be used advantageously in combination with MS analysis to aid in identification of the products. Possible structures for a bark beetle pheromone were narrowed down in this way when nonane was identified as a product (see Fig. 3). The hydrogenolysis data plus the data from spectral probes then allowed its structure to be determined (Silverstein *et al.*, 1968).

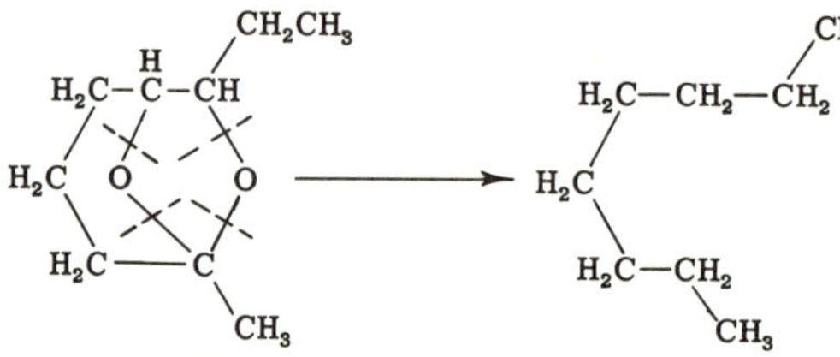

FIG. 3. Production of nonane in the carbon-skeleton chromatography of brevicomin.

A natural partner of carbon-skeleton chromatography in identification is the Kovats retention index system (Section II,F,1a). Carbon-skeleton chromatography gives the parent hydrocarbon and/or its next lower homolog, and this can be used as a starting point in the application of the retention index system. The chemist can add the index increments of the various functional-group possibilities to the retention index of the carbon skeleton for comparison with the retention index of the original compound, and thereby he can frequently identify the compound or its functional groups, or at least reduce the number of possibilities (Beroza and Inscoe, 1969).

Another interesting reaction that is conducted within the GC pathway utilizes short lengths of tubing ("reaction or subtraction loops") containing a reactive chemical that subtracts compounds with a particular functional group (Bierl *et al.*, 1969). In effect, it provides a means of performing functional group analysis on small amounts of compound by simply injecting the compound and determining whether it reacts (if it does, the group is present, and its peak disappears from the chromatogram), or whether it passes unchanged, in which case the target functional group is absent. "Subtraction" of a GC peak after passage through a loop containing zinc oxide indicates the presence of a carboxylic acid, while boric acid subtracts primary and secondary alcohols, benzidine subtracts carbonyl compounds, *o*-dianisidine subtracts aldehydes, and phosphoric acid subtracts epoxides. Regnier and Huang (1970) subtracted compounds with oxygen-containing functional groups with lithium borohydride or lithium aluminum hydride; sodium borohydride could be used to differentiate between aldehydes and ketones, and sodium trimethoxyborohydride to distinguish ketones from most esters.

In studying the constituents of Dufour's glands of ants, Morgan and Wadhams (1972), injected glands by a solid injection technique; injection of the glands together with concentrated sulfuric acid or bromine eliminated many of the peaks previously formed, thus indicating the presence of unsaturated compounds.

4. ELECTROANTENNOGRAM METHOD FOR FINDING SEX ATTRACTANTS

Because the sex lures of many of the Lepidoptera are long-chain unsaturated acetates, alcohols, and aldehydes (in the C_9 to C_{18} range), it has been possible to use EAG bioassays of standard series of monounsaturated compounds to determine structures of sex attractants (Roelofs and Comeau, 1971a). The simple chemical reactions under Preliminary Probes (Section II,D) are used with a crude extract to determine the presence of unsaturation and the nature of functional groups in the target compound. GC retention times of the sex attractant are then determined by injecting the crude attractant on polar and nonpolar columns, and the GC effluent is collected at 1-minute intervals in 12-inch-long glass capillary tubes; the collection time of the tube giving maximum EAG response when 1 ml of air is puffed through it into an air stream passing over the antenna defines the retention time. From the retention times on the different columns, the number of carbon atoms and unsaturated sites may be estimated. The EAG responses of the appropriate series of compounds, e.g., *E* and *Z* C_{12} aldehydes, are tested, and the compound with the maximum response is presumed to be the sex attractant. Should the compound contain two double bonds, as indicated by GC retention times, there will be two maxima when the C_{12} monounsaturated aldehydes are tested (assuming all members are available). Figure 4 illustrates responses obtained with the almond moth. Maximum responses were obtained with (*E*)-12- and (*Z*)-9-tetradecen-1-ol acetates, which suggested that the natural attractant is (*Z,E*)-9,12-tetradecadien-1-ol acetate (Roelofs and Comeau, 1971a). As with all procedures, the presumed structure for the sex attractant must be verified by synthesizing it and its isomers and demonstrating that the compound does indeed attract the target insect in the field. It is further desirable to check the effect of

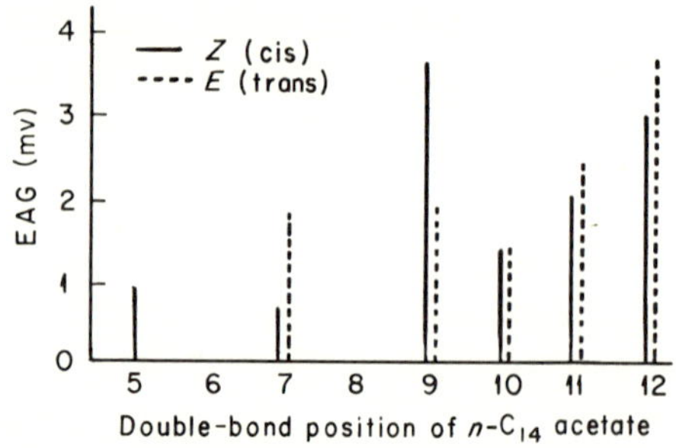

FIG. 4. EAG responses of the male almond moth to tetradecenyl acetates (Beroza, 1972; see also Roelofs and Comeau, 1971a).

adding several concentrations of each isomer to the primary isomer of the pheromone to determine whether attraction can be improved.

A major advantage of the EAG procedure is that only 50–200 female tips and a small supply of male antennae are needed. Disadvantages of the procedure are the need for many model compounds and the fact that, for the most part, only the structures of lepidopteran pheromones have been amenable to this approach because they are straight-chain compounds with primary functional groups in most cases.

Even if a synthetic compound found by EAG tests is attractive in the field, it does not necessarily follow that it is the natural pheromone or one of its attractive constituents, because chemicals closely related to the natural pheromone are often attractive. In a recent study (Sarmiento *et al.*, 1972) fifty-five chemicals closely related to the gypsy moth sex lure, *cis*-7,8-epoxy-2-methyloctadecane, were tested. Two of the compounds gave both strong EAG responses and good attraction of males in the field. They were also very similar in structure to disparlure, so their GC retention times would be very difficult to distinguish from those of disparlure on the packed GC columns normally used. Verification by some other procedure is therefore necessary before a chemical can be accepted as the natural sex lure of the species. In one such confirmational procedure that can be used with very little extract, a computerized search of data from GC–MS of a partially purified extract equivalent to only forty-five female tips was shown to produce a mass spectrum that matched that of the synthetic compound (Section II,F,2a).

III. TRACE ANALYSIS

The minute amounts of pheromone emitted, as for example by the female gypsy moth, cause the insects no difficulty in detection of the material, owing to the exceptionally high sensitivity of the male's olfactory apparatus. However, determination of pheromones or other behavior-controlling chemicals released by or found in insects does pose problems for the analyst because of the low levels at which they occur. Up to the present, this area of endeavor has received only sporadic attention. The problem is complicated by the fact that secondary chemicals, sometimes present in amounts much less than the major pheromonal component, play an important role in insect communication systems, yet the identity of all of the pheromonal ingredients may not be known.

In addition to the measurement of pheromone emission rates (Section III,A) and pheromone content of insects, trace analyses come into play in studies of metabolic pathways of behavior-controlling chemicals or determination of residues (e.g., of repellents on the skin or on treated cloth). Radiolabeled and isotopically labeled compounds (Section III,B) are often useful in such studies.

With few exceptions, the behavior chemicals produced by insects or effective with insects consist of carbon and hydrogen, usually oxygen, and sometimes nitrogen. After appropriate separation and cleanup, the determination of these materials is generally by GC with a flame-ionization detector. If a compound can be derivatized with a group determined by one of the relatively highly selective detectors (electron capture, flame photometric, or thermionic), a considerable increase in analytical sensitivity may be achieved. Kuwahara and Casida (1973) have described such a procedure; the alcohol moiety of a sex pheromone, (*Z,E*)-9,12-tetradecadien-1-ol acetate, as well as the free alcohol, was determined as the trichloroacetate by electron-capture GC. Possible interference from nontarget compounds must be taken into account.

In the analysis of trace constituents, several techniques that utilize GC–MS have been advanced. The most sensitive procedure is single-ion monitoring during the GC run; the MS does not repeatedly scan hundreds of mass values but is fixed on a single mass value. The technique has been criticized for lack of specificity, since the designated ion may arise from other ingredients in a biological sample emerging with the compound being determined. While this lack of specificity makes quantification unreliable at this time, the qualitative results obtained can often be helpful.

Another procedure that retains much of the high sensitivity of single-ion monitoring is mass fragmentography (Hammar *et al.*, 1968); the technique involves the monitoring during GC of several selected fragments in the mass spectrum of the target compound. If a greater number of ions are monitored (compared to single-ion monitoring), the specificity of the analysis is increased. Sensitivity in the picogram range is claimed. Mass fragmentography is not practical with magnetic mass spectrometers due to hysteresis problems but it is easily accomplished with quadrupole instruments by varying electrical fields within the quadrupole structure. Any of the mass fragments in a spectrum may be scanned. Quantitative analysis has been based on the observed relationship between compound concentration and the intensity ratio of several selected fragment ions, utilizing a procedure analogous to the GC internal standard method. At this stage, the reliability of mass fragmentography has not been fully established, and caution dictates that these results should be confirmed by other procedures. One such procedure has involved the addition of a known amount of an isotopically labeled standard, which can be recognized and quantified by the mass spectrometer by selection of the proper masses to scan. Since no other means are generally available for determining picogram amounts of compounds containing only carbon, hydrogen, and oxygen, mass fragmentography should be explored to assess and improve its reliability in the determination of behavior-controlling chemicals.

A. Pheromone Emission Rates

Determinations of pheromone emission rates with individual insects or small groups of insects have been made using cold-trap condensation (Sower *et al.,* 1971; Browne *et al.,* 1974), adsorption on paper (Nordlund and Brady, 1974), trapping in solvent (Richerson and Cameron, 1974), or on GC packing (Byrne *et al.,* 1975; Rudinsky *et al.,* 1973); the pheromone was measured by bioassay or by GC. In those studies in which a parallel determination of recovery from a synthetic sample was made under the same experimental conditions, losses were usually quite high.

A major source of error in the determination of these air-borne compounds in the low concentrations at which they occur is loss of material by adsorption on the walls of the containing vessels. Inefficient collection of the volatile material from an air-stream also causes losses (see Section II,C,5).

Determinations of the rate of release of pheromone from wicks, planchets, or formulations have been made by weight loss (Farkas *et al.,* 1974; Smith *et al.,* 1974), by measurement of residual radioactivity of radiolabeled material (Kuhr *et al.,* 1972), or by GC determination of residual pheromone (McKibben *et al.,* 1972; Fitzgerald *et al.,* 1973; Rudinsky *et al.,* 1974). Although helpful for estimates of release rates, none of these determinations can give an accurate measure of the rate of volatilization, since hygroscopicity may cause difficulties and material may be lost by decomposition or degradation as well as by volatilization.

For materials to be used in the field for insect control, it is necessary that the compound be given off at an adequate rate over an extended time. Because of the many variables to be considered in the development of a satisfactory formulation and the difficulty and expense of suitably controlled field tests, a laboratory method for measurement and comparison of emission rates from formulated materials is needed. Beroza *et al.* (1975) have devised an apparatus for holding samples of wicks or formulations under controlled conditions of air-flow and temperature. The samples are periodically removed for determination of pheromone emission rates; for these analyses, air at a controlled rate is passed over the sample, and the material is collected in hexane for subsequent GC analysis. It is thus possible to compare the performance of several samples over an extended period of time.

B. Radiolabeled Materials

Trace analyses can often be facilitated by the use of isotopically labeled materials. For example, in juvenile hormone studies, deuterium-labeled JH was used with GC–MS to follow the progress of purification and obtain an estimate of the amount of natural JH in single insects (Bieber *et al.,* 1972), and JH studies on biosynthetic and metabolic pathways have been made with compounds labeled with ^{14}C and tritium (e.g., Ajami and Riddiford, 1973; Gill *et al.,* 1974; Metzler *et al.,* 1971). In the area of insect behavior-controlling chemicals, the use of radiolabelled materials is not yet as widespread, but there has been an increasing number of applications reported since Schneider *et al.* (1968) first used tritium-labeled bombykol to study the olfactory threshold of the silkworm.

A semiautomatic scraper for precoated TLC plates (Kasang *et al.,* 1971) that facilitates work with minute amounts of material has proven most useful with radiolabeled pheromones; e.g., Kasang and co-workers have used TLC extensively to follow the metabolism of the sex pheromones on the antennae of the silkworm moth and of the gypsy moth with extraordinarily low levels of the tritiated pheromones (Kasang *et al.,* 1974a,b; Kasang and Weiss, 1974).

Labeled materials have played a part in a variety of investigations with pheromones. To locate the position of unsaturation in an allenic compound presumed to be a sex attractant of a beetle, Horler (1970) used tritiation followed by stepwise degradation of the resulting saturated acid. One of the double bonds in the sex attractant of the Indian meal moth was located by ozonolysis of the ^{14}C-labeled pheromone (Dahm *et al.,* 1971b). The movement of a honeybee pheromone from one part of a bee to another was studied with tritium-labeled 9-oxo-2-

decenoic acid (Butler *et al.*, 1974). ^{14}C-Valine was injected to explore the biosynthetic pathway of a beetle defensive secretion (Benn *et al.*, 1973). Though not directly related to analysis, radiolabeling with ^{32}P was used to tag boll weevils before placing them in a cotton field; monitoring with a portable Geiger counter provided an indication of the number of females attracted to each male and consequently a measure of male pheromone production (McGovern *et al.*, 1973).

IV. ANALYSIS OF SYNTHETIC MATERIALS AND FORMULATIONS

A. Analysis and Specifications for the Unformulated Chemical

Analysis of a synthetic behavior-controlling chemical usually requires a different approach from that needed in the identification or the analysis of trace levels of these compounds. With the synthetic materials, the presence or absence of specific impurities becomes the crucial consideration. Thus, some synthetic pheromones require a minor amount of an opposite geometrical isomer for best attractancy (e.g., Beroza *et al.*, 1973c; Klun *et al.*, 1973; Roelofs and Cardé, 1974), while the activity of similar pheromones for other insect species is not affected by the presence of the opposite isomer other than as the result of dilution (e.g., Read *et al.*, 1974; McLaughlin *et al.*, 1975). In many other cases, the opposite isomer of a pheromone is found to be strongly inhibitory (e.g., Smith *et al.*, 1974).

In setting up specifications for a synthetic pheromone, consideration will have to be given to possible synthetic routes and the impurities that might be encountered in material from each route. Once it is established (usually by bioassay of known mixtures) which possible ingredients are desirable, harmful, or inert, quantitative analyses will have to be devised and permissible limits set for the crucial ingredients to assure procurement of material giving optimum performance. For example, in preparing specifications for the gypsy moth sex attractant disparlure,

cis-7,8-epoxy-2-methyloctadecane $CH_3(CH_2)_9—\underset{\diagdown_{O}\diagup}{CH—CH}(CH_2)_4CH(CH_3)_2$,

it was considered desirable to restrict the amount of the olefin intermediact, (*Z*)-2-methyl-7-octadecene [$CH_3(CH_2)_9CH{=}CH(CH_2)_4—CH(CH_3)_2$] to 0.5% or less because this intermediate is known to diminish the attraction of disparlure (Cardé *et al.*, 1973). On the other hand, since the *trans* isomer of the epoxide has no inhibitory effect (Bierl *et al.*, 1970), at least 15% can be tolerated. Alcohol intermediates have been shown to inhibit attractancy of several acetate pheromones including those of the oriental fruit moth and the cabbage looper (Beroza *et al.*, 1973b; Tumlinson *et al.*, 1972a); in such cases complete acetylation is essential, and an analysis to show the absence of the free alcohol is needed. Preparation of the proper geometrical isomers, either pure or in the proper proportion, is crucial in many instances; analysis for such isomers can become very difficult when doubly unsaturated pheromones are involved; e.g., the *E,E* and *E,Z* isomers of the pink bollworm pheromone inhibit the attraction of the 1:1 *Z,E* + *Z,Z* pheromone (Bierl *et al.*, 1974); the *Z,Z* and *E,E* isomers are readily determined, but determination of the *Z,E* and *E,Z* isomers was found to require the use of capillary column chromatography. Other inhibitors of sex pheromone attraction may be double-bonded positional isomers or saturated analogs [e.g., tetradecyl acetate for the pink bollworm (Bierl *et al.*, 1974].

In the preparation of the specifications for procurement of synthetic material, a balance will have to be struck between what is thought to give the best performance and what can be obtained for a realistic price. If the inclusion of a small amount of an inhibitory intermediate or degradation product causes only minor reduction of activity, it may be far less expensive to use somewhat more of a slightly less pure product than to apply stringent limits on impurity concentrations and obtain a slightly more active product at considerably greater expense. It is well for the buyer to supply the vendor with a standard sample of the desired chemical(s) and specify that the properties of the submitted product should agree with those of the standard; e.g., agreement should be required between the infrared spectra, the gas chromatograms (usually to within a few percent of peak height of the standard, with peak width about the same and with impurities below a given level), and physical properties such as density, refractive index, and boiling point. In addition, specialized tests to demonstrate that the concentrations of detrimental impurities are within acceptable limits should be specified.

Methods for separating and determining geometrical isomers and other functional groups, particularly instrumental analysis, which have already been discussed in Section II, are applicable. Since the final judge of acceptability is the insect itself, it would be well to include a bioassay evaluation in the specifications. Efficacy equal to the standard is the usual criterion of acceptability. In this regard, it is desirable for the chemical manufacturer to submit a sample of his product (obtained in a small-scale synthesis) for approval in such a bioassay before venturing into large-scale production.

B. Analysis of Formulations

Appropriate formulation of a material can make the difference between success and failure. After a satisfactory formulation has been worked out, the analyst will have to check different batches for the crucial ingredients to be certain that their amounts are within acceptable limits; decisions must be made as to which chemicals are to be determined. Analysis of formulations can be troublesome (in requiring extensive cleanup) and less accurate than one might like because of the presence of such nonactive but essential ingredients as stickers, thickeners, emulsifying or wetting agents, antioxidants, sun-screen compounds, keepers (Beroza *et al.,* 1971b), and the like. Not only may these materials in themselves interfere with analyses, but additional interference may arise from them as a result of aging. There are no general rules that can be applied, owing to the diversity of active materials and additives that may be encountered in a formulation. Discussions on formulation analysis in Volumes I (Beckman, 1963) and VI (Pasarela, 1972) of this treatise contain helpful information. Measurement of emission rates of formulations, already discussed (Section III,A), is a valid recourse. If a small experiment will serve, use of radiolabeled or isotopically labeled compounds can be useful (See Section III,B). The possibility of subjecting formulations exposed under practical circumstances to bioassay should also be considered, even though quantification is limited in many instances simply to orders of magnitude. Indeed, chemical analyses without confirmation of activity by bioassay can be highly misleading at least at the outset of a project.

REFERENCES

Acree, F., Jr., Turner, R. B., Gouck, H. K., Beroza, M., and Smith, N. (1968). *Science* **161,** 1346.

Ajami, A. M., and Riddiford, L. M. (1973). *J. Insect Physiol.* **19,** 635.

Andersson, C.-O., Bergström, G., Kullenberg, B., and Ställberg-Stenhagen, S. (1967). *Ark. Kemi* **26,** 191.

Arn, H., Schwartz, C., Limacher, H., and Mani, E. (1974). *Experientia* **30,** 1142.

Baggiolini, M., and Dewald, B. (1967). *J. Chromatogr.* **30,** 256.

Barber, I. A., McGovern, T. P., Beroza, M., Hoyt, C. P., and Walker, A. (1971). *J. Econ. Entomol.* **64,** 1041.

Bayer, E., and Anders, F. (1959). *Naturwissenschaften* **46,** 380.

Beckman, H. F. (1963). *In* "Analytical Methods for Pesticides, Plant Growth Regulators, and Food Additives" (G. Zweig, ed.), Vol. 1, pp. 7–23. Academic Press, New York.

Bedoukian, P. Z. (1970). *Advan. Chemoreception* **1,** 19.

Beevor, P. S., Campion, D. G., Moorhouse, J. E., and Nesbitt, B. F. (1973). *Bull. Entomol. Res.* **62,** 439.

Bell, W. J., Burns, R. E., and Barth, R. H. (1974). *Behav. Biol.* **10,** 419.

Benn, M. H., Lencucha, A., Maxie, S., and Telang, S. A. (1973). *J. Insect Physiol.* **19,** 2173.

Bentley, K. W., and Kirby, B. W., eds. (1972). "Elucidation of Organic Structure by Physical and Chemical Methods," 2nd ed., Parts 1–3. Wiley (Interscience), New York.

Benz, G., and von Salis, G. (1973). *Experientia* **29,** 729.

Berger, R. S. (1966). *Ann. Entomol. Soc. Amer.* **59,** 767.
Berger, R. S. (1972). *Science* **177,** 704.
Bergström, G. (1973). *Chem. Scripta* **4,** 135.
Bergström, G., and Löfqvist, J. (1968). *J. Insect Physiol.* **14,** 995.
Bergström, G., and Löfqvist, J. (1970). *J. Insect Physiol.* **16,** 2353.
Bergström, G., and Lundgren, L. (1973). *Zoon, Suppl.* **1,** 67.
Bergström, G., and Svensson, B. G. (1973). *Zoon, Suppl.* **1,** 61.
Bergström, G., Kullenberg, B., Ställberg-Stenhagen, S., and Stenhagen, E. (1968). *Ark. Kemi* **28,** 453.
Bergström, G., Kullenberg, B., and Ställberg-Stenhagen, S. (1973). *Chem. Scripta* **4,** 174.
Beroza, M. (1962). *Anal. Chem.* **34,** 1801.
Beroza, M. (1964). *J. Gas Chromatogr.* **2,** 330.
Beroza, M. (1970a). *In* "Column Chromatography; Fifth International Symposium on Separation Methods, Lausanne 1969" (E. Sz. Kovats, ed.) [*Chimia,* Supplement], pp. 92–103. Säurländer AG, Aarau, Switzerland.
Beroza, M. (1970b). *Accounts Chem. Res.* **3,** 33.
Beroza, M. (1972). *Toxicol. Environ. Chem. Rev.* **1,** 109.
Beroza, M., and Acree, F., Jr. (1964). *J. Ass. Offic. Agr. Chem.* **47,** 1.
Beroza, M., and Bierl, B. A. (1966). *Anal. Chem.* **38,** 1976.
Beroza, M., and Bierl, B. A. (1967). *Anal. Chem.* **39,** 1131.
Beroza, M., and Bierl, B. A. (1969). *Mikrochim. Acta* p. 720.
Beroza, M., and Inscoe, M. N. (1969). *In* "Ancillary Techniques of Gas Chromatography" (L. S. Ettre and W. H. McFadden, eds.), pp. 89–144. Wiley (Interscience), New York.
Beroza, M., and Sarmiento, R. (1963). *Anal. Chem.* **35,** 1353.
Beroza, M., and Sarmiento, R. (1964). *Anal. Chem.* **36,** 1744.
Beroza, M., and Sarmiento, R. (1966). *Anal. Chem.* **38,** 1042.
Beroza, M., Alexander, B. H., Steiner, L. F., Mitchell, W. C., and Miyashita, D. H. (1960). *Science* **131,** 1044.
Beroza, M., Green, N., Gertler, S. I., Steiner, L. F., and Miyashita, D. H. (1961). *J. Agr. Food Chem.* **9,** 361.
Beroza, M., Staten, R. T., and Bierl, B. A. (1971a). *J. Econ. Entomol.* **64,** 580.
Beroza, M., Bierl, B. A., Tardif, J. G. R., Cook, D. A., and Paszek, E. C. (1971b). *J. Econ. Entomol.* **64,** 1499.
Beroza, M., Punjabi, A. A., and Bierl, B. A. (1973a). *J. Econ. Entomol.* **66,** 1215.
Beroza, M., Gentry, C. R., Blythe, J. L., and Muschik, G. M. (1973b). *J. Econ. Entomol.* **66,** 1307.
Beroza, M., Muschik, G. M., and Gentry, C. R. (1973c). *Nature (London), New Biol.* **244,** 149.
Beroza, M., Katagiri, K., Iwata, Z., Ishizuka, H., Suzuki, S., and Bierl, B. A. (1973d). *Environ. Entomol.* **2,** 966.
Beroza, M., Bierl, B. A., and Moffitt, H. R. (1974). *Science* **183,** 89.
Beroza, M., Bierl, B. A., James, P., and DeVilbiss, D. (1975). *J. Econ. Entomol.* **68,** 369.
Bevan, C. W. L., Birch, A. J., and Caswell, H. (1961). *J. Chem. Soc., London* p. 488.
Bieber, M. A., Sweeley, C. C., Faulkner, D. J., and Peterson, M. R. (1972). *Anal. Biochem.* **47,** 264.
Bierl, B. A., and Beroza, M. (1974). *J. Amer. Oil Chem. Soc.* **51,** 466.
Bierl, B. A., Beroza, M., and Ruth, J. M. (1968). *J. Gas Chromatogr.* **6,** 286.
Bierl, B. A., Beroza, M., and Ashton, W. T. (1969). *Mikrochim. Acta* p. 637.
Bierl, B. A., Beroza, M., and Collier, W. W. (1970). *Science* **170,** 87.

Bierl, B. A., Beroza, M., and Aldridge, M. H. (1971). *Anal. Chem.* **43,** 636.
Bierl, B. A., Beroza, M., and Collier, C. W. (1972). *J. Econ. Entomol.* **65,** 659.
Bierl, B. A., Beroza, M., Staten, R. T., Sonnet, P. E., and Adler, V. E. (1974). *J. Econ. Entomol.* **67,** 211.
Birch, A. J., Brown, W. V., Corrie, J. E. T., and Moore, B. P. (1972). *J. Chem. Soc., Perkin Trans.* **1,** 2653.
Blackwood, J. E., Gladys, C. L., Petrarca, A. E., Powell, W. H., and Rush, J. E. (1968). *J. Chem. Documentation* **8,** 30.
Block, E. F., and McChesney, J. D. (1974). *J. Insect Physiol.* **20,** 1683.
Blum, M. S. (1970). *In* "Chemicals Controlling Insect Behavior" (M. Beroza, ed.), pp. 61–94. Academic Press, New York.
Blum, M. S., Warter, S. L., Monroe, R. S., and Chidester, J. C. (1963). *J. Insect. Physiol.* **9,** 881.
Blum, M. S., Wallace, J. B., and Fales, H. M. (1973). *Insect Biochem.* **3,** 353.
Boch, R., Shearer, D. A., and Stone, B. C. (1962). *Nature (London)* **195,** 1018.
Borden, J. H., Brownlee, R. G., and Silverstein, R. M. (1968). *Can. Entomol.* **100,** 629.
Bowers, W. S., Nault, L. R., Webb, R. E., and Dutky, S. R. (1972). *Science* **177,** 1121.
Brady, U. E. (1973). *Life Sci.* **13,** 227.
Brady, U. E., and Ganyard, M. C., Jr. (1972). *Ann. Entomol. Soc. Amer.* **65,** 898.
Brady, U. E., Tumlinson, J. H., III, Brownlee, R. G., and Silverstein, R. M. (1971). *Science* **171,** 802.
Brand, J. M., Blum, M. S., Fales, H. M., and MacConnell, J. G. (1972). *Toxican* **10,** 259.
Brand, J. M., Blum, M. S., Fales, H. M., and Pasteels, J. M. (1973). *J. Insect Physiol.* **19,** 369.
Brossut, R., Dubois, P., and Rigaud, J. (1974). *J. Insect Physiol.* **20,** 529.
Browne, L. E., Birch, M. C., and Wood, D. L. (1974). *J. Insect Physiol.* **20,** 183.
Brownlee, R. G., and Silverstein, R. M. (1968). *Anal. Chem.* **40,** 2077.
Budzikiewicz, H., Djerassi, C., and Williams, D. H. (1967). "Mass Spectrometry of Organic Compounds." Holden-Day, San Francisco, California.
Burkholder, W. E., and Dicke, R. J. (1966). *J. Econ. Entomol.* **59,** 540.
Buser, H.-R., and Arn, H. (1975). *J. Chromatogr.* **106,** 83.
Butenandt, A., Bechmann, R., Stamm, D., and Hecker, E. (1959). *Z. Naturforsch. B.* **14,** 283.
Butler, C. G., Callow, R. K., and Johnston, N. C. (1961). *Proc. Roy. Soc., Ser. B* **155,** 417.
Butler, C. G., Callow, R. K., Greenway, A. R., and Simpson, J. (1974). *Entomol. Exp. Appl.* **17,** 112.
Butler, L. I., McDonough, L. M., Tamaki, G., and Landis, B. J. (1972). *J. Econ. Entomol.* **65,** 1174.
Byrne, K. J., Swigar, A. A., Silverstein, R. M., Borden, J. H., and Stokkink, E. (1974). *J. Insect Physiol.* **20,** 1895.
Byrne, K. J., Gore, W. E., Pearce, G. T., and Silverstein, R. M. (1975). *J. Chem. Ecol.* **1,** 1.
Cardé, R. T., Roelofs, W. L., and Doane, C. C. (1973). *Nature (London)* **241,** 474.
Cargill, D. I. (1962). *Analyst (London)* **87,** 865.
Carlson, D. A., Mayer, M. S., Silhacek, D. L., James, J. D., Beroza, M., and Bierl, B. A. (1971). *Science* **174,** 76.
Cavill, G. W. K., and Hinterberger, H. (1960). *Aust. J. Chem.* **13,** 514.
Cavill, G. W. K., and Hinterberger, H. (1962). *Proc. Int. Congr. Entomol. 11th, 1960* Vol. 3, p. 53.
Cavill, G. W. K., and Houghton, E. (1974). *J. Insect Physiol.* **20,** 2049.

Chow, Y.-S., Lu, F.-M., Peng, C.-T., and Cheng, P.-C. (1972). *Bull. Inst. Zool., Acad. Sinica* **11,** 1.
Comeau, A., and Roelofs, W. L. (1973). *Entomol. Exp. Appl.* **16,** 191.
Copius Peereboom, J. W., and Beekes, H. W. (1962). *J. Chromatogr.* **9,** 316.
Crewe, R. M., and Blum, M. S. (1970). *J. Insect Physiol.* **16,** 141.
Crewe, R. M., Brand, J. M., and Fletcher, D. J. C. (1969). *Ann. Entomol. Soc. Amer.* **62,** 1212.
Crewe, R. M., Brand, J. M., Fletcher, D. J. C., and Eggers, S. H. (1970). *J. Ga. Entomol. Soc.* **5,** 42.
Cronin, D. A., and Gilbert, J. (1973). *J. Chromatogr.* **87,** 387.
Curtius, H.-C., and Müller, M. (1968). *J. Chromatogr.* **32,** 222.
Dahm, K. H., and Röller, H. (1970). *Life Sci., Part II,* **9,** 1397.
Dahm, K. H., Meyer, D., Finn, W. E., Reinhold, V., and Röller, H. (1971a). *Naturwissenschaften* **58,** 265.
Dahm, K. H., Richter, I., Meyer, D., and Röller, H. (1971b). *Life Sci., Part II* **10,** 531.
Davis, H. G., Eddy, G. W., McGovern, T. P., and Beroza, M. (1967). *J. Med. Entomol.* **4,** 275.
Davis, H. G., Eddy, G. W., McGovern, T. P., and Beroza, M. (1969). *J. Econ. Entomol.* **62,** 1245.
Davis, H. G., Peterson, R. J., Rogoff, W. M., McGovern, T. P., and Beroza, M. (1972). *Environ. Entomol.* **1,** 673.
Duffield, R. M., Blum, M. S., and Brand, J. M. (1974). *Ann. Entomol. Soc. Amer.* **67,** 821.
Edwards, L. J., Siddall, J. B., Dunham, L. L., Uden, P., and Kislow, C. J. (1973). *Nature (London)* **241,** 126.
Elgamal, M. H. A., and Fayez, M. B. E. (1967). *Fresenius' Z. Anal. Chem.* **226,** 408.
Ettre, L. S. (1964). *Anal. Chem.* **36**(8), 31A.
Fales, H. M., Blum, M. S., Crewe, R. M., and Brand, J. M. (1972a). *J. Insect Physiol.* **18,** 1077.
Fales, H. M., Milne, G. W., Pisano, J. J., Brewer, H. B., Jr., Blum, M. S., MacConnell, J. G., Brand, J., and Law, N. (1972b). *Recent Progr. Horm. Res.* **28,** 591.
Farkas, S. R., Shorey, H. H., and Gaston, L. K. (1974). *Environ. Entomol.* **3,** 876.
Fitzgerald, T. D., St. Clair, A. D., Daterman, G. E., and Smith, R. G. (1973). *Environ. Entomol.* **2,** 607.
Friedman, L., and Miller, J. G. (1971). *Science* **172,** 1044.
Fukui, H., Matsumura, F., Ma, M. C., and Burkholder, W. E. (1974). *Tetrahedron Lett.* p. 3563.
Gary, N. E. (1962). *Science* **136,** 773.
Gaston, L. K., Fukuto, T. R., and Shorey, H. H. (1966). *Ann. Entomol. Soc. Amer.* **59,** 1062.
Gertler, S. I., Steiner, L. F., Mitchell, W. C., and Barthel, W. F. (1958). *J. Agr. Food Chem.* **6,** 592.
Ghent, R. L. (1961). Ph.D. Thesis (cited in Blum, 1970).
Gill, S. S., Hammock, B. D., and Casida, J. E. (1974). *J. Agr. Food Chem.* **22,** 386.
Girard, J. E., Hendry, L. B., and Snetsinger, R. (1974). *Mushroom J.* **13,** 29.
Granges, J., and Baggiolini, M., (1971). *Rev. Suisse Vitic. Arboric.* **3,** 93.
Green, N., Jacobson, M., and Keller, J. C. (1969). *Experientia* **25,** 682.
Halstead, D. G. H. (1973). *J. Stored Prod. Res.* **9,** 109.
Hammar, C.-G., Holmstedt, B., and Ryhage, R. (1968). *Anal. Biochem.* **25,** 532.
Hathaway, D. O., McGovern, T. P., Beroza, M., Moffitt, H. R., McDonough, L. M., and Butt, B. A. (1974). *Environ. Entomol.* **3,** 522.

Heath, R. R., Tumlinson, J. H., Doolittle, R. E., and Proveaux, A. T. (1975). *J. Chromatogr. Sci.* **13,** 380.

Hedin, P. A., Hardee, D. D., Thompson, A. C., and Gueldner, R. C. (1974). *J. Insect Physiol.* **20,** 1707.

Heftmann, E., ed. (1974). "Chromatography," 3rd ed. Van Nostrand-Reinhold, Princeton, New Jersey.

Henderson, H. E., Warren, F. L., Augustyn, O. P. H., Burger, B. V., Schneider, D. F., Boshoff, P. R., Spies, H. S. C., and Geertsema, H. (1973). *J. Insect Physiol.* **19,** 1257.

Hendry, L. B., Jugovich, J., Roman, L., Anderson, M. E., and Mumma, R. O. (1974). *Experientia* **30,** 886.

Henzell, R. F., and Lowe, M. D. (1970). *Science* **168,** 1005.

Hill, A., Cardé, R., Comeau, A., Bode, W., and Roelofs, W. (1974). *Environ. Entomol.* **3,** 249.

Hill, A. S., Cardé, R. T., Kido, H., and Roelofs, W. L. (1975). *J. Chem. Ecol.* **1,** 215.

Hirsch, J., and Ahrens, E. H., Jr. (1958). *J. Biol. Chem.* **233,** 311.

Holloway, P. J., and Challen, S. B. (1966). *J. Chromatogr.* **25,** 336.

Horler, D. F. (1970). *J. Chem. Soc., C* p. 859.

Hornstein, I., and Crowe, P. F. (1962). *Anal. Chem.* **34,** 1354.

Houghton, E. (1974). *J. Chromatogr.* **90,** 57.

Houx, N. W. H., Voerman, S., and Jongen, W. M. F. (1974). *J. Chromatogr.* **96,** 25.

Hummel, H., and Karlson, P. (1968). *Hoppe-Seyler's Z. Physiol. Chem.* **349,** 725.

Hummel, H. E., Gaston, L. K., Shorey, H. H., Kaae, R. S., Byrne, K. J., and Silverstein, R. M. (1973). *Science* **181,** 873.

Huwyler, S., Grob, K., and Viscontini, M. (1973). *Helv. Chim. Acta* **56,** 976.

Issenberg, P., Kobayashi, A., and Mysliwy, T. J. (1969). *J. Agr. Food Chem.* **17,** 1377.

Iwaki, S., Marumo, S., Saito, T., Yamada, M., and Katagiri, K. (1974). *J. Amer. Chem. Soc.* **96,** 7842.

Jacobson, M. (1972). "Insect Sex Pheromones." Academic Press, New York.

Jacobson, M., Lilly, C. E., and Harding, C. (1968). *Science* **159,** 208.

Jacobson, M., Redfern, R. E., Jones, W. A., and Aldridge, M. H. (1970). *Science* **170,** 542.

Jacobson, M., Ohinata, K., Chambers, D. L., Jones, W. A., and Fujimoto, M. S. (1973). *J. Med. Chem.* **16,** 248.

Jones, P. A., Coppel, H. C., and Casida, J. E. (1965). *J. Econ. Entomol.* **58,** 465.

Jones, R. L., Burton, R. L., Bowman, M. C., and Beroza, M. (1970). *Science* **168,** 856.

Jones, R. L., Lewis, W. J., Bowman, M. C., Beroza, M., and Bierl, B. A. (1971). *Science* **173,** 842.

Jones, R. L., Lewis, W. J., Beroza, M., Bierl, B. A., and Sparks, A. N. (1973). *Environ. Entomol.* **2,** 593.

Kaae, R. S., Shorey, H. H., and Gaston, L. K. (1973a). *Science* **179,** 487.

Kaae, R. S., Shorey, H. H., McFarland, S. U., and Gaston, L. K. (1973b). *Ann. Entomol. Soc. Amer.* **66,** 444.

Karger, B. L., Snyder, L. R., and Horvath, C. (1973). "An Introduction to Separation Science." Wiley (Interscience), New York.

Karlson, P., Lüscher, M., and Hummel, H. (1968). *J. Insect Physiol.* **14,** 1763.

Kasang, G., and Weiss, N. (1974). *J. Chromatogr.* **92,** 401.

Kasang, G., Göldner, G., and Weiss, N. (1971). *J. Chromatogr.* **59,** 393.

Kasang, G., Knauer, B., and Beroza, M. (1974a). *Experientia* **30,** 147.

Kasang, G., Schneider, D., and Beroza, M. (1974b). *Naturwissenschaften* **61,** 130.

Kawano, Y., Mitchell, W. C., and Matsumoto, H. (1968). *J. Econ. Entomol.* **61,** 986.

King, S. T. (1973). *J. Agr. Food Chem.* **21,** 526.

Kinzer, G. W., Fentiman, A. F., Jr., Page, T. F., Jr., Foltz, R. L., Vite, J. P., and Pitman, G. B. (1969). *Nature (London)* **221,** 477.
Kirchner, J. G. (1967). "Thin-layer Chromatography," Vol. XII, Technique of Organic Chemistry. Wiley (Interscience), New York.
Kirkland, J. J., ed. (1971). "Modern Practice of Liquid Chromatography." Wiley (Interscience), New York.
Klun, J. A. (1968). *J. Econ. Entomol.* **61,** 484.
Klun, J. A., and Brindley, T. A. (1970). *J. Econ. Entomol.* **63,** 779.
Klun, J. A., and Robinson, J. F. (1971). *Ann. Entomol. Soc. Amer.* **64,** 1083.
Klun, J. A., and Robinson, J. F. (1972). *Ann. Entomol. Soc. Amer.* **65,** 1337.
Klun, J. A., Chapman, O. L., Mattes, K. C., Wojtkowski, P. W., Beroza, M., and Sonnet, P. E. (1973). *Science* **181,** 661.
Kovats, E. (1965). *Advan. Chromatogr.* **1,** 229.
Kuhr, R. J., Comeau, A., and Roelofs, W. L. (1972). *Environ. Entomol.* **1,** 625.
Kuwahara, Y., and Casida, J. E. (1973). *Agr. Biol. Chem.* **37,** 681.
Kuwahara, Y., Hara, H., Ishii, S., and Fukami, H. (1971a). *Agr. Biol. Chem.* **35,** 447.
Kuwahara, Y., Kitamura, C., Takahashi, S., Hara, H., Ishii, S., and Fukami, H. (1971b). *Science* **171,** 801.
Lange, R., and Hoffmann, D. (1972). *Naturwissenschaften* **59,** 217.
Levinson, H. Z., Levinson, A. R., Müller, B., and Steinbrecht, R. A. (1974). *J. Insect Physiol.* **20,** 1231.
Leyrer, R. L., and Monroe, R. E. (1973). *J. Insect Physiol.* **19,** 2267.
Libbey, L. M., Morgan, M. E., Putnam, T. B., and Rudinsky, J. A. (1974). *J. Insect Physiol.* **20,** 1667.
McDonough, L. M., and George, D. A. (1970). *J. Chromatogr. Sci.* **8,** 158.
McGovern, T. P., Fiori, B., Beroza, M., and Ingangi, J. C. (1970a). *J. Econ. Entomol.* **63,** 168.
McGovern, T. P., Beroza, M., Schwartz, P. H., Hamilton, D. W., Ingangi, J. C., and Ladd, T. L. (1970b). *J. Econ. Entomol.* **63,** 276.
McGovern, T. P., Beroza, M., Ladd, T. L., Jr., Ingangi, J. C., and Jurimas, J. P. (1970c). *J. Econ. Entomol.* **63,** 1727.
McGovern, W. L., Cross, W. H., Leggett, J. E., and Mitlin, N. (1973). *J. Econ. Entomol.* **66,** 809.
McGurk, D. J., Frost, J., Eisenbraun, E. J., Vick, K., Drew, W. A., and Young, J. (1966). *J. Insect Physiol.* **12,** 1435.
McKibben, G. H., Gueldner, R. C., Hedin, P. A., Hardee, D. D., and Davich, T. P. (1972). *J. Econ. Entomol.* **65,** 1512.
McLaughlin, J. R., Mitchell, E. R., Chambers, D. L., and Tumlinson, J. H. (1974). *Environ. Entomol.* **3,** 677.
McLaughlin, J. R., Mitchell, E. R., Beroza, M., and Bierl, B. A. (1975). *J. Ga. Entomol. Soc.* **10,** 338.
Maddison, P. A., Beroza, M., and McGovern, T. P. (1973). *J. Econ. Entomol.* **66,** 591.
Maini, S. (1973). *Inform. Fitopatol.* **23,** (9), 11.
Malins, D. C., and Mangold, H. K. (1960). *J. Amer. Oil Chem. Soc.* **37,** 576.
Mangold, H. K. (1959). *Fette, Seifen, Anstrichm.* **61,** 877.
Mathis, C. (1965). *Ann. Pharm. Fr.* **23,** 331.
Matsumura, F., Coppel, H. C., and Tai, A. (1968). *Nature (London)* **219,** 963.
Matsumura, F., Tai, A., and Coppel, H. C. (1969). *J. Econ. Entomol.* **62,** 599.
Matsumura, F., Jewett, D. M., and Coppel, H. C. (1972). *J. Econ. Entomol.* **65,** 600.

Meijer, G. M., Ritter, F. J., Persoons, C. J., Minks, A. K., and Voerman, S. (1972). *Science,* **175,** 1469.

Meinwald, J., and Meinwald, Y. C. (1966). *J. Amer. Chem. Soc.* **88,** 1305.

Metzler, M., Dahm, K. H., Meyer, D., and Röller, H. (1971). *Z. Naturforsch. B* **26,** 1270.

Miller, J. M., and Kirchner, J. G. (1953). *Anal. Chem.* **25,** 1107.

Minks, A. K., and Voerman, S. (1973). *Entomol. Exp. Appl.* **16,** 541.

Minyard, J. P., Tumlinson, J. H., Thompson, A. C., and Hedin, P. A. (1967). *J. Chromatogr.* **29,** 88.

Mitchell, E. R., and Tumlinson, J. H. (1973). *Ann. Entomol. Soc. Amer.* **66,** 917.

Mitchell, E. R., Webb, J. C., Baumhover, A. H., Hines, R. W., Stanley, J. W., Endris, R. G., Linquist, D. A., and Masuda, S. (1972). *Environ. Entomol.* **1,** 365.

Mitchell, E. R., Copeland, W. W., Sparks, A. N., and Sekul, A. A. (1974). *Environ. Entomol.* **3,** 778.

Miwa, T. K., Mikolajczak, K. L., Earle, F. R., and Wolff, I. A. (1960). *Anal. Chem.* **32,** 1739.

Mody, N. V., Miles, D. H., Neel, W. W., Hedin, P. A., Thompson, A. C., and Gueldner, R. C. (1973). *J. Insect Physiol.* **19,** 2063.

Moore, B. P. (1966). *Nature (London)* **211,** 746.

Moore, B. P. (1968). *J. Insect Physiol.* **14,** 33.

Moore, B. P., and Brown, W. V. (1971). *J. Chromatogr.* **60,** 157.

Moorhouse, J. E., Yeadon, R., Beevor, P. S., and Nesbitt, B. F. (1969). *Nature (London)* **223,** 1174.

Morgan, E. D., and Wadhams, L. J. (1972). *J. Insect Physiol.* **18,** 1125.

Moser, J. C., Brownlee, R. G., and Silverstein, R. M. (1968). *J. Insect Physiol.* **14,** 529.

Nesbitt, B. F., Beevor, P. S., Cole, R. A., Lester, R., and Poppi, R. G. (1973a). *Nature (London) New Biol.* **244,** 208.

Nesbitt, B. F., Beevor, P. S., Cole, R. A., Lester, R., and Poppi, R. G. (1973b). *Tetrahedron Lett.* p. 4669.

Neumark, S., Teich, I., and Green, N. (1973). *Environ. Lett.* **5,** 1.

Neumark, S., Jacobson, M., and Teich, I. (1974). *Environ. Lett.* **7,** 21.

Nolte, D. J., Eggers, S. H., and May, I. R. (1973). *J. Insect Physiol.* **19,** 1547.

Nordlund, D. A., and Brady, U. E. (1974). *Environ. Entomol.* **3,** 797.

Pasarela, N. R. (1972). *In* "Analytical Methods for Pesticides, Plant Growth Regulators, and Food Additives" (G. Zweig and J. Sherma, eds.), Vol. 6, pp. 107–131. Academic Press, New York.

Peacock, J. W., Silverstein, R. M., Lincoln, A. C., and Simeone, J. B. (1973). *Environ. Entomol.* **2,** 355.

Pearce, G. T., Gore, W. E., Silverstein, R. M., Peacock, J. W., Cuthbert, R., Lanier, G. N., and Simeone, J. B. (1975). *J. Chem. Ecol.* **1,** 115.

Persoons, C. J., Minks, A. K., Voerman, S., Roelofs, W. L., and Ritter, F. J. (1974). *J. Insect Physiol.* **20,** 1181.

Pitman, G. B., and Vité, J. P. (1969). *Can. Entomol.* **101,** 143.

Pitman, G. B., and Vité, J. P. (1970). *Ann. Entomol. Soc. Amer.* **63,** 661.

Pitman, G. B., and Vité, J. P. (1971). *J. Econ. Entomol.* **64,** 402.

Prasad, R. S., Gupta, A. S., and Dev, S. (1974). *J. Chromatogr.* **92,** 450.

Read, J. S., Warren, F. L., and Hewitt, P. H. (1968). *Chem. Commun.* p. 792.

Read, J. S., Hewitt, P. H., Warren, F. L., and Myberg, A. C. (1974). *J. Insect Physiol.* **20,** 441.

Regnier, F. E. (1972). *In* "Biochemical Applications of Mass Spectrometry" (G. R. Waller, ed.), pp. 723–734. Wiley (Interscience), New York.

Regnier, F. E., and Huang, J. C. (1970). *J. Chromatogr. Sci.* **8,** 267.
Regnier, F. E., and Wilson, E. O. (1968). *J. Insect Physiol.* **14,** 955.
Regnier, F. E., and Wilson, E. O. (1969). *J. Insect Physiol.* **15,** 893.
Regnier, F. E., and Wilson, E. O. (1971). *Science* **172,** 267.
Renwick, J. A., and Vité, J. P. (1972). *J. Insect Physiol.* **18,** 1215.
Richerson, J. V., and Cameron, E. A. (1974). *Environ. Entomol.* **3,** 475.
Riddiford, L. M. (1967). *Science* **158,** 139.
Riess, J. (1965). *J. Chromatogr.* **19,** 527.
Riley, R. G., Silverstein, R. M., Carroll, B., and Carroll, R. (1974a). *J. Insect Physiol.* **20,** 651.
Riley, R. G., Silverstein, R. M., and Moser, J. C. (1974b). *J. Insect Physiol.* **20,** 1629.
Ritter, F. J., and Coenen-Saraber, C. M. A. (1969). *Entomol. Exp. Appl.* **12,** 611.
Ritter, F. J., Rotgans, I. E. M., Talman, E., Verwiel, P. E. J., and Stein, F. (1973a). *Experientia* **29,** 530.
Ritter, F. J., van Oosten, A. M., and Persoons, C. J. (1973b). Paper presented at *7th Int. Congr. Int. Union Study Soc. Insects.*
Rodin, J. O., Silverstein, R. M., Burkholder, W. E., and Gorman, J. E. (1969). *Science* **165,** 904.
Roelofs, W. L., and Arn, H. (1968). *Nature (London)* **219,** 513.
Roelofs, W. L., and Cardé, R. T. (1971). *Science* **171,** 684.
Roelofs, W. L., and Cardé, R. T. (1974). *Environ. Entomol.* **3,** 586.
Roelofs, W. L., and Comeau, A. (1969). *Science* **165,** 398.
Roelofs, W. L., and Comeau, A. (1970). *J. Econ. Entomol.* **63,** 969.
Roelofs, W. L., and Comeau, A. (1971a). *In* "Pesticide Chemistry, Vol. 3; Chemical Releasers in Insects" (A. S. Tahori, ed.), pp. 91–112. Gordon & Breach, New York.
Roelofs, W. L., and Comeau, A. (1971b). *J. Insect Physiol.* **17,** 435.
Roelofs, W. L., and Feng, K.-C. (1967). *Ann. Entomol. Soc. Amer.* **60,** 1199.
Roelofs, W. L., and Tette, J. P. (1970). *Nature (London)* **226,** 1172.
Roelofs, W. L., Comeau, A., and Selle, R. (1969). *Nature (London)* **224,** 723.
Roelofs, W. L., Cardé, R., Benz, G., and von Salis, G. (1971a). *Experientia* **27,** 1438.
Roelofs, W., Comeau, A., Hill, A., and Milicevic, G. (1971b). *Science* **174,** 297.
Roelofs, W. L., Tette, J. P., Taschenberg, E. F., and Comeau, A. (1971c). *J. Insect Physiol.* **17,** 2235.
Roelofs, W., Kochansky, J., Cardé, R., Arn, H., and Rauscher, S. (1973). *Mitt. Schweiz. Entomol. Ges.* **46,** 71.
Roelofs, W. L., Hill, A. S., Cardé, R. T., and Baker, T. C. (1974a). *Life Sci.* **14,** 1555.
Roelofs, W., Hill, A., Cardé, R., Tette, J., Madsen, H., and Vakenti, J. (1974b). *Environ. Entomol.* **3,** 747.
Röller, H., Biemann, K., Bjerke, J. S., Norgard, D. W., and McShan, W. H. (1968). *Acta Entomol. Bohemoslov.* **65,** 208.
Rudinsky, J. A., Morgan, M., Libbey, L. M., and Michael, R. R. (1973). *Environ. Entomol.* **2,** 505.
Rudinsky, J. A., Morgan, M. E., Libbey, L. M., and Putnam, T. P. (1974). *Environ. Entomol.* **3,** 90.
Rusiecki, W., and Henneberg, M. (1963). *Ann. Pharm. Fr.* **21,** 843.
Russell, G. F., and Hills, J. I. (1971). *Science* **172,** 1043.
Sakurai, K., Matsuda, K., Sugawara, R., and Muto, T. (1973). *Appl. Entomol. Zool.* **8,** 248.
Sanders, C. J., and Lucuik, G. S. (1972). *Can. Forest. Serv. Bi-Mon. Res. Notes* **28,** 10.
Sanders, C. J., Bartell, R. J., and Roelofs, W. L. (1972). *Can. Forest. Serv. Bi-Mon. Res. Notes* **28,** 9.

Sanders, C. J., Daterman, G. E., Shepherd, R. F., and Cerezke, H. (1974). *Can. Entomol.* **106,** 157.

Sannasi, A., and Rajulu, G. S. (1971). *Life Sci., Part II* **10,** 195.

Sarmiento, R., Beroza, M., Bierl, B. A., and Tardif, J. G. R. (1972). *J. Econ. Entomol.* **65,** 665.

Schneider, D. (1957). *Z. Vergl. Physiol.* **40,** 8.

Schneider, D. (1969). *Science* **163,** 1031.

Schneider, D., Kasang, G., and Kaissling, K.-E. (1968). *Naturwissenschaften* **55,** 395.

Schönherr, J. (1972). *Z. Angew. Entomol.* **71,** 260.

Sekul, A. A., and Sparks, A. N. (1976). *USDA Tech. Bull.* (in press).

Shorey, H. H., Gaston, L. K., and Jefferson, R. N. (1968). *Advan. Pest. Contr. Res.* **8,** 57.

Siggia, S., ed. (1972). "Instrumental Methods of Organic Functional Group Analysis." Wiley (Interscience), New York.

Silverstein, R. M. (1970). *In* "Control of Insect Behavior by Natural Products" (D. L. Wood, R. M. Silverstein, and M. Nakajima, eds.), pp. 285–299. Academic Press, New York.

Silverstein, R. M., and Rodin, J. O. (1966). *J. Econ. Entomol.* **59,** 1152.

Silverstein, R. M., Rodin, J. O., and Wood, D. L. (1966). *Science* **154,** 509.

Silverstein, R. M., Rodin, J. O., Burkholder, W. E., and Gorman, J. E. (1967). *Science* **157,** 85.

Silverstein, R. M., Brownlee, R. G., Bellas, T. E., Wood, D. L., and Browne, L. E. (1968). *Science* **159,** 889.

Silverstein, R. M., Bassler, G. C., and Morrill, T. C. (1974). "Spectrometric Identification of Organic Compounds," 3rd ed. Wiley, New York.

Skinner, W. A., Tong, H. C., Maibach, H. I., and Skidmore, D. (1970). *Experientia* **26,** 728.

Smith, J. F. (1960). *Chem. Ind. (London)* p. 1024.

Smith, R. G., Daterman, G. E., Daves, D. G., Jr., McMurtrey, K. D., and Roelofs, W. L. (1974). *J. Insect Physiol.* **20,** 661.

Smith, R. G., Daterman, G. E., and Daves, G. D., Jr. (1975). *Science* **188,** 63.

Smythe, R. V., Coppel, H. C., Lipton, S. H., and Strong, F. M. (1967). *J. Econ. Entomol.* **60,** 228.

Solomon, J. D., Doolittle, R. E., and Beroza, M. (1972). *Ann. Entomol. Soc. Amer.* **65,** 1058.

Sonnet, P. E., and Moser, J. C. (1973). *Environ. Entomol.* **2,** 851.

Sower, L. L., Gaston, L. K., and Shorey, H. H. (1971). *Ann. Entomol. Soc. Amer.* **64,** 1448.

Sower, L. L., Vick, K. W., and Tumlinson, J. H. (1974). *Environ. Entomol.* **3,** 120.

Stahl, E. (1969). "Thin-layer Chromatography," 2nd ed. Springer-Verlag, Berlin and New York.

Ställberg-Stenhagen, S., and Stenhagen, E. (1968). *Sv. Kem. Tidsskr.* **80,** 178.

Staten, R. T., Beroza, M., Bierl, B. A., and Adler, V. E. (1973). *J. Econ. Entomol.* **66,** 1263.

Steinbrecht, R. A. (1964). *Z. Vergl. Physiol.* **48,** 341.

Steiner, L. F. (1952). *J. Econ. Entomol.* **45,** 241.

Steiner, L. F., Rohwer, G. G., Ayers, E. L., and Christenson, L. D. (1961). *J. Econ. Entomol.* **54,** 30.

Steiner, L. F., Mitchell, W. C., Harris, E. J., Kozuma, T. T., and Fujimoto, M. S. (1965). *J. Econ. Entomol.* **58,** 961.

Stenhagen, E., Abrahamsson, S., and McLafferty, F. W., eds. (1969). "Atlas of Mass Spectral Data," Vols. 1–3. Wiley (Interscience), New York.

Stenhagen, E., Abrahamsson, S., and McLafferty, F. W., eds. (1974). "Registry of Mass Spectral Data," Vols. 1–4. Wiley (Interscience), New York.
Storherr, R. W., and Watts, R. R. (1965). *J. Ass. Offic. Agr. Chem.* **48,** 1154.
Storherr, R. W., Murray, E. J., Klein, I., and Rosenberg, L. A. (1967). *J. Ass. Offic. Anal. Chem.* **50,** 605.
Sutherland, O. R. W., and Hutchins, R. F. N. (1972). *Nature (London)* **239,** 170.
Sutherland, O. R. W., and Hutchins, R. F. N. (1973). *J. Insect Physiol.* **19,** 723.
Tamaki, T., and Yushima, T. (1974). *J. Insect. Physiol.* **20,** 1005.
Tamaki, Y., Noguchi, H., Yushima, T., and Hirano, C. (1971a). *Appl. Entomol. Zool.* **6,** 139.
Tamaki, Y., Noguchi, H., Yushima, T., Hirano, C., Honma, K., and Sugawara, H. (1971b). *Konchu* **39,** 338.
Tamaki, Y., Noguchi, H., and Yushima, T. (1973). *Appl. Entomol. Zool.* **8,** 200.
Tashiro, H., Gertler, S. I., Beroza, M., and Green, N. (1964). *J. Econ. Entomol.* **57,** 230.
Trave, R., and Pavan, M. (1956). *Chim. Ind. (Milan)* **38,** 1015.
Tumlinson, J. H., and Heath, R. R. (1976). *J. Chem. Ecol.* **2,** 87.
Tumlinson, J. H., Minyard, J. P., Hedin, P. A., and Thompson, A. C. (1967). *J. Chromatogr.* **29,** 80.
Tumlinson, J. H., Hardee, D. D., Minyard, J. P., Thompson, A. C., Gast, R. T., and Hedin, P. A. (1968). *J. Econ. Entomol.* **61,** 470.
Tumlinson, J. H., Hardee, D. D., Gueldner, R. C., Thompson, A. C., Hedin, P. A., and Minyard, J. P. (1969). *Science* **166,** 1010.
Tumlinson, J. H., Gueldner, R. C., Hardee, D. D., Thompson, A. C., Hedin, P. A., and Minyard, J. P. (1971). *J. Org. Chem.* **36,** 2616.
Tumlinson, J. H., Mitchell, E. R., Browner, S. M., Mayer, M. S., Green, N., Hines, R., and Lindquist, D. A. (1972a). *Environ. Entomol.* **1,** 354.
Tumlinson, J. H., Mitchell, E. R., Browner, S. M., and Lindquist, D. A. (1972b). *Environ. Entomol.* **1,** 466.
Tumlinson, J. H., Moser, J. C., Silverstein, R. M., Brownlee, R. G., and Ruth, J. M. (1972c). *J. Insect Physiol.* **18,** 809.
Tumlinson, J. H., Heath, R. R., and Doolittle, R. E. (1974a). *Anal. Chem.* **46,** 1309.
Tumlinson, J. H., Yonce, C. E., Doolittle, R. E., Heath, R. R., Gentry, C. R., and Mitchell, E. R. (1974b). *Science* **185,** 614.
Tumlinson, J. H., Hendricks, D. E., Mitchell, E. R., Doolittle, R. E., and Brennan, M. M. (1975). *J. Chem. Ecol.* **1,** 203 (submitted for publication)
VandenHeuvel, W. J. A., and Horning, E. C. (1962). *Biochim. Biophys. Acta* **64,** 416.
Vick, K. W., and Sower, L. L. (1973). *J. Econ. Entomol.* **66,** 1258.
Vick, K. W., Su, H. C. F., Sower, L. L., Mahany, P. G., and Drummond, P. C. (1974). *Experientia* **30,** 17.
Vité, J. P., and Pitman, G. B. (1970). *J. Econ. Entomol.* **63,** 1132.
Vité, J. P., and Renwick, J. A. A. (1971). *J. Insect Physiol.* **17,** 1699.
Vité, J. P., and Williamson, D. L. (1970). *J. Insect Physiol.* **16,** 233.
Vité, J. P., Bakke, A., and Hughes, P. R. (1974). *Naturwissenschaften* **61,** 365.
Wearing, C. H., and Hutchins, R. F. N. (1973). *J. Insect Physiol.* **19,** 1251.
Weatherston, J., and Maclean, W. (1974). *Can. Entomol.* **106,** 281.
Weatherston, J., Roelofs, W., Comeau, A., and Sanders, C. J. (1971). *Can. Entomol.* **103,** 1741.
Weatherston, J., Davidson, L. M., and Simonini, D. (1974). *Can. Entomol.* **106,** 781.
Wheeler, J. W., Evans, S. L., Blum, M. S., and Torgerson, R. L. (1975). *Science* **187,** 254.
Wientjens, W. H. J. M., Lakwijk, A. C., and van der Marel, T. (1973). *Experientia* **29,** 658.
Williamson, D. L. (1971). *Ann. Entomol. Soc. Amer.* **64,** 586.

Yamamoto, R. (1963). *J. Econ. Entomol.* **56,** 119.
Young, J. C., Brownlee, R. G., Rodin, J. O., Hildebrand, D. N., Silverstein, R. M., Wood, D. L., Birch, M. C., and Browne, L. E. (1973a). *J. Insect Physiol.* **19,** 1615.
Young, J. C., Silverstein, R. M., and Birch, M. C. (1973b). *J. Insect Physiol.* **19,** 2273.
Yushima, T., Noguchi, H., Tamaki, Y., Fukazawa, N., and Sugino, T. (1973). *Appl. Entomol. Zool.* **8,** 18.

Part II

Insecticides

3

Allethrin*

(*d-trans*-Allethrin)

Joseph Sherma

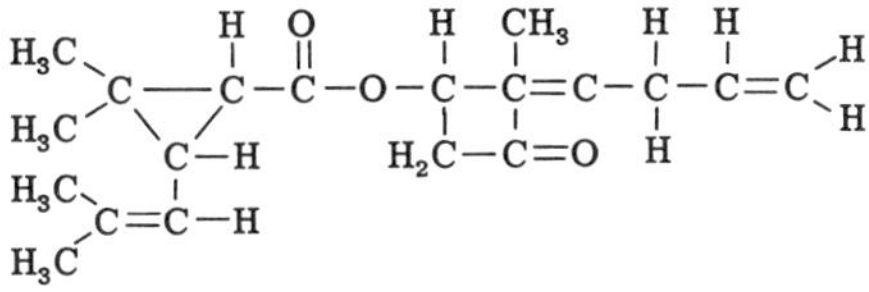

dl-2-Allyl-4-hydroxy-3-methyl-2-cyclopenten-1-one ester of *d-trans*-2,2-dimethyl-3-(2-methylpropenyl)cyclopropanecarboxylic acid

I. GENERAL

See Volume II, pp. 25–28, for general information on mixed *cis*- and *trans*-allethrin. A summary of some of the characteristics of *d-trans*-allethrin (allyl homolog of cinerin I), the topic of this chapter, is given below.

A. Empirical Formula

$C_{19}H_{26}O_3$

B. Alternate Name

Bioallethrin (in areas outside the United States).

C. Characteristics of d-trans® Concentrate, 90%

d-trans-Allethrin (USDA Reg. No. 1021–1060) is available from McLaughlin Gormley King Co., Minneapolis, Minnesota in the form of this concentrate, standardized to 90%, and in the form of intermediates in combination with synergists, other toxicants, and auxiliary materials. Technical material, which varies slightly in purity from batch to batch, is

* See also Volume II, Chapter 2 and Volume VI, Chapter 9.

not available from McLaughlin Gormley King Co. except for research or experimental purposes.

Description: Amber viscous liquid.

Odor: Slight aromatic odor.

Specific gravity: Typical, 0.994.

Weight per gallon: 8.28 lbs.

Refractive index ($n_d^{25°C}$): Typical, 1.500.

Solubility: Miscible in all proportions with acetone, benzene, ethanol, hexane, methanol, toluene, trichloroethane, and refined kerosene. Insoluble in water.

Flash point: Above 150°F, tag open cup.

Free acid chloride: When the sample is titrated to a pH of 5, it shall contain not more than 0.5% chrysanthemum monocarboxylic acid chloride.

Chrysanthemumic acid and chrysanthemic anhydride: Less than 4%.

d-trans-Allethrin content: When assayed in accordance with the AOAC ethylenediamine method, this concentrate shall contain not less than 90% of the ester.

Insecticidal evaluations of *d-trans*-allethrin were reported by Fales *et al.* (1970–1971), and the photodecomposition of allethrin and four related pyrethroids was studied by Ueda *et al.* (1974). Rickett (1973) determined the enantiometric purity of allethrin and related synthetic pyrethroids.

II. FORMULATION ANALYSIS

Volume II contains directions for performing the standard ethylenediamine titrimetric analysis of allethrin formulations. McLaughlin Gormley King Co. has developed a more convenient gas chromatography (GC) procedure (Meinen and Kassera, 1973) for *d-trans*-allethrin concentrate and end product, which will be described below.

A. Principle

d-trans-Allethrin is diluted with acetone containing dibutyl phthalate as internal standard. Ratios of GC peak heights of allethrin and the standard are compared for quantitative determination.

B. Apparatus and Reagents

Gas chromatograph. Use a gas chromatograph equipped with a flame ionization detector and a 4 foot × 4 mm glass column packed with

5% OV-1 on 80–100 mesh Chromosorb W(HP). The column is conditioned before use for 2–3 hours at 275°C with a nitrogen flow of 50 ml/minute. Operating parameters are adjusted to yield retention times of about 4 minutes for the internal standard and 7 minutes for *d-trans*-allethrin and a peak height ⩾ 100 mm for each compound. The column should exhibit at least 200 theoretical plates per foot. Typical conditions are column 165°C, injection port 230°C, detector 230°C; nitrogen carrier gas 125 ml/minute, air 350–400 ml/minute, hydrogen 40–50 ml/minute; sensitivity 10^{-9} A full scale, attenuation 4× for technical material and 1× for formulations; 16 μg allethrin and 8 μg DBP injected (see Fig. 1).

Internal standard solution. 4.0 mg dibutyl phthalate per milligram of acetone.

d-trans-Allethrin standard solutions. (A) Approximately 4 mg/ml

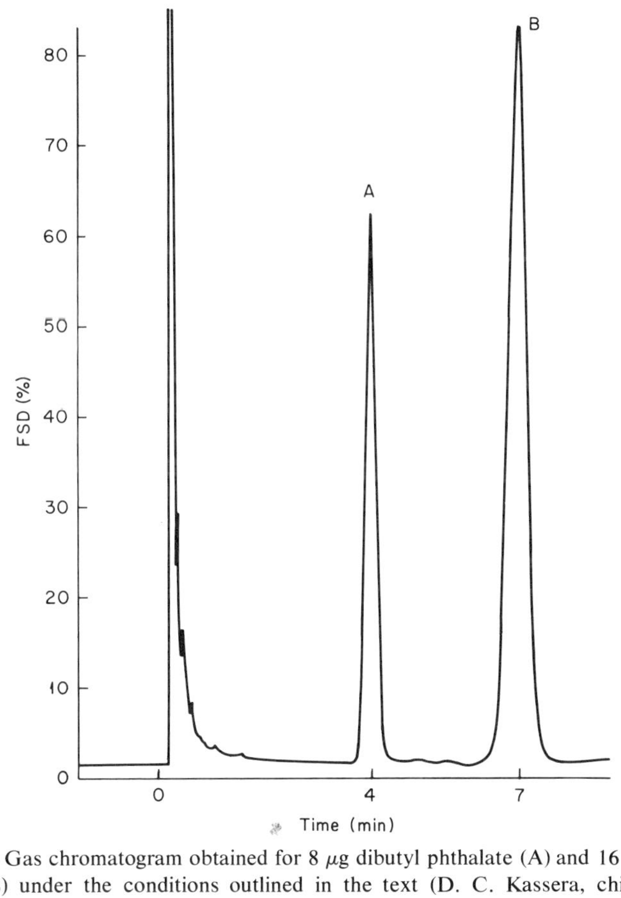

FIG. 1. Gas chromatogram obtained for 8 μg dibutyl phthalate (A) and 16 μg *d-trans*-allethrin (B) under the conditions outlined in the text (D. C. Kassera, chief chemist, McLaughlin Gormley King Co., personal communication, 1974).

for determining technical material: Weigh accurately about 1.0 gm allethrin standard into a 50-ml volumetric flask and dilute to the line with acetone. Pipette 20 ml of this solution into a 100-ml volumetric flask, add 50 ml internal standard solution by pipette, and dilute to volume with acetone. (B) Approximately 1 mg/ml for determining formulations: Dilute 25 ml of solution A to 100 ml with acetone.

C. Sample Preparation

Technical material. Accurately weigh a sample containing about 1.0 gm *d-trans*-allethrin into a 50-ml volumetric flask and dilute to volume with acetone. Pipette a 20-ml portion into a 100-ml volumetric flask, add 50 ml internal standard solution by pipette, and dilute to volume with acetone.

Formulations. Accurately weigh a sample containing about 200 mg *d-trans*-allethrin into a 50-ml volumetric flask, add 25 ml internal standard solution by pipette, and dilute with acetone. Further dilute 25 ml of this solution to 100 ml with acetone in a volumetric flask.

D. Gas Chromatography

Technical material. Inject approximately 3 μl portions of standard solution A until the ratio of peak heights for allethrin and DBP agrees within ±1% or less for successive injections. Repeat with the sample solution, followed by duplicate injections of standard solution. If the peak height ratios differ by more than ±1% from the previous standard injections, repeat the series of injections.

Formulations. Proceed as above using standard solution B, repeating standard injections after each series of three sample injections. If peak height ratios differ by more than ±1.5% from previous standard injections, repeat injections.

E. Calculations

Technical material. Calculate peak height ratios for duplicate standard injections before and after sample injections and average these four values. Calculate and average the peak height ratios for the sample injections and use the following equation

$$\%\ d\text{-}trans\text{-allethrin} = \frac{W'PR}{WR'}$$

where W' and W are the respective standard and sample weights in grams, P is the percentage purity of the standard, and R' and R are the respective peak height ratios of the standard and sample.

Formulations. Calculate average values for all standard and sample peak height ratios and use the equation

$$\%\ d\text{-}trans\text{-allethrin} = \frac{2W'PR}{WR'}$$

where W' is the grams standard in the final dilution.

F. Applications

The method is applicable to both technical *d-trans*-allethrin and various formulations. It is not applicable to formulations containing large amounts of MGK Repellent 874 (2-hydroxyethyl-*n*-octyl sulfide).

G. Results and Discussion

A collaborative study of this GC procedure was described by Meinen and Kassera (1973). The resolution of the GC column was found to be critical in obtaining good results, so that the minimum requirement of 200 theoretical plates per foot should be observed

$$N = \text{theoretical plates per foot} = 16L^2/M^2F$$

where L = retention of the allethrin peak in millimeters, M = peak baseline width in millimeters, F = length of column in feet. Statistical evaluation of accuracy and precision of the collaborative study was favorable, and the method was adopted by the AOAC as official first action.

III. RESIDUE ANALYSIS

See chapter 35 on pyrethrum in Volume II for methods applicable to allethrin.

Acknowledgment

The author is grateful to D. C. Kassera for his help and interest and for supplying the information upon which the chapter is based.

References

Fales, J. H., Bodenstein, O. F., Waters, R. M., and Fields, E. S. (1970). *Soap Chem. Spec.* **46,** 78.

Fales, J. H., Bodenstein, O. F., Waters, R. M., and Fields, E. S. (1971). *Soap Chem. Spec.* **47,** 64.

Meinen, V. J., and Kassera, D. C. (1973). *J. Ass. Offic. Anal. Chem.* **56,** 562.

Rickett, F. E. (1973). *Analyst* **98,** 687.

Ueda, K., Gaughan, G., and Casida, J. E. (1974). *J. Agr. Food Chem.* **22,** 212.

4

Formothion

M. Wisson, C. van Hoek, and H. H. Sauer

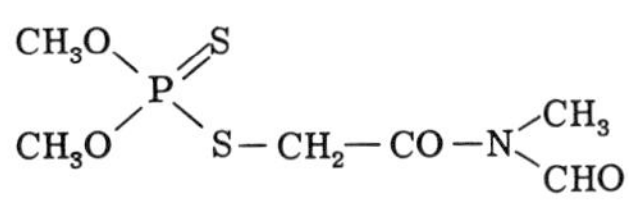

O,O-Dimethyl-*S*-(*N*-methyl-*N*-Formylcarbamoylmethyl) Phosphorodithioate

I. GENERAL

A. Empirical Formula

$C_6H_{12}NO_4PS_2$ (Mol. wt. 257.27).

B. Alternative Names

S-[2-(Formylmethylamino)-2-oxoethyl]*O,O*-dimethyl phosphorodithioate; *O,O*-dimethyl-*S*-(*N*-formyl-2-mercapto-*N*-methylacetamide) phosphorodithioate; Anthio® is the registered trademark of Sandoz Ltd., Basel, Switzerland.

C. Source of Analytical Standard

Sandoz Ltd., Agrochemical Research Division, Basel, Switzerland.

D. Biological Properties

1. INSECTICIDAL ACTIVITY

Formothion is an organophosphorous insecticide with systemic and contact action against a wide range of sucking, mining, and some chewing pests on all main crops. Full action is evident soon after application with a moderate persistence (Dujardin, 1963; Wood and Tyson, 1965; Berg and Riggenbach, 1966).

The metabolism of formothion in the plant leads to approximately equal amounts of dimethoate, which is degraded consequently to its P=O metabolite dimethoxon, and *O,O*-dimethyldithiophosphorylacetic acid and bis(*O,O*-dimethylthiophosphoryl) disulfide (Sauer, 1972). These two products of low toxicity are shown to act as synergists upon dimethoate (Bassand and Klotzsche, 1970).

2. TOXICITY

Formothion is one of the least toxic systemic organophosphates. In rats, the acute oral LD_{50} is 365–535 mg/kg, and the acute dermal LD_{50} ranges between 400 and >2800 mg/kg according to type and concentration of formothion. The LC_{50} is 45 mg formothion per 10 liters of air per minute on inhalation lasting for 4 hours. After 24 months' feeding in rats and dogs, the no-effect level was found to be 80 ppm.

E. History

Formothion was discovered and developed by Sandoz Ltd. (1959). It has been registered for use in vegetables, orchards, and industrial crops in many countries.

F. Physical Properties

1. Formothion is an odorless, yellowish, viscous liquid or crystalline mass. It has a melting point of 25–26°C and a vapor pressure of 8.5×10^{-6} Torr at 20°C. It cannot be distilled without decomposition.

2. The density d_4^{20} is 1.361, the refractive index n_D^{20} is 1.5541.

3. It is miscible with most organic solvents except aliphatic hydrocarbons, in which it is practically insoluble. Its water solubility is 2600 ppm at 24°C.

G. Chemical Properties

Formothion is fairly unstable in its pure form, but stable in nonpolar solvents. It is hydrolyzed in the presence of water to dimethoate and *O,O*-dimethyldithiophosphorylacetic acid. Hydrolysis occurs faster under alkaline than acidic conditions.

H. Formulations

Anthio® insecticide is available as emulsifiable liquid concentrate containing 25, 33, and 40% active ingredient, and as ultra low volume (ULV) liquid concentrates containing 35 and 70% active ingredient.

II. ANALYSIS

A. Formulations

1. REVIEW OF METHODS

Formothion in technical products and liquid formulations can be assayed by two methods, which are both approved and recommended by the Collaborative International Pesticides Analytical Council Ltd. (CIPAC) after extensive work of collaborative analytical laboratories.

The first method utilizes gas chromatography as a rapid, accurate, and selective method for analyzing formothion. Formulations are diluted with toluene and injected, preferably by an automatic injector, to the gas chromatograph equipped with a flame ionization detector or flame photometric detector. Quantitation is achieved using the internal standard technique (ethion and/or triphenyl phosphate as standards) by comparing the peak area ratios of the unknown sample with the corresponding peak area ratios of a reference solution containing known amounts of formothion and internal standard.

The second method is based on the separation from by-products and formulation additives by paper chromatography, followed by combustion of the chromatogram spots according to Schöniger (1955, 1956). The content of formothion is determined colorimetrically by the phosphorus molybdenum blue method described by Faderl (1962). This procedure has shown to be an accurate and selective method for the determination of formothion. Other materials present, such as solvents, emulsifiers, dye stuffs, or other additives do not interfere. By-products containing phosphorus, e.g., dimethoate or other breakdown products, are found to be separated by paper chromatography using appropriate solvent systems. Furthermore, this procedure does not need a formothion standard.

Both methods yield identical and satisfactory results. They were tested by a CIPAC collaborative study (Sauer and Bosshardt, 1974) and accepted as approved CIPAC methods, the gas–liquid chromatography (GLC) method being the referee method. In comparing the two methods, the GLC method provides better reproducibility, and the phosphorus combustion is certainly the more complicated method needing skilled manual performance. However, the second method is of great versatility, being applicable to all organophosphorus compounds and independent of analytical organophosphorus ester standards.

2. RECOMMENDED METHOD—GLC METHOD

See Association of Official Analytical Chemists (1975).

3. RECOMMENDED METHOD—PHOSPHORUS DETERMINATION

See Sauer and Bosshardt (1974).

B. Residues

1. REVIEW OF METHODS

Gas chromatography with the alkali flame ionization detector or flame photometric detector has been shown to be the best method for determining formothion residues because of their high selectivity, sensitivity, and precision. For confirmatory purposes, it is recommended to use paper chromatography.

Metabolism studies and toxicological investigations have shown that plant material and soil treated with formothion preparations must be analyzed not only for the active ingredient itself, but also for its toxicologically important metabolites, dimethoate and dimethoxon.

2. RECOMMENDED METHOD—GLC METHOD

a. *Principle*

Residues of formothion and its metabolites are extracted with acetonitrile containing acetic acid and purified by two cleanup steps utilizing partition between water and chloroform, as well as adsorption on a Florisil charcoal column. Quantitative determination is achieved by GLC equipped with an alkali flame ionization detector (AFID) and/or flame photometric detector (FDP) using the external calibration method.

b. *Reagents*

Celite 545 (see Food and Drug Administration, 1971).

Acetonitrile, technical, redistilled from P_2O_5/H_3PO_4 (see Food and Drug Administration, 1971).

Acetic acid, glacial, Analytical Grade.

Chloroform, redistilled and purified over alumina basic, activity grade I (M. Woelm, Eschwege, Germany or equivalent).

Sodium sulfate, anhydrous, Analytical Grade.

Petroleum ether, as described for chloroform.

Florisil, 60–100 mesh.

Activated charcoal, Norit A (American Norit Co.) or equivalent.

Acetic anhydride, Analytical Grade.

Hexane, Analytical Grade.

Toluene, Analytical Grade.

Formothion analytical standard of known purity.

Dimethoate analytical standard of known purity.

Dimethoxon analytical standard of known purity.

Florisil–charcoal column: Prepare a suspension of 3 gm Florisil in 10 ml of acetonitrile and pour into a chromatographic tube. Allow the liquid to drain to 5 cm above the Florisil. Add 2 gm of a Florisil–charcoal mixture (9 : 1) followed by another 2 gm of florisil. Elute the column finally with 20 ml of acetonitrile containing 0.5% (v/v) acetic anhydride.

c. *Apparatus*

Blender, Waring or equivalent, with 1 liter glass jars.

Shaking machine.

Rotary evaporator equipped with a water bath settled at a maximum temperature at 50°C.

Gas chromatograph, Varian Aerograph 1200 and 2740 or equivalent equipped with an alkali flame ionization detector or flame photometric detector (P mode).

Automatic injector, Hewlett Packard 7670 A or equivalent.

Integrator, Infotronics CRS 208 or equivalent.

Syringe for injection, Hamilton, 10 μl size.

Effluent splitter at column end with ratio 1 : 100 to 1 : 1000 in favor of outlet.

d. *Experimental Procedure*

i. *Extraction Procedure*

Reduce the entire sample to small pieces and mix thoroughly. Place 100 gm of material in a blender jar, add 20 gm of Celite, 200 ml of acetonitrile, and 0.5 ml of acetic acid, and blend for 3 minutes.

Filter with suction through a filter paper placed in a suction funnel, and collect the filtrate. Return the filter cake to the blender jar, add 150 ml acetonitrile and blend for 2 minutes. Transfer the homogenate into a 1 liter screw-cap bottle. Rinse the jar with the aid of 50 ml acetonitrile and add it also to the bottle. Shake the slurry mechanically for 30 minutes and filter into the same filtrate. Evaporate in a rotatory evaporator to a final volume of about 40 ml.

Pour the remaining solution into a 250-ml separatory funnel, rinse the flask successively with 30 ml redistilled water and 90 ml chloroform containing 0.1% (v/v) acetic acid, and transfer both rinsings into the separatory funnel. Shake vigorously, allow to separate and drain the lower phase through anhydrous sodium sulfate into a 500-ml round-bottom flask. Extract the aqueous phase again 4 times with each 70 ml of chloroform containing 0.1% acetic acid. Evaporate the combined organic extracts in a rotatory evaporator.

ii. *Cleanup Procedure*

If the extract is still heavily contaminated, add 40 ml of petroleum ether to the flask and transfer it into a 250-ml separatory funnel. Rinse the flask with 40 ml redistilled water and pour it into the separatory funnel. Shake well and run off the aqueous phase into a second separatory funnel. Repeat the procedure another 3 times with each 40 ml of redistilled water and collect all aqueous phases. Discard the petroleum ether.

Re-extract the aqueous layer 6 times with each 90 ml portions of chloroform containing 0.1% acetic acid. Evaporate the combined organic phases to dryness in a rotatory evaporator. Discard the aqueous phase.

Dissolve the remainder in 10 ml acetonitrile containing 0.5% of acetic anhydride and transfer it onto the Florisil–charcoal column. Rinse several times with each 5 ml of the same solvent mixture, and pour it onto the column just before the proceeding solution has sunk to the adsorbent surface. Complete the elution until the total volume of the eluate reaches 65 ml. Evaporate in a rotatory evaporator to a small volume. Transfer quantitatively to a 10-ml volumetric flask, rinse with the aid of

TABLE I
GLC CONDITIONS

Column	Glass spiral (70 cm × $\frac{1}{8}$ inch)	Glass spiral (100 cm × $\frac{1}{8}$ inch)
Stationary phase	6% Carbowax 20 *M* on Chromosorb W AW DMCS (80–100 mesh)	10% Dexsil 300 on Gaschrom Q (60–80 mesh)
Temperature (°C)		
Column oven	190	150
Injection port	230	180
Detector (AFID)	240	240
Detector (FPD)	180	180
Carrier gas (ml/minute)		
Nitrogen	30	60
Hydrogen (AFID)	40	40
Hydrogen (FPD)	200	200
Air (AFID)	200	200
Air (FPD)	70	70
Oxygen (FPD)	10	10
Retention time (minutes)		
Formothion	8.0	5.2
Dimethoate	6.5	3.1
Dimethoxon	4.2	1.3

hexane or toluene, and dilute to the volume. Quantitate the content by GLC using the external calibration method.

iii. *GLC Determination* (see Table I)

Inject 1.0–10 μl of the solutions containing up to 2 ng (AFID) or up to 50 ng (FPD) of each compound. In order to avoid contamination and flame extinction, bypass the solvent around the detector by using the splitter at column end. Determine the appropriate time interval for splitting by a test chromatogram.

iv. *Calibration*

Weigh accurately 200 mg each of formothion and its metabolites into 20-ml volumetric flasks and dissolve in toluene. Pipette 1 ml of each solution into a 100-ml volumetric flask and fill up to the mark with a hexane–toluene mixture (1 : 1). Dilute this stock solution with hexane to yield standard solutions.

Equilibrate the gas chromatograph, inject each solution, and evaluate by plotting the peak areas versus nanograms injected for each compound. Ascertain that the linearity of the calibration curve ranges for the AFID are at least from 0.2 to 2.0 ng, and from 0.2 to 50 ng using the FPD.

e. *Calculation*

Read the amount of formothion and its metabolites in ng from the calibration curve. Calculate the residues by

$$\text{ppm} = \frac{\text{ng read} \times \text{ml final extract}}{\mu\text{l injected} \times \text{gm of sample}} \times \text{dilution factor}$$

f. *Applicability of the Method*

This procedure has been applied successfully to apples, artichokes, beans, cabbage, carrots, cherries, peel and flesh of citrus fruits, grapes, leek, lettuce, peaches, peas, radish, tea, tobacco, tomatoes, and wheat.

g. *Modification of the Method for Determination of Formothion Residues in Soil*

Mix the entire sample thoroughly and eliminate stones larger than 1 cm diameter. Place 100 gm of the undried samples in a 1-liter screw-cap bottle and add 20 gm Celite, 200 ml acetonitrile, and 0.5 ml acetic acid. Shake mechanically for 30 minutes; then filter with suction through a filter paper and collect the filtrate. Return the filter cake to the bottle with the aid of 200 ml of acetonitrile. Shake the slurry again and filter into the same flask. Then treat the crude extracts as described in Section d above.

h. *Discussion*

As established for analyses of dimethoate residues by Boyd (1972), gas chromatographic determinations of formothion residues using the phosphorus-sensitive detector give excellent results. In general, no interfering substances from plant materials or soil are present as shown by injecting extracts of untreated materials for each sample. Recoveries at 0.1 to 1.0 ppm levels generally have been in the range between 70 and 90% for formothion, 80 and 100% for dimethoate, and 60 and 90% for dimethoxon. The limit of detection has ranged down to 0.02 ppm for formothion and dimethoxon and 0.01 ppm for dimethoate.

3. CONFIRMATORY METHOD—PAPER CHROMATOGRAPHY

a. *Principle*

Cleaned-up sample extracts and reference formothion, dimethoate, and dimethoxon standards are spotted on impregnated paper chromatograms and developed with several solvents. The developed chromatograms are sprayed with appropriate spray reagents. The method is used to obtain additional information on identity of formothion residues determined by GLC.

b. *Reagents*

Chromatographic papers, Whatman no. 1, 16 × 51 cm sheets.
Formamide, Reagent Grade.
Acetone, Analytical Grade.
Toluene, Reagent Grade.
2,2,4-Trimethylpentane, Reagent Grade.
Chloroform, redistilled from all glass apparatus.
Bromine, Analytical Grade.
β-Methylumbelliferone, Reagent Grade.
Ethanol, Reagent Grade.
Ammonia, 0.1 *N* in redistilled water.

β-Methylumbelliferone reagent. Dissolve 0.15 gm β-methylumbelliferone in 100 ml ethanol and add 100 ml redistilled water. Immediately before use, add 10 ml 0.1 *N* ammonia.

Potassium hexaiodoplatinate, (Heraeus GmbH, Hanau, Germany).
Hydroiodic acid, 57%, Reagent Grade.

Potassium hexaiodoplatinate reagent. Dissolve 1 gm potassium hexaiodoplatinate in 20 ml redistilled water, mix with 2 ml hydroiodic acid, and dilute with redistilled water to a final volume of 1000 ml.

Sodium hydroxide, 2 *N* in aqueous methanol (1 : 1).
Hydrochloric acid, 3 *N* in aqueous methanol (1 : 1).
Formothion analytical standard of known purity.

Dimethoate analytical standard of known purity.
Dimethoxon analytical standard of known purity.

c. *Apparatus*

Chromatographic tank and accessories for descending paper chromatography.

Micropipettes, 5 μl, Drummond Microcaps (Drummond Scientific Co., Broomall, Pa., USA) or equivalent.

UV lamp, 366 nm.

d. *Experimental Procedure*

i. *Preparation of the Chromatographic Papers*

With a lead pencil, rule one horizontal line 7.5 cm from the top to mark the origin, and a second 30 cm below the first to mark the advance of the solvent front. At the origin, mark two to four dots, equally spaced, with the extreme dots approximately 3 cm from the edge of the paper.

Charge a dipping trough with a 12% solution of formamide in acetone. Impregnate the paper by holding it in spring clips by two edges and drawing it through the solution as evenly as possible from the origin to the line indicating the advance of the solvent front. Remove it from the solution immediately and hang it up by the top and allow to dry for 15 minutes. Charge the trough with 25 ml of the mobile phase (see Table II).

ii. *Application of the Sample Extracts*

Spot a portion of the sample solution obtained after extraction and cleanup procedure (as described in Sections 2,d,i and 2,d,ii) onto one of the dots at the origin and allow to dry. Repeat as often as necessary to provide a suitable amount by taking into account the GLC results and the lower detection limit. Apply standard solutions containing 1–10 μg of each formothion, dimethoate, and dimethoxon to the other dots on the same paper.

iii. *Development*

Hang the paper in the developing tank with the top dipping about 1.5 cm into the solvent in the trough. Cover the top of the tank with a glass plate. When the solvent front reaches the second line (about 1.5–2 hours), mark the exact position and hang the paper up to dry.

iv. *Visualization*

With β-methylumbelliferone. Expose the dry paper to bromine vapor in a 5-liter wide-mouthed bottle for a few seconds (in the fume cupboard). Expose the paper to the air until it is white again. Spray with β-

TABLE II
R_f VALUES OF FORMOTHION RESIDUES ON PAPER IMPREGNATED WITH FORMAMIDE

Residue	Toluene–2,2,4-trimethylpentane (6:4)	Toluene–chloroform (4:6)
Formothion	0.70	–
Dimethoate	0.15	0.71
Dimethoxon	–	0.22

methylumbelliferone solution and examine under UV in the dark. The compounds appear as dark spots against a light blue fluorescent background. The contrast can be improved by blowing the vapor of 0.1 *N* ammonia over the paper.

With potassium iodoplatinate. Spray the dry paper with potassium iodoplatinate solution. After about 15 minutes, formothion and dimethoate appear as yellow spots on a pink background. To detect dimethoxon, spray the same paper with 2 *N* sodium hydroxide in aqueous methanol and then with 3 *N* hydrochloric acid in aqueous methanol.

v. *Evaluation*

By means of the R_f, coordinate the spots derived from the sample with those derived from the standards on the same paper and make a semiquantitative estimation by comparing the size and intensity of the spots.

vi. *R_f Values*

See Table II.

vii. *Limit of Detection*

The lower limit of detection was established at 0.8 μg for formothion, 0.3 μg for dimethoate, and 0.5 μg for dimethoxon.

e. *Discussion*

The described procedure follows closely the guidelines of paper chromatography as given by the Food and Drug Administration (1968). If necessary the chromatographic spots can be determined quantitatively by the phosphorus combustion in accordance with the procedure described for formulation analysis.

REFERENCES

Association of Official Analytical Chemists. (1975). "Methods of Analysis," 12th ed., p. 119. Washington, D.C.

Bassand, D., and Klotzsche, C. (1970). *Proc. Int. Plant. Protect. Congr., 7th, 1970,* p. 703–704.

Berg, W., and Riggenbach, A. (1966). *World Crops* **18,** 56–59.

Boyd, J. E. (1972). *In* "Analytical Methods for Pesticides, Plant Growth Regulators, and Food Additives" (G. Zweig and J. Sherma, eds.), Vol. 6, pp. 362–375 and literature cited therein. Academic Press, New York.

Dujardin, P. J. (1963). *Def. Veg.* **98,** 6–12.

Food and Drug Administration. (1968). "Pesticide Analytical Manual," Vol. I, p. 511. FDA, Washington, D.C.

Food and Drug Administration. (1971). "Pesticide Analytical Manual," Vol. I, p. 121. FDA, Washington, D.C.

Faderl, N. (1962). *Mitt. Geb. Lebensmittelunters. Hyg.* **53,** 154–175.

Sandoz Ltd. (1959). Swiss Patent 379,833.

Sauer, H. H. (1972). *J. Agr. Food Chem.* **20,** 578–583.

Sauer, H. H., and Bosshardt, H. P. (1974). *J. Ass. Offic. Anal. Chem.* **57,** 771–777.

Schöniger, W. (1955). *Mikrochim. Acta* **1,** 123–129.

Schöniger, W. (1956). *Mikrochim. Acta* **2,** 869–876.

Wood, C. H. P., and Tyson, D. (1965). *Proc. Brit. Insectic. Fungic. Conf., 3rd, 1965,* p. 407–416.

5

Mecarbam

V. P. LYNCH

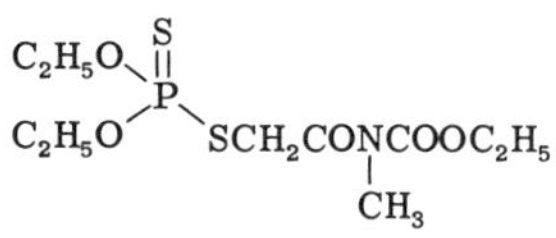

S-(*N*-Ethoxycarbonyl-*N*-methylcarbamoylmethyl)diethyl phosphorothiolothionate

I. GENERAL

A. Empirical Formula

$C_{10}H_{20}NO_5PS_2$ (mol. wt. 329).

B. Alternative Name

Murfotox® is the registered trade mark of Murphy Chemical Ltd.

C. Source of Analytical Standard

Murphy Chemical Ltd., Wheathampstead, St. Albans, Hertfordshire, United Kingdom.

D. Biological Properties

Mecarbam possesses insecticidal and acaricidal properties. It is active against members of the order Hemiptera and against leaf, stem, and root mining larvae of the orders Diptera and Lepidoptera. Its principal use has been in the control of scale insects on citrus and deciduous fruits; leaf hoppers in paddy rice; fruit flies in citrus fruits, cherries, and olives; and root flies in vegetable crops. Mecarbam has also shown promise in the control of leaf miner on coffee. Favorable results have been obtained in the control of pasture pests, particularly the earth mite and Lucerne flea.

The acute oral LD_{50} for rats is 36 mg/kg and 106 mg/kg for mice. The no-effect level in rats lies between 0.85 and 1.64 mg/kg/day.

E. History

Mecarbam was first described by Pianka in 1961.

F. Physical Properties

Mecarbam is a colorless liquid at room temperature. It is practically insoluble in water, soluble to about 5% in aliphatic hydrocarbons, and miscible with alcohols and aromatic hydrocarbons. The purity of the technical material is about 94%.

G. Chemical Properties

1. METHOD OF SYNTHESIS

Mecarbam is prepared by the reaction of a salt of *O*,*O*-diethyldithiophosphoric acid with *N*-chloracetyl-*N*-methylethyl carbamate.

2. CHEMICAL REACTIONS

Mecarbam is slowly hydrolyzed in aqueous acid solutions. It is rapidly hydrolyzed in the presence of alkali.

H. Formulations

Mecarbam is available as a 90, 80, 60, or 50% emulsifiable concentrate and also as a 25% wettable powder.

II. ANALYSIS

A. Analysis of Technical Material and of Formulations

RECOMMENDED METHOD

a. *Principle*

Mecarbam is determined by a gas–liquid chromatography procedure using a flame ionization detector and employing di-*n*-butyl phthalate as internal standard.

b. *Reagents*

Mecarbam, pure reference material.
Di-n-butyl phthalate, internal standard.
Dichloromethane.

c. *Apparatus*

Gas chromatograph, a Pye 104 gas chromatograph (or equivalent) fitted with a flame ionization detector.

Column, 5′ × $\frac{1}{4}$″ glass column packed with 5% SE 30 on Chromosorb W (80–100 mesh), AW-DMCS.

Operating conditions, temperature (°C); oven 200, injection port 225.

Carrier gas, nitrogen (O_2-free) 38 ml/minute. The hydrogen and air flow rates are adjusted to give optimum response.

Integrator, Infotronic Digital Integrator CRS 204 (or equivalent).

Recorder, Vitatron 10 mV (or equivalent).

d. *Procedure*

i. *Calibration of Instrument*

Weigh accurately approximately 0.38 gm of mecarbam standard into a 10-ml stoppered tube; to this add approximately 0.1 gm (accurately weighed) of di-*n*-butyl phthalate and 8 ml of dichloromethane and shake to dissolve. Inject a suitable volume (ca. 0.3 μl) to give an 80% full-scale deflection on the recorder. Carry out the experiments in triplicate.

Calculate the response factor *R* of mecarbam versus di-*n*-butyl phthalate. (Under the conditions described, di-*n*-butyl phthalate and mecarbam elute at about 8 and 12 minutes, respectively).

$$R = \frac{\text{peak area internal standard} \times \text{weight mecarbam}}{\text{peak area mecarbam} \times \text{weight internal standard}}$$

This value has been found to be about 2.7.

ii. *Analysis of Technical Mecarbam and Liquid Formulations*

Accurately weigh a sample to contain approximately 0.38 gm mecarbam and to this add about 0.1 gm (accurately weighed) di-*n*-butyl phthalate. Dissolve the mixture in 8 ml dichloromethane and inject a portion (0.3 μl) into the gas chromatograph. Calculate the mecarbam content as follows:

$$\%\ \text{mecarbam} = \frac{\text{peak area of mecarbam} \times \text{weight internal standard} \times \text{R} \times 100}{\text{peak area internal standard} \times \text{weight of mecarbam}}$$

iii. *Analysis of Wettable Powder Formulation*

Accurately weigh a sample to contain about 0.38 gm of mecarbam into a Soxhlet extractor and extract the mecarbam using about 75 ml dichloromethane for 30 minutes. Cool the flask and concentrate the extract to about 8 ml and add to this solution 0.1 gm di-*n*-butyl phthalate (accurately weighed). Inject a portion (0.3 μl) into the chromatograph and calculate the mecarbam content as above.

B. Residue Analysis

RECOMMENDED METHOD

a. *Principle*

Mecarbam is extracted from crops using hexane. The extract is subjected to a cleanup involving solvent partition and/or column chromatography on Florisil. Mecarbam is determined on a gas–liquid chromatograph fitted with an alkali-flame ionization (thermionic) detector.

b. *Apparatus*

Gas chromatograph. A Pye 104 model 154 isothermal gas chromatograph (or equivalent) fitted with a thermionic detector and a $5' \times \frac{1}{4}''$ glass column packed with 8% silicone fluid MS 200/12500 on Chromosorb W (AW-DMCS) 60–80 mesh.

Operating conditions. Temperature (°C): oven 200, injection port 210.

Carrier gas. Nitrogen (O_2 free) 35 ml/minute flow rate. The flow rates of the air and hydrogen and the position of the cesium bromide tip are adjusted to give optimum response. Under the conditions described, mecarbam has a retention time of about 16 minutes.

Integrator. Infotronic digital CRS 204 (or equivalent).

Recorder. Vitatron 1 mV (or equivalent).

Food cutter. Hobart model 8142 (or equivalent).

Homogenizer. A Silverson multipurpose emulsifier model L2/air.

Sintered glass Büchner funnel. 10 cm diameter, porosity 3.

Rotary-film evaporator.

Chromatographic column. A glass tube 50 cm long, 2.1 cm i.d., fitted with a tap at the lower end and a ground glass joint at the upper end to which is attached a 100-ml vessel serving as a reservoir.

c. *Reagents*

Hexane, Reagent Grade, redistilled.

Acetonitrile, Mallinckodt nanograde quality.

Sodium sulfate, anhydrous, **neutral.**

Charcoal, Darco G.60 (Honeywill-Atlas Ltd).

Florisil, 60–100 mesh.

Hexane–diethyl ether mixture (1 : 1).

d. *Procedure*

i. *Preparation of Chromatographic Column*

Prepare a slurry of the Florisil (30 gm) in hexane (30 ml). Place a cotton wool support in the chromatographic column, add 40 ml of hexane, and transfer the slurry to the column. Set the flow rate to 6 ml/minute. When the adsorbent height is constant, place a small cotton wool pad on top.

ii. *Preparation of Standard Curve*

Prepare solutions of mecarbam in hexane to cover the range between 0.1 and 20 μg/ml. Inject 1-μl portions of these solutions into the gas chromatograph and plot peak area (or peak height) against nanograms of mecarbam.

iii. *Extraction and Cleanup Procedure for Crops*

Weigh 100 gm of a finely chopped sample into the extraction jar. Add 200 ml hexane and homogenize for 3 minutes. Add sufficient sodium sulfate to break any emulsion formed and blend the mixture for another minute. Decant the supernatant liquid onto a sintered glass funnel containing a 1-cm layer of sodium sulfate. Reextract the residue with a further 200 ml hexane. Finally, transfer the residue from the extraction jar to the funnel and wash with about 50 ml hexane.

Concentrate the extract to 100 ml on a rotary-film evaporator at about 40°C. Transfer a 50-ml portion to a 250-ml separating funnel and extract with 100 ml of acetonitrile saturated with hexane. Run the lower layer into a 500-ml separating funnel. Repeat the extraction with a further 100 ml acetonitrile and bulk the acetonitrile extracts. Back-extract the latter twice with 50 ml hexane saturated with acetonitrile. Discard the hexane phases. Transfer the acetonitrile extract to a 500-ml conical flask, add 1 gm Darco G 60, and shake for 5 minutes. Filter the solution through a fine filter paper and evaporate to incipient dryness on a rotary-film evaporator. Remove the final traces of solvent under a gentle stream of nitrogen. Dissolve the residue in 5 ml hexane and transfer to the Florisil column with minimum hexane washings (3 × 3 ml). Elute the column with the (1 : 1) ether–hexane mixture at a rate of 6 ml/minute. Discard the first 120 ml and collect the following 150 ml. (*Note*: As a preliminary, it is essential that the elution pattern of mecarbam be checked for each batch of Florisil). Remove the solvent on a rotary-film evaporator and dissolve the residue in a suitable volume of hexane and inject 1 μl into the gas chromatograph. Refer to the standard calibration graph.

iv. *Extraction Procedure for Olive Oil*

Place 10 ml of the oil in a 250-ml separating funnel. Add 40 ml hexane and extract with 100 ml acetonitrile saturated with hexane. Run the lower layer into a 500-ml separating funnel. Repeat the extraction with a further 100 ml of acetonitrile and combine the acetonitrile extracts. Back-extract the latter twice with 50 ml hexane saturated with acetonitrile. Discard the hexane phases. Transfer the acetonitrile extract to a 500-ml conical flask, add 1 gm Darco G 60 and shake for 5 minutes. Filter the solution through a fine filter paper and evaporate to incipient dryness on a rotary-film evaporator. Remove the final traces of solvent under a gentle stream of nitrogen. Dissolve the residue in 5 ml hexane and transfer to the Florisil column and follow the procedure described in Section iii above.

v. *Applicability of Method*

The method has been successfully applied to determining mecarbam residues in citrus, olives, olive oil, pears, and lucerne. By employing a mixture of acetone–hexane (1:4) as extracting solvent instead of hexane, the method was applicable to grapes and to sugar beet roots. The method is sensitive to about 0.01 ppm. The recovery of mecarbam at the 0.1–2.0 ppm level ranged from 70–100%, except in the case of pears, where the recovery was between 98 and 100%.

e. *Discussion*

At the time of its introduction, only the nonspecific, but sensitive, biological method of analysis using *Drosophila melanogaster* was available for determining residues of mecarbam. Subsequently, a chemical method involving a thin-layer chromatographic separation followed by a wet oxidation of mecarbam was developed. The resulting phosphate was converted to phosphomolybdic acid and determined spectrophotometrically.

The more recently developed method using GLC originally employed the electron capture detector. The latter gave good results but became readily contaminated when oily crops were being analyzed. As a result, it has been replaced by the alkali flame ionization detector, which has given equally satisfactory results without the problem of contamination.

Reference

Pianka, M. (1961). *Chem. Ind.* (*London*) p. 324.

6

Methidathion

D. O. EBERLE AND R. SUTER

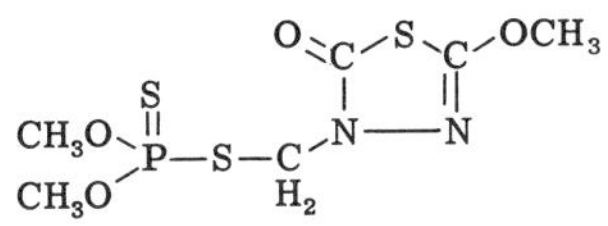

O,O-Dimethyl-*S*-[2-methoxy-1,3,4-thiadiazol-5-(4*H*)-onyl-(4)-methyl] dithiophosphate

I. GENERAL

A. Empirical Formula

$C_6H_{11}N_2O_4PS_3$ (Mol. wt. 302.3).

B. Alternative Names

GS 13005. The names Ultracide® and Supracide® are trademarks of CIBA-GEIGY. Methidathion is the common name.

C. Source of Analytical Standard

Agrochemicals Division, CIBA-GEIGY Ltd., CH-4002 Basel, Switzerland, or Agrochemicals Division, CIBA-GEIGY Corporation, Greensboro, North Carolina.

D. Biological Properties

1. INSECTICIDAL ACTIVITY

Methidathion was introduced in 1964. It is a broad-spectrum contact and stomach poison insecticide and acaricide. It has a foliar penetrative action against hidden insects. The main fields of application are against chewing and sucking pests in stone and pome fruits, grapes, citrus, alfalfa, cotton, potatoes, beet crops, hops, cereals, vegetables, olives, and sugar cane. Methidathion is registered for use in many countries. The rate of application varies according to pest and ranges from 30–60 gm a.i./100 liters or 250–1000 gm/ha. Depending upon crop species, rate of application, and weather, effective control of pests ranges from 1 to 3 weeks. The chemistry and insecticidal performance of methidathion was presented by Grob *et al.* (1965), Grob (1967), and Rufenacht (1968). Wilson and Oliver (1969) described the acaricidal effect of methidathion in woody ornamentals, whereas Campbell *et al.* (1971), Furr and Laster (1971), and Summers *et al.* (1971) studied its effectiveness against wire worms in soil, spider mites on cotton, and alfalfa weevil. Mount *et al.* (1970) found good mosquito control with methidathion when applied as an ultra-low volume aerial spray.

2. TOXICITY

Methidathion is characterized by a relatively high acute oral toxicity (LD_{50} of 25–54 mg/kg for the rat, up to 89 mg/kg for the rabbit and hen) and by a moderate dermal toxicity to the rat (LD_{50} of 150 mg/kg for the active ingredient, up to 400 mg/kg for formulated products.)

In medium- and long-term dietary feeding studies with rats and dogs, the no-effect level is considered to be 0.2 mg of methidathion/kg/day. Methidathion behaves as a characteristic organophosphate inhibitor of cholinesterase, and its clinical effects are typical of anticholinesterase compounds.

E. History

Methidathion was discovered by K. Rüfenacht in 1961 and is disclosed and described in the Swiss Patent 392,521 assigned to J. R. Geigy AG. (Priority date 5, October, 1961.)

F. Physical Properties

Methidathion is colorless, noncombustible, nonexplosive, and noncorrosive.

The melting point of the substance is 39–40°C, and the vapor pressure 1.0×10^{-6} mm Hg at 20°C. Solubility in water is 240 ppm at 20°C. It is readily soluble in methanol, acetone, and benzene. The half-life of methidathion at 25°C in acid and neutral aqueous media is approximately 1 month. Hydrolytic cleavage occurs primarily at the C—S bond to generate dimethyl-dithiophosphoric acid and R—CH_2OH.* RCH_2OH further degrades to RH.* In alkaline media, the hydrolysis rate is a function of hydroxyl ion concentration. The half-lives at 25°C at pH 9.2 and pH 13.0 are 6 days and 0.5 hours, respectively. Alkaline

*R =

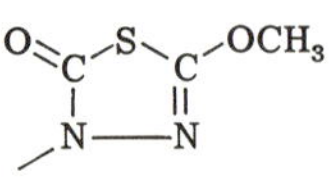

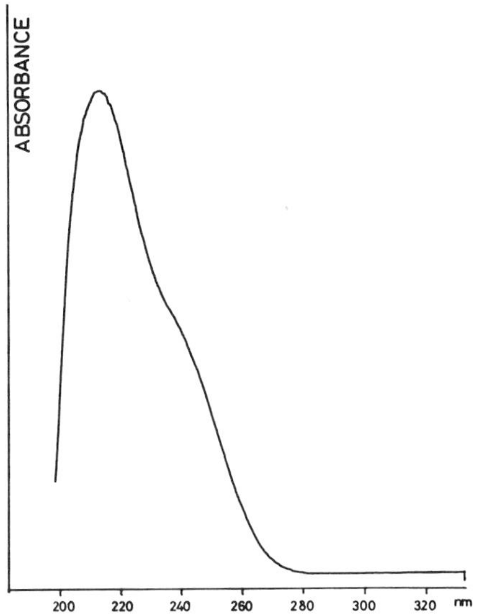

Fig. 1. UV Spectrum of methidathion. Concentration 25 μg/ml in methanol. Scan 10 nm/cm. Instrument: Beckman DBG.

cleavage occurs primarily at the P—S bond to yield RCH_2SH* and a dimethyl monothiophosphate salt. UV, IR, and NMR spectra are shown in Figs. 1–3.

G. Chemical Properties

1. METHOD OF SYNTHESIS

The synthesis of methidathion includes several reaction steps. The heterocyclic ring is synthesized by reacting potassium methyl xanthate (I) with hydrazine hydrate to form thiocarbazic acid *O*-methyl ester (II). Reaction of (II) with phosgene yields 2-methoxy-1,3,4-thiadiazol-5-(4*H*) one (III). The heterocyclic compound (III) reacts with aqueous formaldehyde to give 2-methoxy-4-hydroxymethyl-1,3,4-thiadiazol-5-(4*H*)one (IV), which can be converted to 2-methoxy-4-chloromethyl-1,3,4-thiadiazol-5-(4*H*)one (VI) by reaction with thionyl chloride. The chloromethyl compound (VI) readily reacts with the potassium salt of *O,O*-dimethyldithiophosphoric acid (V) to form methidathion (VII). Alternatively, the insecticide can be obtained directly by reaction of compound (IV) with (V) in strong mineral acid or by condensation of a mixture of (III), (V), and formaldehyde under acid conditions.

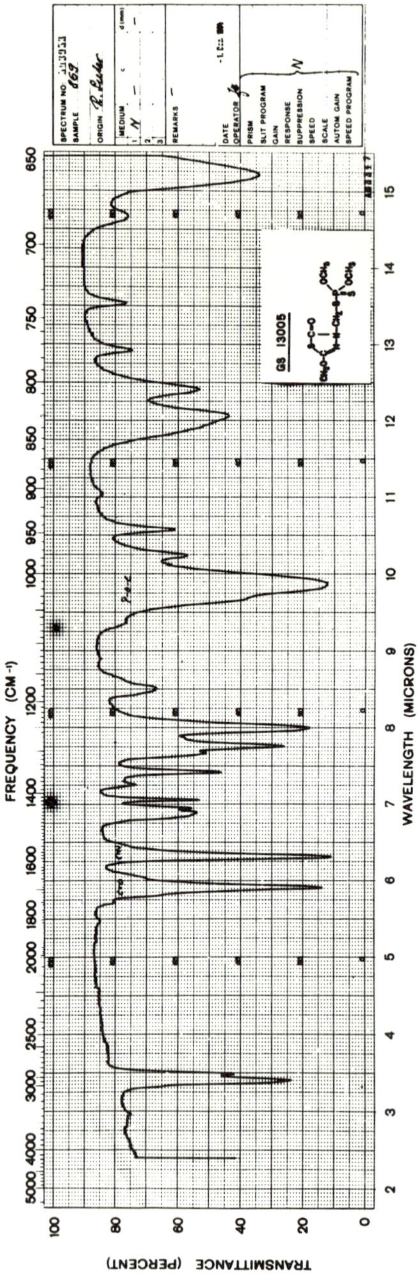

FIG. 2. IR Spectrum of methidathion. Preparation technique: Nujol mull. Instrument: PE 221 with NaCl prism. Characteristic bonds: 5,9 μ (C═O), 6,3 μ (C═N—), 9,8 μ (P—O—C).

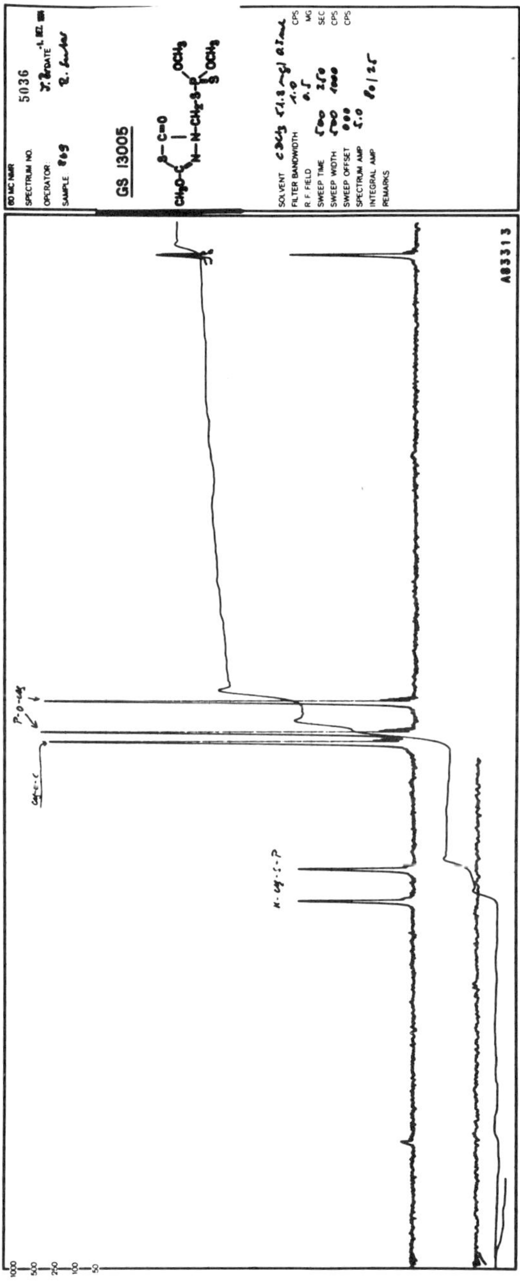

FIG. 3. NMR spectrum of methidathion. Instrument: Varian A 60. Concentration about 10% in $CDCl_3$. Characteristic signals: $\tau = 5.2$; doublet, 2 H (N—CH_2—S—P); $\tau = 6.0$, singulet, 3 H (CH_3—O—C); $\tau = 6.2$, doublet, 3 H (P—O—CH_3).

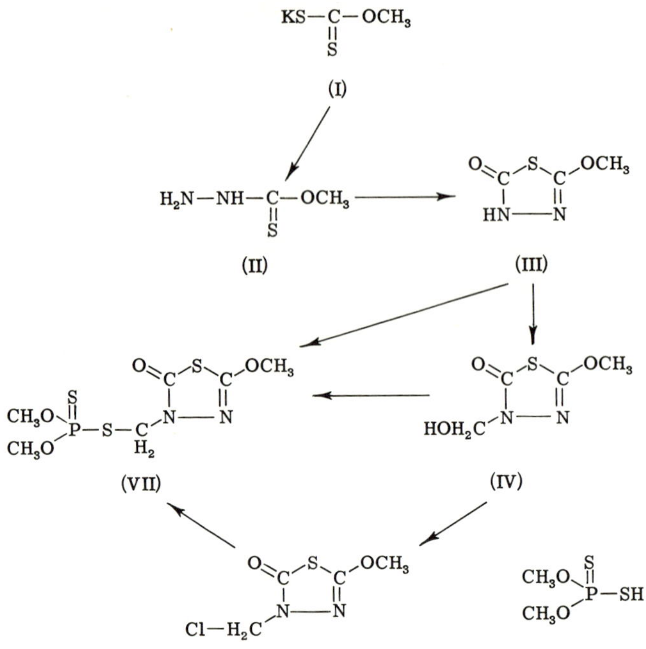

2. COMPOSITION OF TECHNICAL PRODUCT

Technical grade methidathion contains different by-products, which can be determined by thin-layer chromatography, depending on the route of synthesis. The following products are of interest: The oxo analogue of methidathion with the code number GS 13007, being monothiophosphoric acid *O,O*-dimethyl-*S*-[(2-methoxy-1,3,4-thiadiazol-5-(4*H*)-one-4-yl)-methyl] ester, and the dimeric products (VIII), (IX), (X), and (XI).

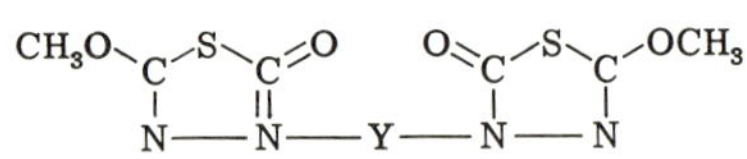

VIII: Y = $-CH_2-$
IX: Y = $-CH_2O-CH_2-$
X: Y = $-CH_2-S-CH_2-$
XI: Y = $-CH_2-S-S-CH_2-$

H. Formulations

Methidathion is formulated as 20% and 40% wettable powder and as 20% and 40% emulsifiable concentrate.

II. ANALYSIS

A. Formulation Analysis

1. RECOMMENDED METHOD

a. *Principle*

Methidathion is determined by gas chromatography (GC) using GS 13006 = *O,O*-diethyl-*S*-[2-methoxy-1,3,4-thiadiazol-5-(4*H*)-onyl-(4)-methyl] dithiophosphate as internal standard. Liquid samples are diluted to an appropriate concentration with toluene, while wettable powders are dispersed and diluted with toluene.

b. *Reagents*

Toluene, Reagent Grade.

GS 13006 = *O,O,*-diethyl-*S*-[2-methoxy-1,3,4-thiadiazol-5-(4*H*)-onyl-(4)-methyl] dithiophosphate = internal standard, purity greater than 97%.

Methidathion, analytical standard of known purity.

c. *Apparatus*

Gas chromatograph, e.g., Hewlett-Packard 5750 B, equipped with a flame ionization detector.

Electronic integrator, e.g., Hewlett-Packard 3370 B.

Column: Pyrex glass, 1.8 m × 2 mm i.d., packed with 3% silicone OV-101 on Gas-Chrom Q, 80–100 mesh.

d. *Experimental Procedure*

For solid formulations and emulsifiable concentrates, weigh an amount of sample equivalent to approximately 0.120 gm of methidathion and approximately 0.120 gm of internal standard (GS 13006) to nearest 0.1 gm into a glass stoppered 100-ml volumetric flask, dilute to volume with toluene and shake vigorously. Prepare a reference standard solution with the analytical standard methidathion in a similar manner.

Analyze the reference solution by GC with the following operating parameters:

Column oven temperature	190°C
Injection port temperature	250°C
Detector temperature	250°C
Carrier gas (nitrogen)	30 ml/minute
Hydrogen and oxygen flows	as recommended for detector
Sample	1.5 μl

Under these conditions, the approximate retention times for methidathion and GS 13006 are 10.5 and 14.3 minutes, respectively.

When successive injections of the reference solution give area ratios reproducible to ±1% relative, make duplicate injections of the sample solution, bracketing each sample with injections of the reference standard solution.

e. *Calculation*

Divide the area of the methidathion peak by that of the internal standard to determine the ratio *R*. Calculate the average *R* for the sample injections and the average *R* for the standard injections.

$$\%\ \text{Methidathion} = \frac{\text{average } R \text{ (sample)}}{\text{average } R \text{ (standard)}} \times \frac{\text{weight (standard)}}{\text{weight (sample)}} \times P$$

$P = \%$ purity of methidathion standard

2. DISCUSSION OF METHOD

Results with a relative standard deviation of less than ±1% can be obtained with the method described.

B. Residue Analysis

1. REVIEW OF PERTINENT LITERATURE

The metabolism of methidathion in rats, bean plants, apples, and soil was studied by Esser and Müller (1966), Esser *et al.* (1967, 1968), and Dupuis *et al.* (1971). Bull (1968) described the metabolic fate of methidathion in cotton plants, and Cassidy *et al.* (1969a) performed a metabolic study in alfalfa.

The uptake, translocation and excretion of methidathion by ruminants was extensively studied by Cassidy *et al.* (1969b), by Polan *et al.* (1969a,c), and by Polan and Chandler (1971). Polan *et al.* (1969b) published a degradation study with methidathion in hay, silage and rumen fluid. Analytical methods based on gas and thin-layer chromatography for studying the dissipation of residues of methidathion and its major metabolites GS 13007, i.e., *O,O*-dimethyl-*S*-[2-methoxy-1,3,4-thiadiazol-5-(4*H*)-onyl-(4)-methyl] thiophosphate, and GS 12956, i.e., 2-methoxy-1,3,4-thiadiazol-5-(4*H*)-one in various commodities have been published by Eberle (1967), Eberle *et al.* (1967), Nelson (1967), Mattson *et al.* (1969), Armbrust and Wilson (1970), Fahey *et al.* (1970), and Eberle and Hörmann (1971). The fate of methidathion in soils was studied by Getzin and Rosefield (1968) and by Getzin (1970).

A flame photometric detection method with extraction and cleanup procedure was described by Wisman and Young (1969) and applied by

Young (1970) to the analysis of methidathion residues in egg yolk and poultry feed.

2. RECOMMENDED METHOD FOR METHIDATHION

a. *Principle*

Crop and soil samples are extracted with methanol and the insecticide partitioned into methylene chloride. Further clean up is obtained by alumina column chromatography. The final determination of methidathion is achieved by gas chromatography using a nitrogen- or phosphorus-specific detector.

b. *Apparatus and Reagents*

Chromatographic columns, glass, 18 × 200 mm.

Mixer, Bauknecht model KU 21.

Centrifuge, Sorvall model SS4, 20,000 rpm.

Soil sieving machine, Werner & Pfleiderer, Stuttgart, Germany, Model Express.

Gas chromatograph, dual column Aerograph Model 1740 with Honeywell 1 mV recorder and the following detector systems:

(1) Sodium thermionic phosphorus detector.

(2) Coulson electrolytic conductivity detection system model 90-80-60 with CIC pyrolyzer model 12, CIC model 80 electrolytic conductivity detector cell, CIC model 90 DC bridge, and CIC model 60 water-recycling system. The quartz reduction tube is packed with a nickel catalyst and a strontium hydroxide acid absorber is placed in the exit end of the reduction tube.

GLC columns: (1) 3′ × $\frac{1}{8}$″ glass tubing packed with 3% OV-17 on 80–100 mesh Chromosorb G, combined with the phosphorus detector; (2) 4′ × $\frac{1}{4}$″ glass tube packed with 2% neopentylglycol succinate (NPGS) on 80–100 mesh Chromosorb G, column effluent attached to the electrolytic conductivity detection system.

Alumina oxide, Woelm basic activity grade I. Prepare activity grade V by thoroughly mixing 100 gm grade I with 15 ml of water and allowing to stand overnight in a sealed bottle.

Florisil, Floridin Co., Pittsburgh, Pa. Activate each new batch by heating 5 hours at 130°C. Store at 130°C in a glass-stoppered bottle.

c. *Extraction and Cleanup*

Chop representative 800–1000 gm subsamples of freshly harvested or deep-frozen crops in a Bauknecht household mixer. Transfer a 200 gm portion to an Omni-Mixer cup, add 400 ml methanol, and blend 2 minutes at high speed. Transfer the macerate to a 2-liter wide-neck bottle and shake mechanically 2–4 hours (extract A).

For soil analysis, remove stones, sieve a representative sample (2–4 kg) mechanically through a 4 mm screen, and extract by shaking a 200 gm subsample mechanically 12 hours with 400 ml methanol (extract B). Determine the water content of soil by drying a sample portion at room temperature and base all results on weight of dried soil.

Fortify untreated check samples with known amounts of methidathion at different levels before extraction and carry out the same analytical procedure as for field-treated samples. Deep freeze portions of field and fortified samples.

Filter extract A or B through a fritted glass funnel using gentle suction; wash the residue remaining in the funnel with two 50-ml portions of methanol, transfer the filtrate to a 1 liter graduated cylinder, and record the volume. Transfer half of the methanolic solution, corresponding to 100 gm of test material, to a 1-liter separatory funnel, add 400 ml water and 10 ml saturated aqueous sodium chloride solution, and shake vigorously with two 100-ml portions of methylene chloride. After the layers have separated, combine the methylene chloride extracts, dry with anhydrous sodium sulfate, and evaporate to dryness with a rotary vacuum evaporator (extract C).

For analysis of water, extract 500 ml by vigorously shaking with 200 ml of chloroform in a 1-liter separatory funnel. Let the mixture stand 1 hour until the layers separate. Drain off the lower phase and centrifuge at 10,000 rpm. Transfer the entire contents of the centrifuge tube to a 250-ml separatory funnel, pass the chloroform phase through a cotton plug into a 300-ml round-bottom flask, and evaporate to dryness with a rotary vacuum evaporator (extract D).

Samples are now ready for cleanup by column chromatography. Fill a chromatographic column with *n*-hexane and 25 gm alumina grade V to produce a packed column at least 7 cm high. Dissolve extract C or D in 2 ml benzene, transfer to the column, and wash the column twice with 25-ml portions of hexane. When the last of the hexane has just entered the adsorbent, begin collecting the eluate in a 250-ml flask and continue elution with 100 ml 5% ethyl ether in hexane at about 5 ml/minute. Collect the eluate and evaporate to dryness.

Add a known volume of hexane–ethyl ether (1 : 1) to the residue remaining after evaporation so that a 5 μl injection will produce about half-scale recorded deflection. Standardize the gas chromatographic instrument for detection of methidathion as follows:

Instruments	Conditions
Injection port	250°C
Chart speed	$\frac{1}{2}$″/minute

i. *Sodium Thermionic Phosphorus Detection Unit*

Carrier gas (nitrogen)	30 ml/minute
Column oven	225°C
Hydrogen	14 ml/minute
Air	120 ml/minute
Sensitivity	4×10^{-10} AFS
Elution time	4.5 minutes
Response to methidathion	1 cm^2/0.2 ng

ii. *Electrolytic Conductivity Detection Unit*

Carrier gas (helium)	70 ml/minute
Column oven	230°C
Combustion furnace	800°C
Hydrogen	40 ml/minute
Purge helium	10 ml/minute
Attenuation	2
Elution time	4 minutes
Response to methidathion	1 cm^2/10 ng
Voltage	30 V

d. *Efficiency of Extraction*

The extraction efficiencies of petroleum ether, benzene, and methanol were compared. Field-treated apples were extracted with benzene and petroleum ether and with methanol as above. The results of the analyses showed that all three solvents extract the same quantity of methidathion and therefore are equally suitable for the extraction of the parent compound from apples and similar fruits.

3. RECOMMENDED METHOD FOR GS 13007

a. *Principle*

The metabolite is extracted from crops by methanol and determined by thin-layer chromatography, using flyhead cholinesterase inhibition.

b. *Sample Preparation*

Subsample and chop material as described for methidathion. Transfer a 20-gm subsample to an Omni-Mixer cup and blend 5 minutes with 40 ml methanol. Fortify check samples before blending. Filter the material through a G-1 fritted glass funnel under vacuum. (No additional solvent washing is necessary.) Transfer the filtrate to a graduated cylinder and record the volume. (The total volume is about 50 ml from crops and about 40 ml from soil.) No further cleanup of these extracts is required.

c. *Apparatus and Reagents*

Thin-layer chromatographic equipment.

Spotting pipettes, Drummond Microcaps, 5, 2, and 1 μl (Drummond Scientific Co., Broomall, Pa.).

Centrifuge, Sorvall model SS4, 20,000 rpm.

Fly-head homogenate, The fly-head homogenate is prepared as follows: 50 flies, 6 days old, of a susceptible strain are anesthesized with CO_2 and beheaded; the heads are ground 10 minutes in a centrifuge tube with 12 ml buffer (pH 7.2). The mixture is centrifuged 5 minutes at 500 rpm and the solution is kept in an icewater bath until use.

1-Naphthyl acetate, Fluka A.G., Buchs, Switzerland.

Buffer solution, pH 7.2. Dissolve 12.5 gm $Na_2HPO_4 \cdot 7H_2O$ and 2.72 gm KH_2PO_4 in 1 liter of water.

Sodium tetraborate buffer salt solution, pH 9.2 (E. Merck A.G., Darmstadt, Germany).

Azoene Fast Blue RR salt, Reagent Grade (Rohner A.G., Pratteln, Switzerland).

Substrate solution, Dissolve 25 mg 1-naphthyl acetate in 10 ml absolute ethyl alcohol and 50 mg Azoene Fast Blue RR salt in 20 ml tetraborate buffer plus 20 ml of distilled water. Combine and mix the solutions.

Fly-head homogenate and substrate solution must be prepared freshly before use because of rapid decomposition.

d. *Thin-Layer Chromatography*

Coat 20 × 20 cm glass plates with a 250-μm layer of silica gel G. Activate plates 30 minutes at 110°C in a drying oven and store in a desiccator until use. Spot a 12 μl crop extract and a 10-μl soil extract (equivalent to 5 mg of material) on the TLC plate, using 10 and 2 μl disposable pipettes (Microcaps). Also spot standards of GS 13007 in methanol. Prepare standard solutions freshly before use. Let spots dry at room temperature. Develop the plate for 10 cm with benzene–ethyl acetate–chloroform (40:40:20), air dry, and spray with fly-head homogenate. Keep the plate at room temperature 15 minutes and spray the substrate solution onto plate. After about 10 minutes, GS 13007 begins to appear as white spots on a pink background. If there is not enough contrast between the spots and the background after 1 hour, the plates may be sprayed again with fresh substrate solution. Maximum spot intensity develops after about 2 hours. GS 13007 has an R_f value of about 0.40 under the conditions used.

Despite efforts to standardize the fly-head preparation, some day-to-day variation in the color and intensity of the background is noticeable.

Other solvent systems have been used successfully to develop GS 13007 thin-layer chromatograms using this detection system; however, best results have been obtained on crop samples using the described procedure.

4. RECOMMENDED METHOD FOR GS 12956 [2-methoxy-1,3,4-thiadiazol-5-(4*H*)-one]

a. *Principle*

The metabolite is extracted from crops with methanol and determined on thin-layer chromatography plates by spraying plates with silver nitrate followed by UV irradiation.

b. *Sample Preparation, Extraction, and Cleanup*

Subsample, chop, and extract crops and soil as described for methidathion. Filter extract A or B through a fritted glass funnel as above and concentrate half of the stripping solution, corresponding to 100 gm of material, to about 20 ml with a rotary vacuum evaporator. Add 15 ml of saturated aqueous sodium chloride solution, transfer to a 200-ml separatory funnel, and extract 3 times with 50-ml portions of chloroform. Combine the chloroform layers and evaporate to dryness.

Fill a chromatographic column with chloroform and pour in 25 gm Florisil. Drain off the supernatant chloroform. Transfer the crop extract, dissolved in 2 ml chloroform, to the column and wash first with 50 ml chloroform; this eluate contains methidathion and may be further cleaned on alumina as described above. Elute GS 12956 with 100 ml methanol–chloroform (1:9) at a flow rate of 5 ml/minute. Collect the eluate and concentrate to about 5 ml. Quantitatively transfer the concentrate with ethyl ether to a graduated test tube and evaporate the solvent at room temperature in a stream of nitrogen. Redissolve the residue in 0.1 ml methanol.

c. *Thin-Layer Chromatography*

Spot 10 μl of crop extract (equivalent to 10 gm of material) on a TLC plate with standard spots of GS 12956 dissolved in methanol. Develop the plate to 10 cm with acetone–toluene (1:9). Dry the chromatogram with a hair dryer and spray with 0.1 *N* aqueous silver nitrate solution. Irradiate the wet plate with unfiltered shortwave UV light (254 nm) for 15 minutes.

5. DISCUSSION

The gas chromatographic response to methidathion with the electrolytic conductivity detector was 1 cm^2 for 10 ng, i.e., 0.01 μg was de-

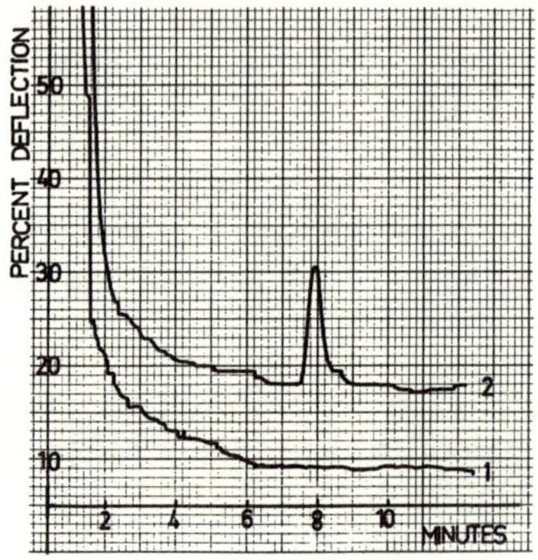

FIG. 4. Gas chromatogram with sodium thermionic phosphorus detector (sensitivity 4×10^{-10} Amp). Response to methidathion in apple extracts. 1 = control; 2 = control plus 0.02 ppm.

tectable in a 1-gm portion of crops and soil (0.01 ppm). Injection of a 100-gm portion of water yielded a sensitivity of 0.0001 ppm (0.1 ppb). The response with the thermionic phosphorus detector was 1 cm^2/0.2 ng, i.e., 0.01 ppm for crops and soil (injection of a 2-gm portion).

The systems were standardized by determining the peak areas obtained from known amounts of methidathion by triangulation. The amount of sample injected was chosen so that the insecticide determined was bracketed by the standards injected. The values so obtained were used to calculate recoveries from the fortified samples. Recoveries found

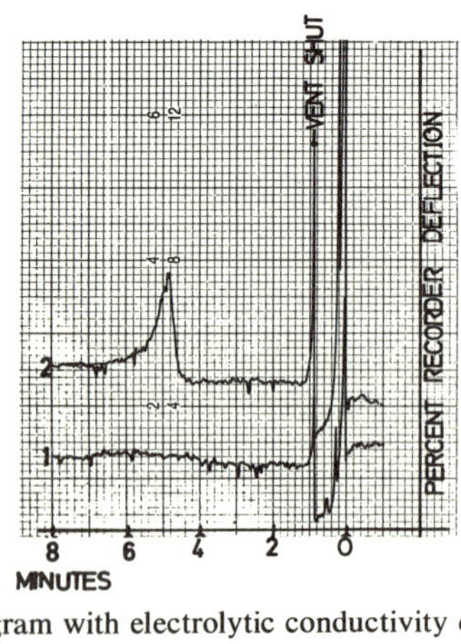

FIG. 5. Gas chromatogram with electrolytic conductivity detection of methidathion in hop extracts. 1 = control; 2 = control plus 0.02 ppm.

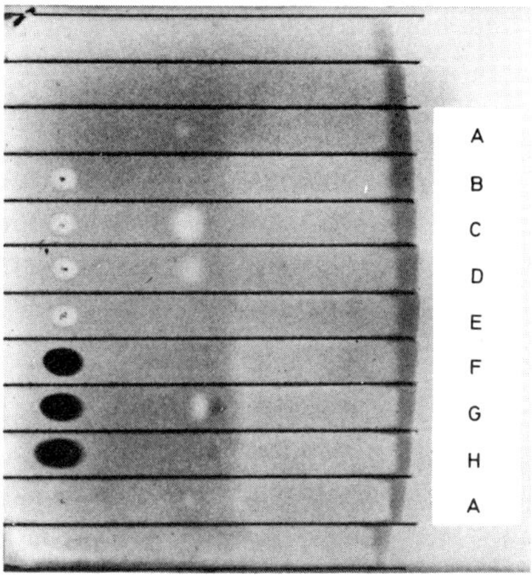

FIG. 6. Thin-layer chromatogram of extracts of field-treated fruits showing the dissipation of GS 13007 ; 5-mg portions spotted. A = standard, 50 pg GS 13007; B = apples, control; C = apples, 2-day sample; D = apples, 21-day sample; E = apples, 42-day sample; F = grapes, control; G = grapes, 21-day sample; H = grapes, 42-day sample.

for apples, cherries, hops, and soil were satisfactory at all fortification levels (0.01, 0.05, 0.1, 0.5, and 1 ppm) and ranged from 70 to 105%, generally close to 90%. Recoveries for water were 75 and 85% at 0.2 and 1 ppb, respectively. Typical gas chromatograms are shown in Figs. 4

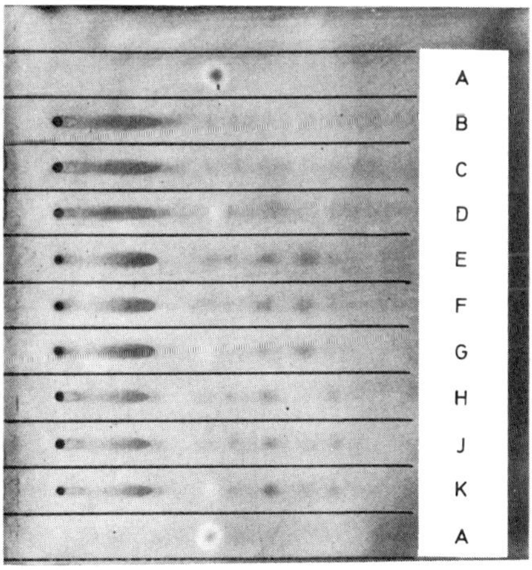

FIG. 7. Thin-layer chromatogram of extracts of field-treated fruits, 10-gm portions spotted. A = standard, 1 μg GS 12956; B = apples, control; C = apples, 21 day sample; D = apples, 4-day sample; E = grapes, control; F = grapes, 21-day sample; G = grapes, 4-day sample; H = prunes, control; J = prunes, 21-day sample; K = prunes, 4-day sample.

and 5 for apples and hops, fortified before extraction with 0.02 ppm methidathion.

With the TLC conditions used, GS 13007 and GS 12956 have R_f values of 0.40 and 0.50, respectively. The minimum detectable amount is 50 pg for GS 13007, so that residues of 0.01 ppm can be detected in a 5 mg subsample of crops, whereas 100 ng of GS 12956 is detectable in a 10-gm subsample (0.01 ppm). Use of petroleum ether as an extraction solvent, as described by Young (1970), yielded considerably lower recoveries of GS 13007 from both fortified and field-treated samples of apples and grapes.

Typical thin-layer chromatograms obtained from the analysis of field-treated apples, grapes, and prunes, showing the dissipation of GS 13007, are presented in Fig. 6. The chromatogram shows that trace amounts of GS 13007 are present after treatment but disappear rapidly and are below 0.01 ppm in all cases 42 days after treatment.

Figure 7 is a thin-layer chromatogram of apple, grape, and prune extracts obtained as described above for the determination of GS 12956. Only the samples picked 4 days after application of methidathion possibly contain traces of GS 12956 at the maximum level of 0.02 ppm, whereas all other chromatograms of treated samples (C, F, I) and untreated samples (B, E, H) are identical showing the absence of GS 12956.

References

Armbrust, E. J., and Wilson, M. C. (1970). *J. Econ. Entomol.* **63,** 189–192.

Bull, D. L. (1968). *J. Agr. Food. Chem.* **16,** 610–616.

Campbell, M. V., Mount, D. A., Heming, B. S., and Raleigh, N. C. (1971). *J. Econ. Entomol.* **64,** 41–44.

Cassidy, J. E., Ryskiewich, D. P., and Murphy, R. T. (1969a). *J. Agr. Food Chem.* **17,** 558.

Cassidy, J. E., Murphy, R. T., Mattson, A. M., and Kahrs, R. A. (1969b). *J. Agr. Food Chem.* **17,** 571–575.

Dupuis, G., Mücke, W., and Esser, H. O. (1971). *J. Econ. Entomol.* **64,** 588–597.

Eberle, D. O. (1967). *Pap., Pestic. Congr., 6th, 1967* Abstract BV-6, pp. 506–561.

Eberle, D. O., and Hörmann, W. D. (1971). *J. Ass. Offic. Agr. Chem.* **54,** 150–159.

Eberle, D. O., Delley, R. G., Székely, G. G., and Stammbach, K. H. (1967). *J. Agr. Food Chem.* **15,** 213–216.

Esser, H. O., and Müller, P. W. (1966). *Experientia* **22,** 36–42.

Esser, H. O., Dupuis, G., Mücke, W., and Müller, P. W. (1967). *Proc. Int. Pestic. Congr., 6th, 1967* Abstract B III, 1–13, pp. 199–201.

Esser, H. O., Mucke, W., and Alt, K. O. (1968). *Helv. Chim. Acta* **51,** 513–517.

Fahey, J. E., Wilson, M. C., and Armbrust, E. J. (1970). *J. Econ. Entomol.* **63,** 589–591.

Furr, R. E., and Laster, M. L. (1971). *J. Econ. Entomol.* **64,** 339–40.

Geigy, J. R., AG. (1961). Swiss Patent 392,521.

Getzin, L. W. (1970). *Bull. Environ. Contam. Toxicol.* **5,** 104–110.

Getzin, L. W., and Rosefield, J. (1968). *J. Agr. Food Chem.* **16,** 598–601.

Grob, H. (1967). *Extr. Phytiatr.-Phytopharm., France* pp. 1–10.

Grob, H., Gasser, R., and Ruzette, M. E. (1965). *Proc. Brit. Insectic. Fungic. Conf., 3rd, 1965* pp. 451–477.

Mattson, A. M., Kahrs, R. A., and Murphy, R. T. (1969). *J. Agr. Food Chem.* **17,** 565–570.

Mount, G. A., Lowe, R. E., Baldwin, K. F., Pierre, N. W., and Savage, K. E. (1970). *Mosquito News* 30, 442–446.

Nelson, L. L. (1967). *J. Econ. Entomol.* **60,** 879–880.

Polan, C. E., and Chandler, P. T. (1971). *J. Dairy Sci.* **54,** 847–853.

Polan, C. E., Huber, J. T., Young, R. W., and Osborne, J. C. (1969a). *J. Agr. Food Chem.* **17,** 857–859.

Polan, C. E., Sandy, R. A., and Huber J. T. (1969). *J. Dairy Sci.* **52,** 1296–1299.

Polan, C. E., Miller, C. N., and Sandy, R. A. (1969c). *J. Dairy Sci.* **52,** 1384–1389.

Rüfenacht, K. (1968). *Helv. Chim. Acta* **51,** 518–526.

Summers, C. G., Byrne, H. D., and Pimentel, D. (1971). *J. Econ. Entomol.* **64,** 478–480.

Wilson, N. L., and Oliver, A. D. (1969). *J. Econ. Entomol.* **62,** 1400–1401.

Wisman, E. L., and Young, R. W. (1969). *Poultry Sci.* **49,** 83–85.

Young, R. W. (1970). *J. Agr. Food Chem.* **18,** 164–167.

7

Phenthoate

B. Bazzi

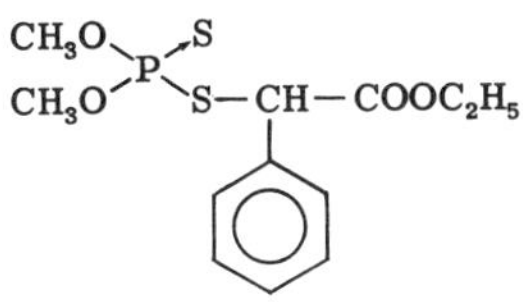

Ethylmercaptophenyl acetate (*O*,*O*-dimethyl phosphorodithioate)

I. GENERAL

A. Empirical Formula

$C_{12}H_{17}O_4PS_2$ (Mol. wt. 320.3).

B. Alternative Names

Cidial® (trademark of Montedison Company), Elsan® (trademark of Nissan Chem. Ind.).

C. Source of Analytical Standard

Agricultural Division, Montedison Company, Milan, Italy.

D. Biological Properties

Phenthoate is an insecticide that works by contact, and stomach poisoning and has no systemic action. Applied at dosage rates of 0.5–1 kg a.i./ha, it protects cotton, rice, citrus, fruit, vegetables, and other crops from Lepidoptera, jassids, aphids, and soft scales. It is also effective against mosquito larvae and adults.

The acute oral LD_{50} of the technical product for rats is 300–400 mg/kg, for mice 350–400 mg/kg. The application of 4800 mg/kg for 4 hours to the skin of rats caused no toxic symptoms. Tests on rats over an 88 week period showed that the highest nontoxic dose level of the technical product is slightly below 0.5 mg/kg/day. The acute oral LD_{50} for hares is 72 mg/kg, for pheasants 218 mg/kg, for quails 300 mg/kg.

TABLE I
PHYSICAL PROPERTIES OF PHENTHOATE

Property	Phenthoate: Pure	Phenthoate: Technical
Physical state	Crystal	Oily liquid
Color	Colorless	Reddish yellow
Odor	Slightly aromatic	Aromatic
Specific gravity (d_4^{20})	1.226	ca. 1.23
Melting point	17.5 ± 0.5°C	–
Refractive index, n_D^{20}	1.5550	–
Flash point (Cleveland)	–	165–170°C
Vapor pressure	4×10^{-5}mmHg at 40°C	–
Solubility at 20°C	11 ppm in water	Completely miscible with alcohols, aromatic hydrocarbon solvents, chlorinated hydrocarbon solvents, acetone, cyclohexanone, acetonitrile, methylcellosolve, carbon disulfide; soluble: 12% in *n*-hexane, 17% in ligroin, 20% in diethylglycol, over 10% in light petroleum

The LD_{50} for bees is 0.12 μg per bee. In tests on fish, the LC_{100} (4-day exposure) for *Carassius auratus* is 4.5 ppm, for *Phoxinus laevis* 3.4 ppm, for *Gambusia affinis* 0.3 ppm, for *Lebistes reticulatus* 2–3 ppm.

E. History

Phenthoate was introduced in 1961 by Montedison Company under the code number L561 and the protection of BP 834,814; USP 2,947,662; and others.

F. Physical Properties

Important physical properties of Phenthoate are given in Table I.

G. Chemical Properties

1. METHOD OF SYNTHESIS

Phenthoate is produced on reaction of sodium *0,0*-dimethyl phosphorodithioate with ethyl α-phenylbromoacetate. Alternatively, sodium *O,O*-dimethyl phosphorodithioate may be reacted with α-phenylbromoacetic acid, thus obtaining α-*O,O*-dimethyl phosphorodithioate–α-phenylacetic acid, which may be esterified with ethyl alcohol.

2. CHEMICAL REACTIONS

The compound is stable in acid and neutral media, but in buffer solution at pH 9.7 about 25% is hydrolized after 20 days. Phenthoate can be converted by oxidizing agents to its oxygen analogue.

H. Formulations

Phenthoate is available as ultra low-volume spray, emulsifiable concentrates, wettable powders, and dusts for agricultural use and for mosquito control.

II. ANALYSIS

A. Formulation Analysis

1. REVIEW OF METHODS

Thin-layer chromatographic (TLC) and gas–liquid chromatographic (GLC) methods have been used to analyze phenthoate in technical products and in its formulations. In the TLC procedures, the active ingredient is separated from impurities on a chromatographic plate and then analyzed; or as $PdCl_2$–phenthoate yellow complex (Nissan Chemical Industries, Ltd., 1965), or by estimation of phosphorus as ammonium phosphomolybdate (Bazzi *et al.,* 1965) or as phosphovanadomolybdate (Nissan Chemical Industries, Ltd., 1972). In the GLC procedures (Bazzi *et al.,* 1969; Nissan Chemical Industries, Ltd., 1972), both methods use a thermal conductivity detector. The GLC procedure is normally preferred, but in some cases (e.g., where analytical standard is not available), the TLC method could be the procedure of choice.

2. RECOMMENDED METHOD—GLC PROCEDURE

a. *Principle*

Phenthoate or a convenient amount of an insecticidal preparation is diluted to appropriate volume with toluene, together with di-*n*-butyl phthalate as an internal standard, and analyzed by gas chromatography using a thermal conductivity detector (Bazzi *et al.,* 1969).

b. *Reagents*

Insecticide standard, Phenthoate Analytical Grade, known purity (Montedison Company, Milan, Italy).

Internal standard, di-*n*-butyl phthalate, Reagent Grade (B.D.H., Poole, England).

GLC stationary phase, SE-30 silicone oil (cat. no. 50350, Applied Science Laboratories).

GLC solid support, Chromosorb W 60–80 mesh acid washed and silanized (cat. no. 08-5216100, Carlo Erba S.p.A., Milan, Italy).

Solvents, Reagent Grade.

c. *Apparatus*

Gas chromatograph, Fractovap model C, type ATC/f (Carlo Erba S.p.A., Milan, Italy), with a thermal conductivity detector or equivalent equipment.

Gas chromatographic column, 70 × 0.5 cm i.d. stainless steel, packed with 5% SE-30 on 60–80 mesh acid washed, silanized Chromosorb W.

Millimetric lens, 8×, 10-mm scale width, and 0.1-mm subdivisions (see section f, below).

d. *Experimental Procedure*

Accurately weigh about 0.65 gm di-*n*-butyl phthalate and about 1 gm pure reference phenthoate into a 50-ml volumetric flask, dissolve in toluene, and dilute the solution to volume (solution 1). In another 50-ml volumetric flask, accurately weigh about 0.65 gm di-*n*-butyl phthalate and an amount of sample equivalent to approximately 1 gm phenthoate, and dilute to volume with toluene (solution 2). Analyze solutions 1 and 2 by gas chromatography with the following operating parameters:

Column temperature	190°C
Injection port temperature	250°C
Helium flow rate	45 ml/minute
Recorder sensitivity	1 mV full scale
Recorder chart speed	75 cm/hour

Adjust the gas chromatographic conditions so that the injections of 3–4 μl will give the peak height between one-third and two-thirds full scale height and half-height base from 0.4 to 1 cm (Fig. 1). Inject 6 portions of solution 1 and 6 portions of solution 2, alternating 3 injections of solution 1 with 3 injections of solution 2.

e. *Calculation*

Calculate the individual correction factor F for each chromatogram from solution 1 as below and determine the mean $\bar{F}$ (for 6 values):

$$F = \frac{A_s}{A_f} \times \frac{W_f}{W_s}$$

where A_s = peak area of internal standard, A_f = peak area of phenthoate,

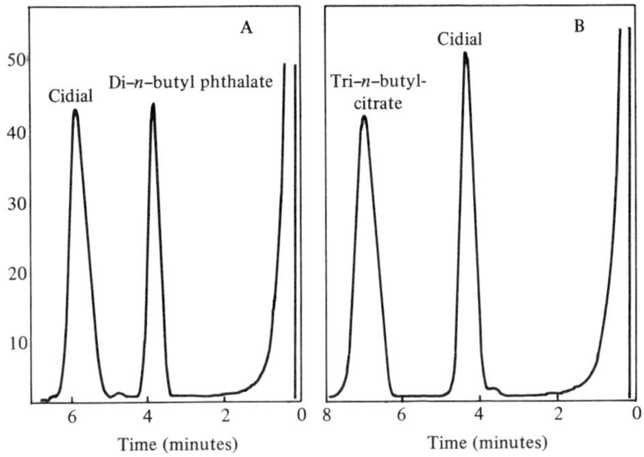

FIG. 1. Gas chromatogram of technical phenthoate. (A) With di-*n*-butyl phthalate: (B) with tri-*n*-butyl citrate (Bazzi *et al.*, 1969).

W_s = weight of internal standard, and W_f = weight of phenthoate reference standard.

From the chromatograms of solution 2, calculate the active ingredient content of the samples as mean of 6 individual values. Calculate each individual value as follows:

$$\% \text{ phenthoate} = \frac{A_f^1}{A_s^1} \times \frac{W_s^1}{W_f^1} \times \bar{F} \times t$$

where A_f^1 = peak area of phenthoate, A_s^1 = peak area of internal standard, W_f^1 = weight of phenthoate sample, W_s^1 = weight of internal standard, $\bar{F}$ = mean of 6 correction factor values, and t = purity of phenthoate analytical standard (percent).

f. *Note on Procedure*

Measurement of the peak area by means of an automatic electronic integrator connected to the gas chromatograph is highly desirable and should result in better precision than specified. In the event that such an instrument is available, the micrometric lens is not required.

3. ALTERNATE METHOD—TLC PROCEDURE

a. *Principle*

Phenthoate, or a convenient amount of an insecticidal preparation, is dissolved in benzene and the solution is spotted on a TLC plate. After development and spraying with the chromogenic reagent, the area containing phenthoate is scraped off, and the pesticide is eluted with methanol. The active ingredient content in the eluate is obtained by determining phosphorus as ammonium phosphomolybdate (Bazzi *et al.*, 1965).

b. *Reagents*

Silica gel G, TLC grade according to Stahl (cat. no. 7731 Merck).

TLC developing solvent: n-hexane, 90 ml; acetone, 10 ml.

Chromogenic reagent. Mix 0.25 gm palladium(ous) chloride with 1 ml concentrated HCl and 5 ml water. Dilute to 100 ml with 95% ethanol.

Ammonium nitrate, 2% water solution.

Molybdate reagent. Dissolve 300 gm of ammonium molybdate in 800 ml boiling water. Add the cool solution dropwise, with stirring, to a solution of 100 gm ammonium sulfate in 1 liter HNO_3 (sp. gr. 1.36)–860 ml of HNO_3 (sp. gr. 1.40) plus 140 ml of water. Let stand in the dark for 3 days, then dilute with water to 2 liters. Filter the reagent before using and store in a dark bottle.

Sulfuric acid, Reagent Grade (sp. gr. 1.84).

Nitric acid, Reagent Grade (sp. gr. 1.40).

Ammonium hydroxide, Reagent Grade (sp. gr. 0.925).

c. *Apparatus*

Chromatographic tank, 22 × 10 × 22 cm.

Glass plates, 20 × 20 cm.

Glass crucible, 10 G 4, with fritted disk, Gooch type, low-form.

Rotary vacuum evaporator, Ascenso (Milan, Italy) or equivalent.

Filtration vacuum apparatus, Fisher filtrator (cat. no. 9-788, Fisher Scientific Company) or equivalent.

Mechanical vacuum pump, capable of vacuum 15 mm Hg.

d. *Preparation of the TLC Plates*

Carefully mix 50 gm silica gel G with 85–90 ml water for 3 minutes in an Erlenmeyer flask; immediately apply 1 mm thick layer to the glass plates and dry then 90 minutes in an oven at 110°C. With a needle, mark a thin line on the plate at 3 cm from the upper edge, and remove 5 mm of silica gel from both sides of the plate.

e. *Procedure*

In a 50-ml volumetric flask, weigh an amount of thoroughly homogenized sample corresponding to 1 gm of phenthoate; dissolve in benzene and dilute to volume. With a controlled pipette, apply 1 ml of pesticide solution in a uniform, thin streak 14 cm long and 3 cm from the bottom edge of the plate. After the solvent has completely evaporated, place the TLC plate in the chromatographic tank containing the solvent system, and let the plate develop until the solvent front has reached the upper scored line. Remove the plate and air dry at least 30 minutes. Then repeat the TLC operation 4 times, each time restoring the initial volume of solvent in the tank. When the plate is completely dried, spray with the chromogenic reagent and let stand at least 2 hours. Phenthoate appears as a bright yellow band at R_f 0.60.

Mark the phenthoate area, allowing a safety margin depending on the presence of other compounds, then remove the silica gel around this zone with a microscope slide and thoroughly clean the glass with a filter paper wetted with methanol.

Scrape off the phenthoate area into a glass crucible mounted on a vacuum assembly and elute with methanol (30–40 ml total) into a 50-ml Kjeldahl flask. After each washing with methanol, release the vacuum, and with a glass rod carefully stir the silica gel in the crucible. Using the rotary evaporator, cautiously evaporate the filtrate at 35°C (maximum) just to dryness. Add 4.5 ml H_2SO_4, followed by 1.5 ml HNO_3, heat gently to white fumes and cool. Add 1 ml HNO_3 and repeat the digestion again to white fumes. After cooling transfer the solution to a 600-ml beaker, carefully rinse the flask with water, add 8 ml HNO_3, dilute to 150 ml, and heat on steam bath. Remove from steam bath and promptly pipette 150 ml of molybdate reagent into the center of the solution, not down beaker walls.

After 3–4 minutes, swirl gently, cover the beaker, and let stand in dark for at least 12 hours. In the meantime, wash a glass crucible with a dilute NH_4OH (1:2) solution, water, warm HNO_3 (1:1), and again with water. Remove the water with two portions of 25 ml ethanol and two of 25 ml acetone, releasing the vacuum after each washing. Suck dry, place the crucible in a vacuum desiccator containing no desiccant and evacuate to 15–20 mm Hg for 30–35 minutes. Release the vacuum and weigh the crucible 5 minutes after removal from the desiccator. Filter the precipitate, wash with 400 ml NH_4NO_3 solution and then with ethanol; wash again twice with alcohol and twice with acetone. Continue as described for weighing the empty crucible.

f. *Calculation*

$$\% \text{ phenthoate} = \frac{A \times 0.014524 \times 10.34 \times 100}{W}$$

where A = weight of ammonium phosphomolybdate (mg), 0.014524 = factor to convert ammonium phosphomolybdate to phosphorus, 10.34 = factor to convert phosphorus to phenthoate, W = weight of sample applied to TLC plate (mg).

g. *Modification of the Method*

Further experiments in the author's laboratory showed that phenthoate can be determined via phosphorus, gravimetrically, as quinoline phosphomolybdate. This method is simple, readily applied, and possesses the same degree of precision obtainable by determining phosphorus as ammonium phosphomolybdate. For this purpose, the eluate from the TLC plate is evaporated to dryness on a rotary evaporator, and the phosphorus of the residue is determined as described for prothoate.

4. DISCUSSION OF THE METHODS

Both of the methods have been applied on technical products and liquid formulations and have given excellent results. The various impurities in the technical phenthoate, as well as the solvents, emulsifiers, wetting agents, etc. present in the formulations do not interfere.

The precision of the GLC method is estimated to be $\pm 1.3\%$, while the one of the TLC method can be better than $\pm 1\%$ if phenthoate is determined via phosphorus as quinoline phosphomolybdate.

B. Residue Analysis

1. REVIEW OF METHODS

Several types of methods are available for the determination of phenthoate residues in crops and animal tissues. Colorimetric methods determine the active ingredient residues after conversion to phosphomolybdate and molybdenum blue. However, these procedures are made reasonably specific by the cleanup and can be sensitive to 0.1–0.2 ppm in apples, olives and olive oil (Bazzi *et al.*, 1964a), oranges, lemons, and tangerines (Bazzi and Canale, 1968). Thin-layer chromatography has been useful for quantitative identification of phenthoate in crops (Bazzi *et al.*, 1964b). The amount of phenthoate is determined by visual comparison of the sample spot with standard spots. Gas chromatography with an electron capture detector was used by Montedison Company (1969) to measure phenthoate residues in crops and in animal tissues. These procedures were validated at levels down to 0.5 ppm in alfalfa and 0.05 ppm in cattle meat. Gunther *et al.* (1972) employed a gas chromatograph equipped with a flame photometric detector, operated in the phosphorus mode, to determine phenthoate residues in oranges and leaves at levels down to 0.5 ppm without cleanup, and 0.1 ppm with a simple cleanup using a Florisil column. A thermionic detector was also tested and found equally satisfactory.

The preferred method, given below, uses gas chromatography with a flame photometric detector to analyze phenthoate and its oxygen analogue residues in maize and cattle meat at levels down to 0.005 ppm for each compound (Bazzi, 1972).

2. RECOMMENDED METHOD

a. *Principle*

An analytical procedure is described for the determination of phenthoate and its oxygen analogue in maize and cattle meat down to levels of 0.005 ppm for both compounds. The method is based on the following steps: extraction of the residues with acetone, cleanup by partition with couples of immiscible liquid phases and by Florex column chromatography, final gas chromatographic determination of the phenthoate and its oxygen analogue with an instrument equipped with a selective phosphorus-sensitive detector.

b. *Reagents*

Solvents, Reagent Grade, redistilled: acetone, benzene, chloroform, methanol, *n*-hexane.

Nitrogen, highly purified.

Anhydrous sodium sulfate, Reagent Grade.

Florex XXS, 60–100 mesh (The Floridin Co, Pittsburgh, Pennsylvania, USA), heated 16 hours at 120°C.

Phenthoate and phenthoate oxygen analogue, Analytical Grade, known purity (Montedison Company, Milan, Italy).

GLC stationary phase, UCCW 982, methylvinyl silicone polymer (cat. no. 8501-3840, Hewlett Packard).

GLC solid support, Gas Chrom Q 80–100 mesh (Applied Science Laboratories Inc., State College, Pennsylvania, USA).

Check all reagents for impurities using the flame photometric detector gas chromatograph before they are used.

c. *Apparatus*

Homogenizer, Omni-Mixer (Sorvall Inc.) or equivalent.

Rotary vacuum evaporator Ascenso (Milan, Italy) or equivalent.

Filtration vacuum apparatus equipped with Buchner funnel.

Filter paper, glass fiber, Whatman GF/C.

Liquid chromatographic column, glass tube, 400 × 12 mm i.d., with Teflon stopcock.

Gas chromatograph, Micro Tek, MT 160 DPH, equipped with a Melpar flame photometric detector with a 526-nm phosphorus filter or equivalent equipment.

Gas chromatographic column, 180 × 0.4 cm i.d., Pyrex glass, packed with 3.8% UCCW 982 on 80–100 mesh Gas Chrom Q. The column is silylated in the vapor phase at 250°C by injecting 20 μl of hexamethyldisilazane five times.

Recorder, Speedomax G, Leeds and Northrup Co., or equivalent.

d. *Experimental Procedure*

i. *Extraction and Preliminary Cleanup*

(a) *Maize.* Homogenize 50 gm of representative sample with 125 ml of acetone and water 90 : 10 mixture in the Omni-Mixer immersed in an icewater bath for 15 minutes. Filter the liquid under vacuum through the Büchner funnel, with filter paper, into a 500-ml separatory funnel, fitted with a Teflon stopcock, containing 265 ml benzene. Wash the pulp three times with 25 ml acetone, filter into the separatory funnel, and then shake for 3 minutes. Let the mixture stand, and discard the separated aqueous phase. Filter the organic layer through 30 gm anhydrous sodium sulfate into a round-bottom flask and evaporate just to dryness with the rotary evaporator.

(b) *Muscle Tissues.* Homogenize 50 gm of representative sample with 125 ml acetone in the Omni-Mixer immersed in an icewater bath, and then follow the procedure described for maize.

(c) *Fatty Tissues.* Homogenize 50 gm of representative sample with 250 ml *n*-hexane in the Omni-Mixer immersed in an icewater bath for 15 minutes, and centrifugate at −20°C for 1 hour. Transfer the liquid phase to a separatory funnel, extract twice with 200-ml portions of acetonitrile, shaking each time for 3 minutes. Combine the acetonitrile extracts into a 500-ml round-bottom flask and concentrate just to dryness on the rotary evaporator.

ii. *Florex Column Chromatography*

Place a glass wool plug into the bottom of the chromatographic tube containing a few milliliters of chloroform and methanol 98 : 2 mixture. Add a slurry of 50 ml solvent mixture and 15 gm Florex to the column and allow it to settle, using a vibrator to displace any trapped air and insure uniform packing. Allow the solvent mixture to percolate through the bed. Pass another 100 ml of mixture under weak nitrogen pressure by adjusting the elution speed to 1.5–2 ml per minute until the level has reached the top of the bed. Dissolve the residue of the preceding cleanup step in 1 ml of solvent mixture and transfer it to the prepared column. Rinse three times with additional 0.3-ml portions of mixture and transfer them to the column. Allow the solvent to percolate through the bed, then elute both compounds with 100 ml solvent mixture. Collect the effluent and concentrate under vacuum to a convenient final volume.

iii. *Gas Chromatography*

Analyze the solution of the chromatographic cleanup by gas chromatography with the following operating parameters:

Column temperature:	225°C
Injection port temperature:	270°C
Detector temperature:	225°C
Carrier gas; Nitrogen flow rate:	70–80 ml/minute
Oxygen flow rate:	20 ml/minute
Air flow rate:	100 ml/minute
Hydrogen flow rate:	150 ml/minute
Recorder sensitivity:	1 mV full scale.

Inject 1–8 μl of the solution into the gas chromatograph and measure the peak heights corresponding to phenthoate and its oxygen analogue. Under these conditions, the retention times are, respectively, about 2 and 1.5 minutes. Drawing off benzene standard solutions, inject 2–16 ng of phenthoate and its oxygen analogue into the gas chromatograph, measure the peak heights and prepare calibration curves plotting nanograms of compound versus peak height.

Calculate the concentration of both compounds as follows:

$$\text{ppm phenthoate (or oxygen analogue)} = W/S$$

where W = weight of phenthoate or its oxygen analogue in nanograms, found in the portion of sample injected (from calibration curve); S = weight of the sample in milligrams, represented by the portion injected.

iv. *Detection Limit*

Considering the several "blanks" examined, the detection limit is about 0.005 ppm for each compound.

v. *Recoveries*

Recovery of phenthoate and its oxygen analogue from the substrates examined, was checked by adding known amount of standard at the blending step. All recoveries were run at the 0.05 ppm level and were in the 80–95% range for phenthoate and in the 75–90% range for oxygen analogue.

3. APPLICABILITY OF RECOMMENDED METHOD

The method has been tested on maize (stems, leaves, male and female inflorescences, cores of the cobs, caryopsides) at different degrees of ripening, and on cattle meat. It has been successfully applied also to some fish samples and to determine phenthoate residues in cottonseed oil. In this latter case, phenthoate is eluted from Florex chromatographic column with *n*-hexane and benzene 20:80 mixture, after 100 ml of *n*-hexane have been allowed to percolate through the column and were then discarded. The average recovery of 12 analyses was 86%, with a standard deviation of 4.7.

References

Bazzi, B. (1972). Report No. 79/1972. Istituto Ricerche Agrarie, Montedison Company, Milan.

Bazzi, B., and Canale, G. (1968). *Contrib. Ist. Ric. Agr. Montedison Company, Milan,* **9,** 53–62.

Bazzi, B., Fabbrini, R., and Canale, G. (1964a). *Pap., Congr. Ass. Int. Exp. Chim., 4th, Athens, 1964* p. 429–433.

Bazzi, B., Santi, R., Canale, G., and Radice, M. (1964b). *Contrib. Ist Ric. Agr. Montedison Company, Milan,* **6,** 49–58.

Bazzi, B., Santi, R., Radice, M., and Fabbrini, R. (1965). *J. Ass. Offic. Agr. Chem.* **48,** 1118–1120.

Bazzi, B., Abbruzzese, L., and Galluzzi, G. (1969). *J. Ass. Offic. Anal. Chem.* **52,** 52–55.

Gunther, F. A., Carman, G. E., Dusch, M. E., and O'Neal, J. R. (1972). Department of Entomology, Citrus Research Center and Agricultural Experiment Station, Riverside, California (personal communication).

Montedison Company. (1969). "Analytical Methods." Istituto Ricerche Agrarie, Milan.

Nissan Chemical Industries, Ltd. (1965). *Pap., 18th Meet. Jap. Chem. Soc.*

Nissan Chemical Industries, Ltd. (1972). "Analysis Methods of Elsan." NCI.

8

Pirimiphos-ethyl

D. J. W. BULLOCK

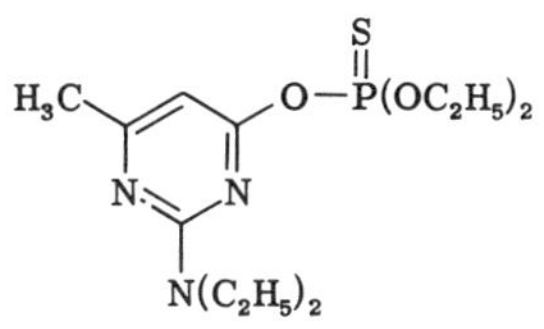

O-2-Diethylamino-6-methylpyrimidin-4-yl
O,O-diethyl phosphorothioate

I. GENERAL

A. Empirical Formula

$C_{13}H_{24}N_3O_3PS$ (Mol. wt. 333.4).

B. Alternative Names

PP211, Primicid®.

C. Source of Analytical Standard

ICI (Plant Protection Division), Yalding, Kent, England.

D. Biological Properties

Pirimiphos-ethyl controls a wide range of soil-dwelling pests, particularly flies and beetles, and shows promise as a replacement for the organochlorine insecticides. It is also effective as a foliar spray against many species of caterpillars, mites, and scale insects. It is relatively nonphytotoxic and may be used as a seed dressing.

The toxicity of pirimiphos-ethyl is relatively low. The acute oral LD_{50} for rats is 140–200 mg/kg, for mice 105 mg/kg, for cats 25–50 mg/kg, for guinea pigs 50–150 mg/kg. The acute dermal LD_{50} for rats is 1000–2000 mg/kg, while the 90-day no-effect level in rat and dog is 0.08 and 0.2 mg/kg/day, respectively. Application of pirimiphos-ethyl to guinea pigs produced no skin sensitization, and no irritation was evident on subacute dermal application of 100 mg/kg to the rabbit. Rats exposed to a vapor concentration of 1 ppm for 6 hours daily over a period of 15 days showed no inhibition of plasma or erythrocyte cholinesterase. Pirimiphos-ethyl fed to mammals is rapidly excreted within 48 hours, mainly in the urine.

Pirimiphos-ethyl is rapidly degraded on leaf surfaces but is more persistent in soil, having a half-life of the order of 40 days, the main degradation product in both cases being the parent hydroxypyrimidine 2-diethylamino-4-hydroxy-6-methylpyrimidine.

E. History

Pirimiphos-ethyl was introduced in 1971 by ICI Ltd. as an insecticide under the code name PP 211 and subsequently under the common name Primicid protected by British Patents 1,019,227 and 1,205,000.

F. Physical Properties

Pure pirimiphos-ethyl is a straw-colored liquid having a density of 1.14 at 20°C and a refractive index of 1.520 (D line) at 25°C. It is soluble in most organic solvents but is practically insoluble in water (1 ppm at 30°C). It is weakly basic and has a volatility of 2.9×10^{-4} mm Hg at 25°C. The ultraviolet spectrum, of pirimiphos-ethyl in methanol shows a strong absorption band at 248 nm ($\epsilon = 21{,}300$) and a weaker band at 300 nm.

G. Chemical Properties

1. METHOD OF SYNTHESIS

Pirimiphos-ethyl (I) is synthesized by reacting 2-diethylamino-4-hydroxy-6-methylpyrimidine with *O,O*-diethyl phosphorochloridothionate in the presence of an acid binding reagent. The parent hydroxypyrimidine is prepared by condensing *N,N*-diethylguanidine with ethyl acetoacetate under basic conditions. The reactions are shown below.

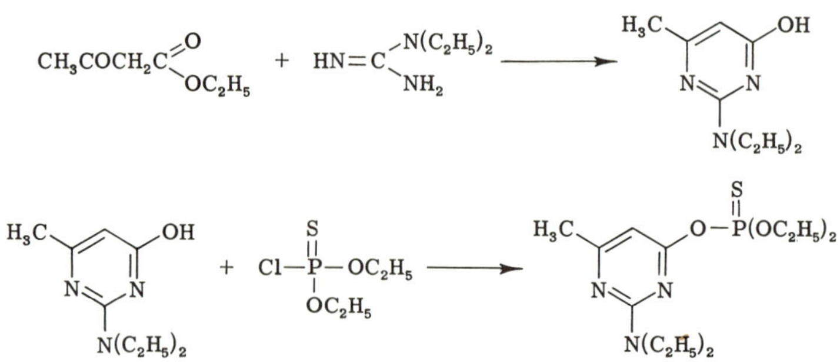

(I)

2. CHEMICAL REACTIONS

Pirimiphos-ethyl is readily hydrolyzed by strong acid and alkali to the parent hydroxypyrimidine 2-diethylamino-4-hydroxy-6-methylpyrimidine (II), hydrolysis being complete after boiling for 30 minutes with 3 *N* hydrochloric acid and no further change being effected on prolonged boiling up to 120 minutes.

Unstabilized technical pirimiphos-ethyl stored at room temperature undergoes a very slow rearrangement to *O,S*-diethyl-(2-diethylamino-6-methylpyrimidin-4-yl-)phosphorothiolate (III), which is hydrolyzed further, yielding eventually 2-diethylamino-4-hydroxy-6-methylpyrimidine (II).

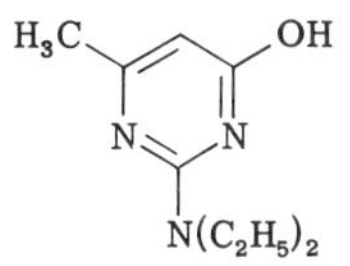

(II)

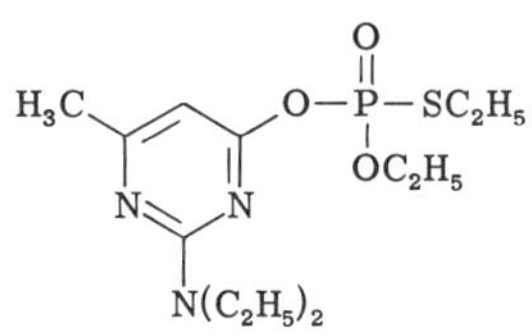

(III)

Other impurities likely to be present in degraded material are 2-diethylamino-6-methyl-4-ethylthiopyrimidine (IV), 2-diethylamino-4-ethoxy-6-methylpyrimidine (V), and traces of *O,O*-diethyl-(2-diethylamino-6-methylpyrimidin-4-yl) phosphate (VI).

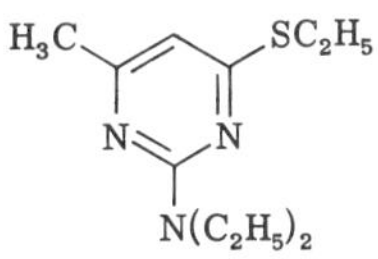

(IV)

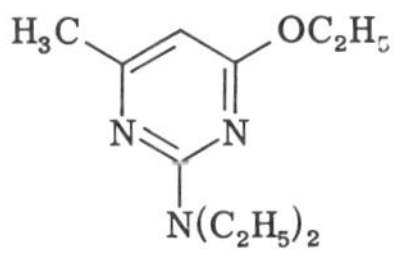

(V)

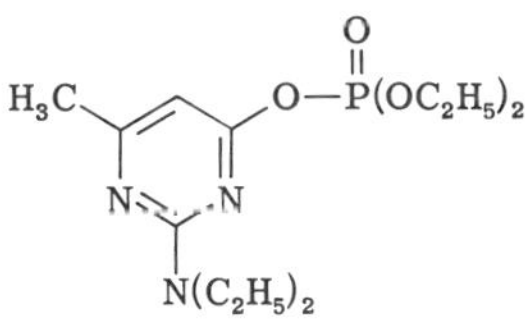

(VI)

These compounds can be resolved by gas–liquid chromatography as shown in Table I, although a polar phase such as fluorosilicone oil (QF-1) or a polyester is necessary for satisfactory resolution of pirimiphos-ethyl oxon (VI) from the photochemically produced *O,O*-diethyl *O*-(2-ethylamino-6-methylpyrimidin-4-yl) phosphorothioate (VII).

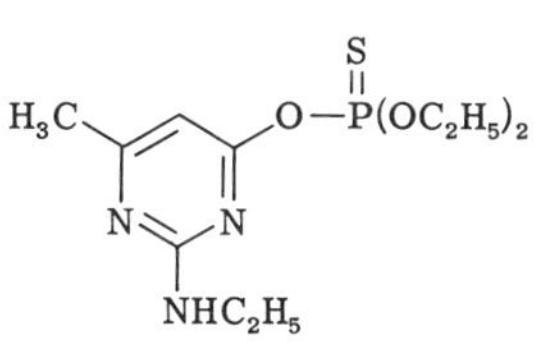

(VII)

TABLE I
RELATIVE RETENTION TIME FOR PIRIMIPHOS-ETHYL DEGRADATION PRODUCTS ON A NONPOLAR COLUMN PACKING[a]

Compound	R_{PE}[b]
Pirimiphos-ethyl (I)	1.00 (~4.7 minutes)
2-Diethylamino-4-hydroxy-6-methylpyrimidine (II)	0.20
O,S-diethyl *O*-(2-diethylamino-6-methylpyrimidin-4-yl) phosphorothiolate (III)	1.45
2-Diethylamino-6-methyl-4-ethylthiopyrimidine (IV)	0.17
2-Diethylamino-4-ethoxy-6-methylpyrimidine (V)	0.13
O,O-Diethyl *O*-(2-diethylamino-6-methylpyrimidin-4-yl) phosphate (VI)	0.90
O,O-Diethyl *O*-(2-ethylamino-6-methylpyrimidin-4-yl) phosphorothionate (VII)	0.90

[a] 5% SE-30 on Chromosorb W-HP (80–100 mesh), column temperature 210°C, column length 270 cm.
[b] PE = pirimiphos-ethyl.

H. Formulation

Formulations include 250- and 500-gm/liter emulsifiable concentrates, 5% and 10% granules, and 20% encapsulated.

II. ANALYSIS

A. Formulation Analysis

1. REVIEW OF METHODS

Three methods are available for the determination of pirimiphos-ethyl in formulated products, namely (i) gas–liquid chromatography (GLC) using a flame ionization detector (Bagness and Sharples, 1974); (ii) thin-layer chromatography (TLC); and (iii) hydrolysis to the parent hydroxypyrimidine followed by ultraviolet spectroscopy. The latter two were developed by Dr. S. H. Yuen of these laboratories.

The GLC procedure is the preferred method followed by quantitative TLC, as these methods are less susceptible to interferences from impurities, degradation products, and formulation ingredients. Good accuracy and reproducibility can be routinely obtained by both methods.

2. RECOMMENDED METHODS

a. *Gas–Liquid Chromatographic Method*

i. *Principle*

The sample is dissolved in acetone or chloroform containing nonadecane as an internal standard. A 1.5–2-μl portion is injected into a gas chromatograph equipped with a flame ionization detector, and the areas of the peaks obtained at the retention times of nonadecane and pirimiphos-ethyl are measured with a digital integrator. These areas are used for calculating the pirimiphos-ethyl content of the sample based on a response factor obtained using pure pirimiphos-ethyl. This method is applicable to stored and fresh samples of:

1. Technical pirimiphos-ethyl
2. Technical pirimiphos-ethyl containing epichlorohydrin stabilizer
3. Ultralow-volume formulations
4. Emulsifiable concentrates and dust formulation.

ii. *Reagents*

Chloroform, redistilled.

Acetone, redistilled.

Nonadecane, pure (Koch-Light Laboratories Ltd., Colnbrook, England)

Pirimiphos-ethyl analytical standard (Plant Protection Division, Yalding, Kent, England); store under refrigeration.

Silicone gum E301 (Phase Separations Ltd., Cheshire, England).

Celite 545 (Phase Separation Ltd., Cheshire, England).

Silyl 8 column conditioner (Phase Separation Ltd., Cheshire, England).

iii. *Apparatus*

Gas chromatograph equipped with a flame ionization detector having on-column injection such as the Pye Series 104.

Digital integrator such as the Kent Chromalog Mark III. Formulations containing mineral oil may show late eluting peaks requiring an integrator capable of measuring peaks on sloping baselines.

iv. *GLC Conditions*

Column,	Glass, 1.5 m × 4 mm i.d.
Column temperature,	220°C
Gas flow rates,	
Nitrogen (carrier gas),	50 ml/minute
Hydrogen,	50 ml/minute
Air,	700 ml/minute
Chart speed,	60 cm/hour
Amplifier attenuation,	50×10^2
Syringe,	5 μl capacity
Injection,	1.5 μl

Column packing: Weigh 250 gm 100–120 mesh Celite 545 into a 2-liter beaker and add concentrated HCl until the Celite is covered by a

2-inch layer of acid. Heat on a boiling water bath in a fume cupboard for 2–3 hours, stirring occasionally. Decant the acid and repeat with fresh acid. Filter off the Celite into a sintered funnel and wash several times with distilled water.

Put the Celite into a clean 2-liter beaker and stir for 15 minutes with about 1 liter of distilled water. Filter off the Celite and repeat the procedure with fresh water until the water remains neutral after 15 minutes stirring. Filter off the Celite and dry in an oven at 130°C overnight. Sieve the Celite through 100- and 120-mesh sieves and use the portion retained on the 120-mesh sieve to prepare the column packing.

Dissolve 5 gm silicone elastomer E 301 in 200 ml chloroform; this is most conveniently done by allowing it to stand overnight in a stoppered flask. Weigh 45 gm of the treated Celite into a large evaporating basin and pour in the silicone elastomer solution. Evaporate the chloroform on a water bath in a fume cupboard. The slurry should be stirred frequently but gently to ensure that the support is evenly coated with the silicone without creating too much fine material. When the bulk of chloroform has evaporated and the slurry is too thick to stir, place the basin in an oven at 100°C to remove the remaining solvent. Sieve the dried material through 100- and 120-mesh sieves, retaining the material on the 120-mesh sieve for the column packing.

Condition the column overnight at 250°C with a carrier gas flow and silanize with Silyl 8-column conditioner before connecting the column to the detector.

v. *Procedure*

(a) *Preparation of Solutions. Internal standard solution:* Weigh accurately about 2.0 gm nonadecane and dissolve in acetone. Dilute to 1 liter with acetone.

Pirimiphos-ethyl standard solution: Weigh accurately about 0.15 gm pure pirimiphos-ethyl and dissolve in exactly 25 ml internal standard solution. Call this solution A.

Sample solution: Weigh accurately sufficient sample to contain about 0.15 gm pirimiphos-ethyl and dissolve in exactly 25 ml internal standard solution. Allow the mixture to stand, until the pirimiphos-ethyl has dissolved. (For encapsulated formulations shake the mixture for 15 seconds every 30 minutes over 3 hours.) Call the supernatant liquid solution B.

(b) *Determination.* Inject a 1.5-μl portion of solution A into the chromatographic column. Use the digital integrator to measure the areas of the peaks due to nonadecane and pirimiphos-ethyl. A potentiometric recorder may be used instead of an integrator, but areas measured as peak height × width at half-height have a poorer precision. Repeat the injection.

Take the mean ratio of these areas to calculate the response factor for pure pirimiphos-ethyl using the formula shown below. This is approximately 2.2.

Take solution B and repeat the procedure as for solution A. Calculate the response factor for the sample.

(c) *Calculations.*

$$_{A}R_{W}^{C} = \frac{\text{weight of pure pirimiphos-ethyl in solution A}}{\text{weight of nonadecane in solution A}}$$

$$_{A}R_{a}^{C} = \frac{\text{peak area of pirimiphos-ethyl in solution A}}{\text{peak area of nonadecane in solution A}}$$

Response factor for pure pirimiphos-ethyl, $F_A = {_A}R_W^C / {_A}R_a^C$.

$$_{B}R_{W}^{C} = \frac{\text{weight of sample in solution B}}{\text{weight of nonadecane in solution B}}$$

$$_{B}R_{a}^{C} = \frac{\text{peak area of pirimiphos-ethyl in solution B}}{\text{peak area of nonadecane in solution B}}$$

Response factor for sample, $F_B = {_B}R_W^C / {_B}R_a^C$. Then percentage pirimiphos-ethyl content (w/w) of sample = $(F_A/F_B) \times 100$.

b. *Thin-Layer Chromatographic Method*

i. *Principle*

A portion of the filtered sample solution is applied to a silica gel chromatoplate. After developing the plate with hexane–acetone (9:1), the plate is examined under ultraviolet light and the area corresponding to pirimiphos-ethyl ($R_f \sim 0.6$) removed. After eluting the pirimiphos-ethyl from the silica gel with methanol, final determination is carried out by measuring the extinction of the methanol solution at 248 nm.

ii. *Reagents and Apparatus*

Thin-layer plates: Make a slurry of 40 gm silica gel GF_{254} (Merck, Darmstadt, Germany) and 80 ml of water by stirring with a glass rod. Coat a 0.3 mm layer on six 20 cm × 20 cm glass plates, then dry at 105°C. Store the prepared plates in a cabinet over an efficient desiccator.

Syringe, 100 μl.

Chromatographic tank, suitable for use with 20 cm × 20 cm plates.

Ultraviolet lamp, 254 nm.

Spectrophotometer, Unicam SP500, Pye-Unicam Ltd., Cambridge, England.

Developing solvent, hexane–acetone 3:1.

Removal and extraction apparatus, suitable for removing and extracting areas of silica gel from TLC plates (Matthews *et al.,* 1962).

iii. *Procedure*

Pour 100 ml of developing solvent into a chromatographic tank and cover with a lid. Weigh a portion of sample to contain 0.25–0.3 gm of pirimiphos-ethyl into a 50-ml volumetric flask, add methanol to the mark, and shake vigorously for 1 minute. Filter the contents through a Whatman no. 40 paper into a 1″ × 6″ stoppered test tube.

Apply 100 μl of the prepared sample solution onto a thin-layer plate as a streak about 5 cm long $1\frac{1}{2}$ cm from the bottom of the plate and from one edge; apply a second portion of sample solution similarly to the other half of the plate. Put the plate in the chromatographic tank and allow the chromatogram to develop until the solvent front has moved to 1 cm from the top of the plate. Dry the plate in a fume cupboard.

Examine the chromatogram under ultraviolet light and mark off the two bands of pirimiphos-ethyl (about R_f 0.6). Remove silica gel in turn from the marked area with the removal and extraction apparatus, then elute with 50 ml methanol into a 50-ml volumetric flask (see Note 1); call these test solutions. Concurrently prepare a blank solution by collecting untreated silica gel from an area corresponding to the treated from the middle of the plate with the removal and extraction apparatus followed by elution with methanol.

Measure the extinctions of the test and blank solutions at 248 nm in a 1-cm silica cell against methanol as reference. The extinction for the latter solution is usually less than 0.01. Then percentage pirimiphos-ethyl content (w/w) of the sample = $39.6\ E/W$ (see Note 2), where E = average extinction of the test solution found (corrected for the blank); W = weight in grams of sample taken.

Note 1: Connect a 50-ml volumetric flask to the B10 joint of the removal and extraction apparatus and a vacuum line to the side tube, then collect silica gel from the marked area using the apparatus as a vacuum cleaner. Reverse the top attachment to serve as a funnel and add methanol in small portions until nearly 50 ml of the eluate have been collected. Detach the volumetric flask, add methanol to the mark and mix.

Note 2: The factor given is a product of the dilution factor and regression coefficient obtained from the calibration graph with a Unicam

SP500 spectrophotometer. This equation is not applicable when a spectrophotometer of other make is used for measuring the extinction. In such a case, a calibration graph ranging from 0 to 0.6 mg of pure pirimiphos-ethyl per 50 ml methanol should be constructed, from which the pirimiphos-ethyl content of the final solution is read off. The percentage pirimiphos-ethyl content of sample is then computed using the expression $50Y/W$, where Y is the weight in milligrams of pirimiphos-ethyl found in the final solution (corrected for the blank) and W is the weight in grams of sample taken.

Note 3: The sintered disk of the TLC extraction and removal apparatus should be cleaned thoroughly with a stiff brush, after each use.

c. *Analysis by Ultraviolet Spectroscopy Following Hydrolysis*

i. *Principle*

A portion of sample is treated with chloroform and the extract percolated through a Florisil column, which removes any initial 2-diethylamino-4-hydroxy-6-methylpyrimidine resulting from degradation of the sample. After evaporation of the solvent, the residual pirimiphos-ethyl is hydrolyzed in boiling 3 *N* hydrochloric acid to yield 2-diethylamino-4-hydroxy-6-methylpyrimidine, which is determined absorptiometrically at 298 nm in methanol.

ii. *Reagents*

Florisil, 100–200 U.S. mesh (Floridin Co.). To prepare the column, transfer 4 gm of Florisil to the chromatographic tube, in which a small cotton–wool pad has been placed near the stopcock end (*N.B.,* rest a loose cotton–wool plug on the top of the column, when a granule sample is analyzed). Gently tap the tube to settle the Florisil, then pass through 15 ml chloroform. Use a new column for each test.

Chloroform and methanol, Analytical Grade should be used. These solvents must be kept in glass bottles; storage in polyethylene or other plastics may cause contamination giving a high background in the final ultraviolet absorptiometry.

Hydrochloric acid, 3 N.

Ammonium acetate buffer solution, pH 6.5. Add with mixing 95 ml glacial acetic acid and 100 ml of concentrated ammonia solution (sp. gr. 0.88) to 1 liter of water, dilute the cooled solution to 2 liters with water and mix.

Note 4: Florisil is used as received, but elution characteristics of each batch must be checked.

iii. *Standard Solutions*

(a) *Standard 2-diethylamino-4-hydroxy-6-methylpyrimidine solutions. Stock solution,* 0.5 mg/ml. Dissolve 0.125 gm pure 2-diethylamino-4-hydroxy-6-methylpyrimidine ($C_9H_{15}N_3O$; mol. wt. 181.2; m.p. 134°C) in methanol and dilute to 250 ml with methanol. Pure 2-diethyl-4-hydroxy-6-methylpyrimidine is obtainable from ICI Plant Protection Division, Yalding, Kent, England.

Working solution, 0.1 mg/ml. Dilute 20.0 ml stock solution to 100 ml with methanol.

Prepare four standards, containing 0.25, 0.5, 0.75, and 1.0 mg of 2-diethylamino-4-hydroxy-6-methylpyrimidine, by diluting in turn 2.5, 5.0, 7.5, and 10.0 ml working solution to 50.0 ml with methanol.

Note 5: Standard 2-diethylamino-4-hydroxy-6-methylpyrimidine solutions are stable under normal laboratory conditions.

iv. *Apparatus*

Chromatographic columns, 30 cm long × 1.3 cm i.d. constricted at the lower end to a 2-mm bore.

Boiling flasks, 250-ml round-bottomed flasks with B24 2-cm necks.

Separating funnels, 250-ml capacity preferably with PTFE stopcocks.

Rotary evaporator.

Spectrophotometer, Unicam SP500 (Pye-Unicam Ltd., Cambridge, England).

v. *Procedure*

(a) *Preparation of Sample Solution. Technical samples and liquid formulations.* Weigh a portion of well-mixed sample containing 0.25–0.3 gm pirimiphos-ethyl into a 100-ml beaker. Add 50 ml chloroform, mix and pass the solution through a Florisil column, collecting the eluate in a 250-ml boiling flask. Wash the beaker twice each with 50 ml chloroform and pass the washings through the column into the boiling flask.

Pumice granules. Weigh a portion of well-mixed sample containing 0.25–0.3 gm pirimiphos-ethyl into a 250-ml beaker, add 50 ml chloroform, and stir thoroughly with a glass rod. Decant the supernatant solution through a Florisil column, collecting the eluate in a 250-ml boiling flask. Wash the pumice in the beaker twice each with 50 ml of chloroform and pass the supernatant solutions through the column into the boiling flask.

(b) *Determination.* Evaporate off the solvent at 40–50°C under reduced pressure, add 50 ml of 3 *N* hydrochloric acid and boil under a reflux condenser for 45 minutes. Wash down the condenser with 100 ml water, transfer the cooled hydrolyzate quantitatively to a 500-ml volumetric flask, make up to the mark with water and mix. Pipette 15.0 ml of the sample solution into a 250-ml separating funnel containing 80 ml buffer solution, add 50 ml chloroform, and shake vigorously for 1 minute. Run the chloroform extract into a 250-ml boiling flask. Repeat the extraction with 50 ml chloroform and combine the extracts in the boiling flask. Evaporate off the solvent at 40–50°C under reduced pressure and dissolve the residue in 50 ml of methanol. Pipette 10.0 ml of the methanol solution into a 50-ml volumetric flask, dilute to the mark with methanol and mix; call this the test solution.

Measure the extinctions of the test solution and four standards, containing 0.25, 0.5, 0.75, and 1.0 mg of 2-diethylamino-4-hydroxy-6-methylpyrimidine at 298 nm in a 1-cm silica cell against methanol as reference. Draw the calibration graph relating to extinctions of standards to amounts of 2-diethylamino-4-hydroxy-6-methylpyrimidine in mg, and read off the 2-diethylamino-4-hydroxy-6-methylpyrimidine content of the test solution; alternatively, compute the 2-diethylamino-4-hydroxy-6-methylpyrimidine content by interpolation. Call this amount Y mg. One mg of 2-diethylamino-4-hydroxy-6-methylpyrimidine is equivalent to 1.84 mg of pirimiphos-ethyl (mol. wt. 333.4). Then percentage pirimiphos-ethyl content (w/w) of sample is

$$\frac{500 \times 50 \times 100 \times 1.84 \times Y}{15 \times 10 \times 1000 \times W} = \frac{30.7Y}{W}$$

where W = weight in grams of sample taken.

d. *General Discussion*

Analysis of technical and formulated pirimiphos-ethyl by GLC, TLC, and UV give results that are generally found to be acceptable, being within the limits of precision for each of the individual methods. GLC is usually the method of choice, particularly for analysis of low purity samples or extensively degraded materials. GLC relative response factors are linear over a wide concentration range and highly reproducible from day to day. There is no evidence of interferences from wetting or dispersing agents. TLC and UV are regarded as confirmatory methods (Bagness and Sharples, 1974).

B. Residue Analysis

1. REVIEW OF METHODS

Residue methods for pirimiphos-ethyl are similar to those described for pirimiphos-methyl in Chapter 9. Crop samples are extracted with

20% hexane in acetone or with methanol, animal tissues are extracted with boiling *n*-hexane or acetonitrile and soil with a mixture of acetone and aqueous ammonium chloride. Cleanup of sample extracts is carried out by using *n*-hexane–acetonitrile partition and/or Florisil column chromatography. Analyses of samples for pirimiphos-ethyl have been largely confined to crop samples and to soil. It is anticipated, however, that all methods described for pirimiphos-methyl will be applicable to pirimiphos-ethyl.

Pirimiphos-ethyl is used mainly as a soil insecticide for control of pests such as carrot fly, onion fly, and banana stem borer, being applied as a seed dressing or as granules. Under these conditions, pirimiphos-ethyl shows limited mobility and little translocation into plants. The main initial degradation product in soil is the hydrolysis product 2-diethylamino-4-hydroxy-6-methylpyrimidine. Pirimiphos-ethyl is, however, occasionally used as a foliar spray, and experiments involving application of pirimiphos-ethyl to leaves under laboratory conditions have shown the formation of small amounts of compounds VI and VII (Section I), which still retain the phosphorus group. For this reason, the analytical method was developed to include these two compounds as well as the parent pirimiphos-ethyl.

Analysis of a large number of crop samples has shown, however,

TABLE II
RELATIVE RETENTION TIMES ON GLC OF ORGANOPHOSPHORUS PESTICIDES[a]

Compound	Retention time (minutes)[b]
Pirimiphos-ethyl	1.0
Compound (VI)	1.5
Compound (VII)	2.3
Parathion	3.0
Dimethoate	1.3
Malathion	2.0
Bromophos	2.0
Diazinon	0.5
Fenithrothion	3.1
Iodofenphos	3.8
Dichlorvos	~0.1
Tetrachlorvinphos	4.5
Azinphos-methyl	12.0
Chlorfenvinphos	3.0

[a] Column: 50 cm, 3% PDEAS, 180°C.
[b] Retention time of pirimiphos-methyl, 2 minutes.

that pirimiphos-ethyl itself invariably represents the total toxic residue. The analysis therefore resolves itself into the determination of residues of the parent compound pirimiphos-ethyl.

Following extraction of the sample, with or without cleanup, determination of the residue level is carried out by injecting a portion into a gas chromatograph equipped with a flame photometric or thermionic detector, using the same conditions and precautions as described for pirimiphos-methyl in Chapter 9. Using polar phases for GLC such as phenyldiethanolamine succinate (PDEAS) or fluorosilicone oil (QF-1), pirimiphos-ethyl and its metabolites (VI and VII) can be readily separated from each other and other common organophosphorus insecticides as shown in Table II. Confirmation of residue identity has been carried out by using a nonpolar column packing such as 3% SE-30 in the GLC analysis. This latter column will also separate pirimiphos-ethyl from pirimiphos-methyl, whereas the retention times are very similar on polar column packings.

An alternate, rarely used, residue method for pirimiphos-ethyl is a colorimetric determination based upon the hydrolysis of pirimiphos-ethyl to H_2S with hydrobromic acid and subsequent determination of the H_2S. While this method is lacking in specificity and sensitivity, it is occasionally of use in laboratories without GLC facilities.

Confirmation of residue identity in cases of apparent high residue

TABLE III

RELATIVE R_f VALUES ON TLC OF ORGANOPHOSPHORUS PESTICIDES[a]

Compound	Relative R_f[b]
Parathion	1.0
Pirimiphos-methyl	1.10
Pirimiphos-ethyl	1.35
Iodofenphos	1.19
Tetrachlorvinphos	0.25
Bromophos	1.25
Malathion	0.60
Diazinon	0.78
Dimethoate	0.02
Chlorfenvinphos	0.18
Azinphos-methyl	0.20

[a] 0.25 mm preprepared silica gel GF_{254} (Merck) layer; developing solvent *n*-hexane–ethyl acetate (4:1); plate allowed to develop for 14 cm in a saturated atmosphere.

[b] R_f parathion, 0.36.

TABLE IV
R_f VALUES FOR PIRIMIPHOS-ETHYL AND METABOLITES ON TLC

Compound	Solvent system: Diethyl ether : *n*-hexane : ethyl alcohol (66:33:10)	Solvent system: Hexane : acetone (4:1)
Pirimiphos-ethyl	0.93	0.55
VI	0.70	0.31
VII	0.80	0.32

can be carried out by thin-layer chromatography (TLC). A portion of the cleaned up extract is applied to silica gel plate, developed in hexane–ethyl acetate (4:1) and sprayed with one of the many reagents available for visualizing organophosphorus pesticides (Volume VII, Chapter 1). Less than 1 μg pirimiphos-ethyl can be detected using a spray reagent such as hydriodic acid–ammonium molybdate (Askew *et al.*, 1974).

Table III shows the TLC properties on silica gel using hexane–ethyl acetate (4:1) as a developing solvent. An alternative solvent is diethyl ether–*n*-hexane–methanol, 66:33:10.

2. RECOMMENDED METHOD

The gas liquid chromatographic method described for pirimiphos-methyl in Chapter 9 can be also applied to pirimiphos-ethyl without modification, although the range of samples to which it has been applied for pirimiphos-ethyl has to date been restricted to crops (leafy and root), oils, stored grain, and soil. The GLC behavior of pirimiphos-ethyl is very similar to pirimiphos-methyl, although it has a longer retention time on nonpolar column packings. As with pirimiphos-methyl, care is needed when concentrating organic solutions containing pirimiphos-ethyl. Water samples should be submitted and stored in aluminium containers. Organic solvents containing pirimiphos-ethyl and its metabolites should not be dried with large quantities of anhydrous sodium sulfate, since losses by absorption can occur.

All comments in Chapter 9, Section III, are applicable also to pirimiphos-ethyl. Table IV gives chromatographic data for pirimiphos-ethyl and metabolites in two confirmatory TLC systems.

REFERENCES

Askew J., Ruzicka, J. H., and Wheals, B. B. (1974). *Analyst* **94,** 275.
Bagness, J. E., and Sharples, W. G. (1974). *Analyst* **99,** 225.
Matthews, J. S., Pereda, V. A. L., and Aguilera, P. A. (1962). *J. Chromatogr.* **9,** 331.

9

Pirimiphos-methyl

D. J. W. BULLOCK

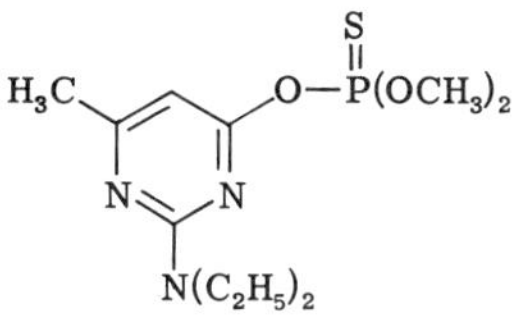

O-(2-Diethylamino-6-methylpyrimidin-4-yl) *O,O*-dimethyl phosphorothioate

I. GENERAL

A. Empirical Formula

$C_{11}H_{20}N_3O_3PS$ (Mol. wt. 305.3).

B. Alternative Names

PP 511, Actellic®, Blex®.

C. Source of Analytical Standard

ICI (Plant Protection Division), Yalding, Kent, England.

D. Biological Properties

Pirimiphos-methyl is a broad-spectrum, fast-acting insecticide and acaricide of low mammalian toxicity. It shows both contact and fumigant activity, controlling many crop pests and particularly pests of stored products, where it shows good eradicant action at low rates of application and gives protection against reinfestation over a long period. Of especial interest is its activity against species resistant to other organophosphorus pesticides.

Pirimiphos-methyl is generally of low toxicity. The acute oral LD_{50} for female rats is 2050 mg/kg, for male mice 1180 mg/kg, for female guinea pigs 1000–2000 mk/kg, for male rabbits 1150–2300 mg/kg, for female cats 575–1150 mg/kg, for hens 30–60 mg/kg, for birds generally in the range 30–400 mg/kg. The acute dermal LD_{50} for rabbits is more than 2000 mg/kg, and pirimiphos-methyl does not cause skin or eye irritation. Rats fed up to 360 ppm in their diet for 90 days showed no abnormalities apart from inhibition of plasma cholinesterase.

It is extensively metabolized in rats and dogs mainly to the parent hydroxypyrimidine 2-diethylamino-4-hydroxy-6-methylpyrimidine and the corresponding *N*-dealkylated derivatives. No accumulation occurs in tissues with repeated doses.

Hens fed up to 32 ppm pirimiphos-methyl in their diet for 7 days showed no detectable residues in flesh or eggs. The 48 hour median tolerance limit for rainbow trout is 0.25 ppm and for the common carp is 1.4 ppm. Pirimiphos-methyl is rapidly degraded in water, mainly to the parent hydroxypyrimidine (50% degradation after one day in sunlight) and shows limited persistence and mobility in soil.

E. History

Pirimiphos-methyl was introduced by ICI in 1970 as an insecticide under the code number PP 511 and subsequently under the common names Actellic® and Blex® protected by British patents 1,019,227 and 1,204,552.

F. Physical Properties

Pure pirimiphos-methyl is a light straw-colored liquid having a density of 1.157 at 30°C and freezing to a white solid at approximately 15°C. It is miscible with or soluble in most organic solvents but practically insoluble in water (5 ppm at 30°C). It is very weakly basic and quite volatile, having a vapor pressure of 1.1×10^{-4} Torr at 30°C. The ultraviolet spectrum shows a strong absorption band at 248 nm ($\epsilon = 20{,}761$ at 248 nm) and a weaker band at 300 nm.

G. Chemical Properties

1. METHOD OF SYNTHESIS

Pirimiphos-methyl (I) is synthesized by reacting 2-diethylamino-4-hydroxy-6-methylpyrimidine with *O,O*-dimethyl phosphorochloridothionate under anhydrous conditions in the presence of an acid binding agent. The parent hydroxypyrimidine is prepared by condensing *N,N*-diethylguanidine with ethyl acetoacetate under basic conditions. The reactions are shown below.

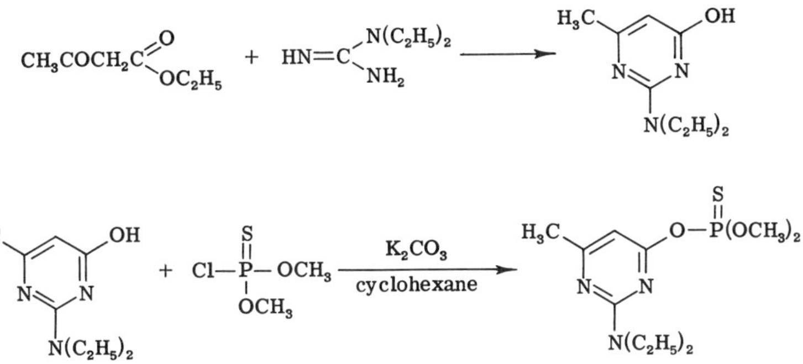

(I)

2. CHEMICAL REACTIONS

Pirimiphos-methyl is readily hydrolyzed by strong acid and alkali to the parent hydroxypyrimidine 2-diethylamino-4-hydroxy-6-methylpyrimidine (II). For example, complete hydrolysis to the parent hydroxypyrimidine is effected by boiling in 3 *N* hydrochloric acid for 30 minutes, no further change being effected on prolonged boiling up to 60 minutes. Unstabilized pirimiphos-methyl stored at room temperature undergoes a very slow rearrangement to *O,S*-dimethyl-*O*-(2-diethylamino-6-methylpyrimidin-4-yl) phosphorothiolate (III). This inactive and unstable compound is readily hydrolyzed by moisture, yielding eventually colorless crystals of the salt 2-diethylamino-4-hydroxy-6-methylpyrimidin-*O,S*-dimethylthiophosphate (IV). This in turn slowly decomposes to give the characteristic odor of methylmercaptan.

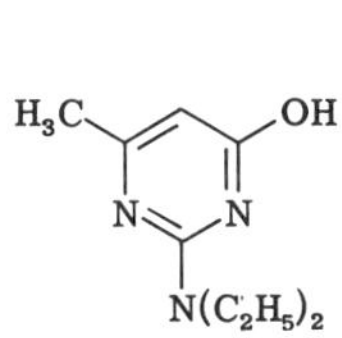

(II)

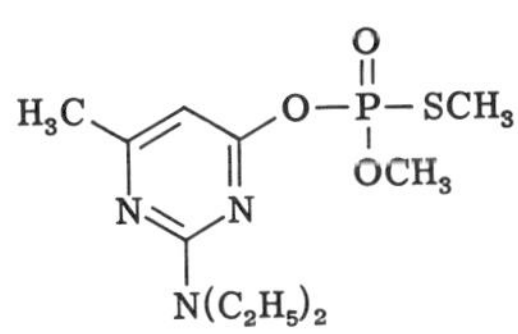

(III)

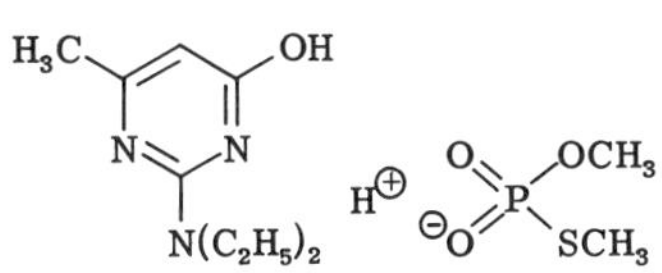

(IV)

Storage of unstabilized pirimiphos-methyl at elevated temperatures leads to a variety of dephosphorylated products, predominant among

TABLE I
RELATIVE RETENTION TIMES FOR PIRIMIPHOS-METHYL DEGRADATION PRODUCTS ON A NONPOLAR COLUMN PACKING[a]

Compound	R_{PM}[b]
Pirimiphos-methyl (I)	1.00 (~4 minutes)
2-Diethylamino-4-hydroxy-6-methylpyrimidine (II)	0.23
O,S-Dimethyl-*O*-(2-dimethylamino-6-methylpyrimidin-4-yl) phosphorothiolate (III)	1.49
2-Diethylamino-4-hydroxy-6-methylpyrimidinium *O,S*-dimethylthiophosphate (IV)	Decomposes to give (II)
2-Diethylamino-4-methyl-6-methylthiopyrimidine (V)	0.18
2-Diethylamino-4-methoxy-6-methylpyrimidine (VI)	0.14
O,O-Dimethyl-*O*-(2-diethylamino-6-methylpyrimidin-4-yl) phosphate (VII)	0.92
O,O-Dimethyl-*O*-(2-ethylamino-6-methyl-pyrimidin-4-yl) phosphorothionate (VIII)	0.92

[a] 5% SE-30 on Chromosorb WHP (80–100 mesh), column temp. 210°C, column length 270 cm.

[b] PM = pirimiphos-methyl.

which are 2-diethylamino-4-methyl-6-methylthiopyrimidine (V) and 2-diethylamino-4-methoxy-6-methylpyrimidine (VI). A small amount of *O,O*-dimethyl *O*-(2-diethylamino-6-methylpyrimidin-4-yl) phosphate (VII) is also formed. These compounds can be resolved by gas–liquid chromatography (GLC) as shown in Table I, although a polar phase such as fluorosilicone oil (QF-I) or a polyester is necessary for satisfactory separation of compound VII from the photochemically produced *O,O*-dimethyl-*O*-(2-ethylamino-6-methylpyrimidin-4-yl) phosphorothionate (VIII).

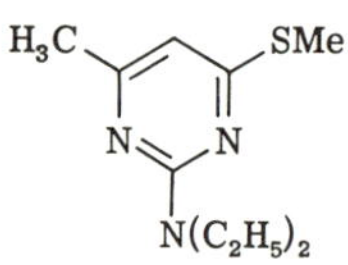

(V)

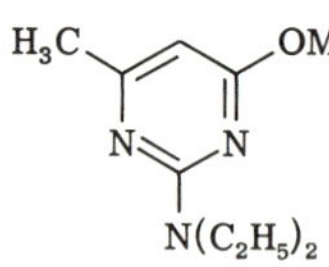

(VI)

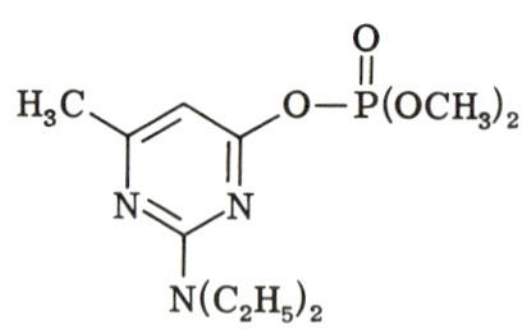

(VII)

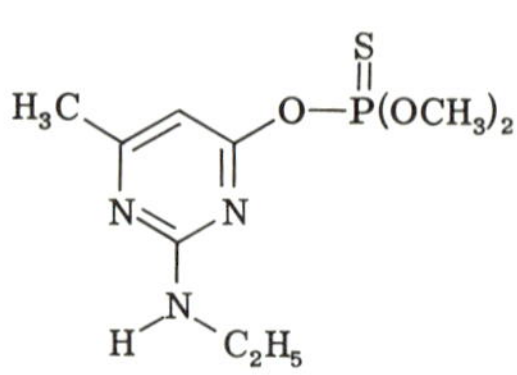

(VIII)

H. Formulation

Pirimiphos-methyl is available as emulsifiable concentrates (8%, 25%, 50% a.i.), dusts (2% a.i.), formulations for ultralow-volume application (10% and 50% a.i.), a solvent-free formulation, and a smoke generator formulation. Trade names include Actellic®, Actellifog®, Silosan®, and Blex®.

II. ANALYSIS

A. Formulation Analysis

1. REVIEW OF METHODS

Three methods are available for the determination of pirimiphos-methyl in formulated products. These include gas–liquid chromatography (GLC) using a flame ionization detector (Bagness and Sharples, 1974); thin-layer chromatography (TLC); and hydrolysis to the parent hydroxypyrimidine followed by ultraviolet spectroscopic determination. The last two were developed by Dr. S. H. Yuen of these laboratories.

The GLC procedure is the preferred method followed by quantitative TLC, as these methods are less susceptible to interferences from impurities, degradation products, and formulation ingredients. Good accuracy and reproducibility can be routinely obtained by both methods.

2. RECOMMENDED METHODS

a. *Gas–Liquid Chromatographic Method*

i. *Principle*

The GC procedure for pirimiphos-methyl is identical to that for pirimiphos-ethyl (Chapter 8), except that a column temperature of 215°C and amplifier attenuation of 20×10^2 are employed, 1 μl of sample is injected, and octadecane is the internal standard.

ii. *Procedure*

(a) *Preparation of Solutions. Internal standard solution:* Weigh accurately about 1.8 gm octadecane and dissolve in 50 ml chloroform. Dilute to 1 liter with chloroform.

Pirimiphos-methyl standard solution: Weigh accurately about 0.35 gm pure pirimiphos-methyl and dissolve in exactly 50 ml internal standard solution. Call this solution A.

Sample solution: Weigh accurately about 0.35 gm technical material (0.70 gm 50% a.i. liquid formulation or 1.40 gm 25% a.i. liquid formulation) and dissolve in exactly 50 ml internal standard solution. Call this solution B.

(b) *Determination.* Inject a 1-μl portion of Solution A into the chromatographic column. Use the digital integrator to measure the areas of the peaks due to octadecane and pirimiphos-methyl. A potentiometric recorder may be used instead of an integrator, but areas measured as peak height $\times$ width at half-height have a poorer precision. Repeat the injection.

Take the mean ratio of these areas to calculate the response factor for pure pirimiphos-methyl using the formula shown below. This is approximately 2.4. Take solution B and repeat the procedure as for solution A. Calculate the response factor for the sample.

(c) *Calculations.* Use the calculation method outlined in Chapter 8, Section II,A,2,a,v,(a), substituting "pirimiphos-methyl" for "pirimiphos-ethyl" and "octadecane" for "nonadecane."

b. *Thin-Layer Chromatographic Method*

Use the same TLC procedure as for formulations of pirimiphos-ethyl (Chapter 8), substituting hexane–acetone (3 : 1) for hexane–acetone (9 : 1). The R_f value of pirimiphos-methyl is $\simeq 0.7$. The percentage of pirimiphos-methyl of the sample $= 34.6E/W$.

c. *Analysis by Ultraviolet Spectroscopy following Hydrolysis*

Follow the procedure exactly as for pirimiphos-ethyl (Chapter 8). One milligram of 2-diethylamino-4 hydroxy-6-methylpyrimidine is equivalent to 1.685 mg of pirimiphos-methyl (mol. wt. 305.3). Then a percentage of pirimiphos-methyl content (w/w) of the sample is equal to

$$\frac{500 \times 50 \times 100 \times 1.685 \times Y}{15 \times 10 \times 1000 \times W} = \frac{28.1\,Y}{W}$$

where W = weight in grams of sample taken.

B. Residue Analysis

1. REVIEW OF METHODS

Two methods are available for the determination of pirimiphos-methyl residues. The first uses a gas–liquid chromatograph equipped with a flame photometric or thermionic (alkali flame) detector and is the method of choice. The second, rarely used, method is a colorimetric determination based upon the hydrolysis of pirimiphos-methyl to H_2S and subsequent determination of this as methylene blue. Thin-layer chromatography on silica gel is used as required for confirmation of residue identity.

Although pirimiphos-methyl is normally considered to be the total toxic residue in crops and stored products, the GLC method has been developed to include the determination of compounds VII and VIII, which are biologically active, and have been detected in very small amounts following application of ^{14}C-labeled pirimiphos-methyl to leaves under laboratory conditions. Analysis of crops sprayed under field conditions, stored grain, and animal tissues show no significant residues of either VII or VIII. Compound VII, pirimiphos-methyl oxon, is particularly difficult to work with, since it is hydrolyzed very easily by atmospheric moisture. Because of the sensitivity of flame photometric and thermionic detectors to pirimiphos-methyl, a limit of detection of 0.01 ppm or lower is generally obtainable without difficulty. Organic solvents containing pirimiphos-methyl residues can be concentrated on a rotary evaporator without loss, provided the operation is carried out carefully and at low temperatures (below 30°C). A mixed solvent of 20% acetone in *n*-hexane has been found satisfactory for extracting weathered residues of pirimiphos-methyl from crops and stored products. Methanol is a good alternative and gives less emulsions in certain cases. Animal tissues are extracted with boiling *n*-hexane or acetonitrile, and milk is extracted with a mixture of acetone and acetonitrile. Extracts of many crops can be injected directly into the gas chromatograph without cleanup. This procedure usually shortens the column lifetime and may entail frequent replacements of injection zone for on-column injection. Analyses of crops by direct injection of uncleaned extracts should always be preceded by injections of control extracts and control extracts fortified by addition of pirimiphos-methyl residues to check that (a) where a flame photometric detector is being used, excessive contamination of the extract with hydrocarbons, lipids or other crop material is not reducing the detector response to pirimiphos-methyl, although interfering peaks may not be apparent on injection of the unfortified control extract, (b) where a thermionic detector is being used, excessive contamination of the extract with crop material is not giving rise to interfering peaks, negative peaks or late-eluting peaks that interfere with subsequent analyses, or subtle enhancement of detector response.

Other signs of unacceptable levels of contamination of the extract before GLC are rising baselines, increased noise levels, decreased column lifetime and detector contamination (particularly for thermionic detectors).

Extracts of oily crops such as palm oil, olive oil, or rapeseed and animal tissues always require some degree of cleanup before analysis. Countercurrent partitioning between *n*-hexane and acetonitrile has been found effective for the removal of oils and fats, while column chromatography on 2% deactivated Florisil can be used for further cleanup if required or to separate pirimiphos-methyl from its metabolites if a suitable GLC column is not available.

Two types of column packings have been used in the GLC determination of pirimiphos-methyl residues. Polar phases such as phenyldiethanolamine succinate give very effective separation of pirimiphos-methyl and its metabolites, although these columns show excessive bleed and rather short lifetimes. Nonpolar phases such as SE-30 or OV1 can also be used to chromatograph pirimiphos-methyl effectively, but on these column packings, as Table I shows, separation of metabolite VII from metabolite VIII is not possible.

Florisil partially deactivated by the addition of 2% water provides excellent cleanup of animal tissues following *n*-hexane–acetonitrile partition if a very clean extract is required. Deactivation with water is used because it gives a superior cleanup, has more reproducible elution characteristics, is less affected by humidity and temperature, and gives better recoveries particularly of pirimiphos-methyl oxon.

Interferences in the GLC residue method by a number of other common crop-protection chemicals has been investigated and shown to be minimal as seen in Table II (Chapter 8), on pirimiphos-ethyl.

For samples showing apparent high residues, confirmation of residue identity can be obtained using thin-layer chromatography (TLC) on silica gel. Table III (Chapter 8) shows the TLC properties of pirimiphos-methyl on silica gel.

2. RECOMMENDED METHOD

a. *Principle*

Crop samples are extracted with 20% acetone in *n*-hexane by high-speed blending. After the plant debris has settled, a sample of the decanted liquid is washed with water and then dried with anhydrous sodium sulfate. A portion of the washed dried extract is injected into a gas chromatograph equipped with a flame photometric or thermionic detector. The detector responses (peak heights) obtained at the retention times of pirimiphos-methyl and its metabolites (VII, VIII) are used to calculate the residues present. Stored products such as grain and other dry produce are extracted with 20% acetone in *n*-hexane following the addition of water or with methanol, animal tissues with boiling *n*-hexane, and milk with acetone–acetonitrile. Oily or fatty extracts are cleaned up by *n*-hexane–acetonitrile countercurrent partition and/or Florisil liquid–solid column chromatography before analyses. The limit of detection in each case is usually better than 0.01 ppm, while recoveries vary from 90% for crop samples to 75–80% for milk and animal tissues.

b. *Reagents*

Acetone, redistilled if necessary.

n-Hexane, redistilled if necessary.

Methanol, redistilled if necessary.

Acetonitrile, redistilled if necessary.

Florisil 60–100 mesh (British Drug Houscs Ltd.).

Celite 545 filter aid (Phase Separations Ltd.).

Phosphate buffer, pH 6.7. Dissolve 136 gm KH_2PO_4 and 174 gm K_2HPO_4 in water and make up to 2 liters.

Sodium sulfate, granular anhydrous, Analar.

Ammonium chloride, Analar.

Chromosorb W-HP 80–100 mesh (Phase Separations Ltd.).

Phenyldiethanolamine succinate (Phase Separations Ltd.).

Fluorosilicone oil QF-1 (Phase Separations Ltd.).

Silicone gum rubber S-30 Ultraphase (Phase Separations Ltd.).

Pirimiphos-methyl and its metabolites, samples of known purity. Store in dark under refrigeration.

Gases for GLC, pure nitrogen (e.g. "white spot" BOC), hydrogen, air, oxygen.

Note 1: Solvents used in the analysis should be examined for possible interfering impurities by concentrating 100 ml down to 0.5 ml and injecting a 5-μl portion into the gas chromatograph. Other reagents used should also be carefully checked for interfering impurities.

c. *Apparatus*

Laboratory chopper used for initial preparation of samplcs, c.g., Hobart model.

High-speed macerator or blender, e.g., Sorvall Omni-Mixcr Homogenizer available from V. A. Howe & Co. Ltd., London, U.K. or I. Sorvall Inc. Norwalk, Connecticut.

Rotary evaporator or Kuderna–Danish concentrator with 3 ball Snyder column.

Water bath capable of maintaining a temperature of 30°C.

Syringes for GLC, e.g., 10 μl Hamilton 701 SN 3".

Separator funnels, 250 ml and 500 ml capacity.

Glass columns for Florisil cleanup 40 cm × 205 cm i.d. fitted with glass stopcocks.

Gas–liquid chromatograph equipped with a flame photometric detector or with a suitable thermionic (alkali flame) detector. The Melpar flame photometric detector is recommended (available from Techmation Ltd., Middlesex, England).

Note 2: Chromatographs fitted with heated injection ports and on-column injection systems are recommended to avoid decomposition of pirimiphos-methyl. Glass columns and easily interchangeable precolumns are also desirable.

Wide-mouth 2-oz glass jars fitted with screw-capped lids.

d. *Experimental procedure*

i. *Gas–liquid Chromatographic Conditions*

The following conditions have been found satisfactory when using a Pye 104 gas chromatograph equipped with glass column and a heated injection port. The manufacturer's manual for the gas chromatograph concerned should be consulted for details concerning amplifier and recorder settings.

Column, 90 cm long × 3 mm i.d. glass.

Column packing, 3% w/w phenyldiethanolamine succinate on Chromosorb W-HP (80–150 mesh).

Alternative column packing, 1:1 mixture of 4% SE-30 (Ultraphase)/6% QF-1 on Chromosorb W-HP (80–100 mesh); 7% QF-1 on Chromosorb W-HP (80–100 mesh).

Confirmatory column packing, 270 cm long × 3 mm i.d. 3% SE-30 (Ultraphase) on Chromosorb W-HP (80–100 mesh).

Column temperatures:

3% PDEAS	180°C (for 90 cm column)
4% SE-30/ 6% QF-l	210°C (for 90 cm column)
7% QF-l	160°C (for 90 cm column)
3% SE-30	220°C (for 270 cm column)

Injection port temperature, 15°C above column temperature.

Carrier gas flow rate, pure dry nitrogen at 100 ml/minute.

Detector, Melpar flame photometric detector with phosphorus filter (526-nm filter). The setting up of the flame photometric detector should be carried out as detailed in the manufacturer's literature. This detector is usually operated at 15–20°C above the column temperature, and with gas flows as follows: hydrogen 200 ml/minute, oxygen 35 ml/minute, air 25 ml/minute. The hydrogen–oxygen gas flow ratio should be adjusted until a satisfactory response is obtained for pirimiphos-methyl and its metabolites.

Retention times. The retention times given below should be regarded as approximate values and will vary slightly according to the age and condition of the column.

3% PDEAS	pirimiphos-methyl	1.9 minutes
	VIII	3.9 minutes
	VII	2.7 minutes
7% QF-1	pirimiphos-methyl	1.0 minutes
	VIII	3.0 minutes
	VII	2.4 minutes
3% SE-30 (270 cm column)	pirimiphos-methyl	4.5 minutes
	VII	3.9 minutes
	VIII	3.95 minutes

Column Preparation. The following method is recommended using the preparation of the 3% PDEAS column packing as an example. Dissolve 0.3 gm phenyldiethanolamine succinate in acetone (100 ml) and add 10 gm Chromosorb W-HP (80–100 mesh) to this solution contained in a round-bottomed flask (250 ml). Remove the majority of the solvent under vacuum on a rotary evaporator until a soft but not dry mass is produced. Pour the packing on to a Petri dish and allow to air dry overnight. Thoroughly mix the packing and then pack the column under vacuum with constant vibration, to within 1 cm of the injection-port end. Close the column with a silanized glass plug. Condition the column by installing it in the gas chromatograph unconnected to the detector and heating it overnight at 15°C above its operating temperature. The phenyldiethanolamine succinate column gives excellent resolution of pirimiphos-methyl and its metabolites from many other organophosphorus insecticides in common use. It does, however, bleed heavily and has a more limited lifetime than the other recommended column packings.

ii. *Extraction*

(a) *Root Crops, Fruit, and Vegetables.*

1. Remove loose soil and roots, etc.

2. Dice or chop the whole sample thoroughly using a laboratory chopper or mill.

Note 3: If the samples have been deep frozen while awaiting analysis, they should be subsampled while only partially thawed to avoid loss of liquid. In some cases, it may be desirable to add solid carbon dioxide to the sample during mincing or chopping.

3. Mix the chopped sample and weigh out a representative portion (100 gm) into a macerating jar. Add 400 ml 20% acetone in *n*-hexane and macerate at high speed for 10 minutes. Allow the plant debris to settle by standing for 30 minutes and then decant a portion of the supernatant liquid (approximately 50 ml) into a separating funnel (250 ml).

4. Add an equal volume (50 ml) of water to the decanted extract in the separating funnel, shake together for 1 minute, and when the two layers have separated, drain off the lower water layer including any interfacial emulsion. Persistent emulsions should be broken by centrifuging a portion of the emulsified layer and/or by agitation with a glass rod.

5. Dry the separated washed *n*-hexane layer by shaking with 1–2 gm anhydrous sodium sulfate and then transfer the dried *n*-hexane extract to a screw-capped, air-tight glass bottle. Keep under refrigeration in the dark until required for analysis, which should be preferably carried out as soon as possible.

Note 4: Mushrooms and similar crops should be extracted with methanol to avoid the persistent emulsions formed on extraction with *n*-hexane–acetone.

A 100-gm subsample is macerated with 400 ml methanol and the macerate filtered through a Whatman no. 1 filter paper (9 cm) on a Büchner funnel. The crop debris is sucked dry and the residuum washed with further methanol until the final volume of the filtrate is 500 ml.

Note 5: Extracts of crops with *n*-hexane–acetone, methanol, or acetonitrile containing pirimiphos-methyl and its metabolites may be concentrated on a rotary evaporator without loss, provided the temperature of the water bath does not exceed 30°C and the evaporation is slow and gentle. A Kuderna–Danish concentrator with three ball Snyder column is recommended where losses are found to occur. The concentration should not proceed beyond the aqueous residue, owing to the possibility of hydrolysis of the pesticide.

(b) *Stored Grain* (*Wheat, Barley, Rice, Peanuts*), *Bread, Bran, and Fine Offal.*

1. Thoroughly mix the whole sample and by quartering take a 100-gm representative subsample. Loaves of bread should be cut up with a knife into 1 inch cubes. When analyzing grain, it is as well to remember that the majority of the residue is on the outside of the grain and is readily lost as fine dust that may be found adhering to plastic bags or containers.

2. Grind the sample to a fine powder in a large mill. The temperature of the sample should not be allowed to rise more than 10°C during grinding, and precautions should be taken to avoid loss of fine dust.

3. Weigh a representative portion (100 gm for grain, 50 gm for bread and bran) into a macerating jar and add 50 ml water followed by 400 ml 20% acetone in *n*-hexane. Blend the mixture on a high-speed macerator for 10 minutes and allow the mixture to stand until a clear

portion of supernatant liquid (~50 ml) may be poured off into a separating funnel (250 ml). A portion of the mixture may be centrifuged at this point if desired.

4. Add an equal volume of water to the portion of extract contained in the separating funnel and shake for 1 minute.

5. Allow the *n*-hexane and water layers to separate and run off the water layer including any interfacial emulsion.

6. Dry the *n*-hexane layer by shaking with 1–2 gm anhydrous sodium sulfate and then transfer the dried *n*-hexane extract to a screw-capped, air-tight glass bottle. Keep under refrigeration in the dark until required for analysis which should be preferably carried out as soon as possible.

Note 6: Grain and Bread samples can also be extracted with methanol as described for mushrooms (see Note 4).

Note 7: Extracts of grain and bread products can be cleaned up by following the *n*-hexane–acetonitrile countercurrent partition method described below for oily or fatty materials.

(c) *Oily or Fatty Crop Samples.* Samples such as olive oil, palm oil, rapeseed oil, etc., must be cleaned up by *n*-hexane–acetonitrile countercurrent partition before a portion of the extract is injected into the gas–liquid chromatograph. Unless this is done, the lifetime of the column packing is very short and the response of the detector may be reduced. In certain cases, however, where the sample contains residues above 1 ppm, sufficient dilution may be possible to avoid clean-up at the discretion of the analyst.

1. For oily or fatty crops (e.g., rapeseed), carry out the procedure as detailed in (a) using 90 ml of the washed *n*-hexane extract (equivalent to a 25 gm crop sample). For samples of oil, dissolve 10 gm in 100 ml *n*-hexane, warming the mixture if necessary. In both cases then proceed as below.

2. Extract the *n*-hexane solution by shaking three times for 1 minute with 100-ml portions of acetonitrile in a separating funnel (500 ml), allowing complete separation of the layers each time and rejecting any interfacial emulsion. The acetonitrile layer is the lower layer.

3. Combine the acetonitrile layers in a separating funnel (500 ml) and wash with *n*-hexane (50 ml) by shaking gently for 1 minute.

4. Separate the layers and wash the *n*-hexane by shaking with acetonitrile (30 ml) in a separating funnel (250 ml).

5. Reject the *n*-hexane layer from (4) and add the acetonitrile wash from (4) to the combined acetonitrile layers from (3).

6. Concentrate the combined acetonitrile layers to 60 ml (±5 ml) on a rotary evaporator, taking great care at this stage to evaporate the excess solvent at a slow and gentle rate and at a temperature not exceeding 30°C.

7. Quantitatively transfer the concentrated acetonitrile extract to a separating funnel (500 ml), washing the concentration flask with 10 ml acetonitrile and adding this to the acetonitrile in the separating funnel.

8. Add 300 ml of 5% aqueous sodium sulfate to the acetonitrile extract and shake gently. Extract this solution twice for 1 minute with 100 ml portions of *n*-hexane.

9. Allow the layers to separate completely by standing, and combine the upper *n*-hexane layers, rejecting any small amounts of interfacial emulsion. Depending on the probable residue levels, this combined *n*-hexane may be dried with anhydrous sodium sulfate before GLC or concentrated gently on a rotary evaporator (temperature below 30°C) to a volume suitable for analysis (usually not below 50 ml).

10. The final *n*-hexane solution after drying should be stored under cool dark conditions until required for analysis as detailed in Section (a) no. 5.

(d) *Milk.*

1. Thoroughly homogenize the milk sample either by vigorous shaking or blending.

2. Take a 50-gm subsample of the homogenized milk, and shake vigorously with 50 ml acetone in a macerating jar for 5 minutes.

3. Add 200 ml acetonitrile to the acetone–milk mixture and blend on a Sorvall Omni mixer for 5 minutes.

4. Filter the precipitated proteins off through a pad of glass wool supported on a large glass funnel, and transfer the filtrate to a separating funnel (500 ml).

5. Add 50 ml *n*-hexane and shake gently for 1 minute. When the two phases have separated, run the bottom acetonitrile layer into a round-bottom flask (500 ml). Reject the top *n*-hexane wash layer.

6. Concentrate the acetonitrile solution down to 60 ± 5 ml on a rotary evaporator. This process must be carried out very carefully to avoid loss of residue by volatilization and hydrolysis. The temperature of evaporation should not exceed 30°C, and the volume of the concentrate should not on any account go below 60 ± 5 ml.

7. Transfer the concentrate quantitatively to a separating funnel (250 ml) with the aid of two 10-ml water washings of the concentrating flask. Extract the aqueous solution by shaking successively with 100-ml and 50-ml portions of *n*-hexane.

8. Combine the *n*-hexane extracts, and dry by shaking with 15 gm

anhydrous sodium sulfate. Filter the solution through a small plug of anhydrous sodium sulfate, finally washing the sodium sulfate with 20 ml *n*-hexane.

9. Concentrate the combined *n*-hexane extracts and washings on a rotary evaporator (temperature $<30°C$) to approximately 7 ml. Transfer the concentrate to a 10-ml volumetric flask and make up to the mark using *n*-hexane washings of the concentration flask.

10. Store the solution under refrigeration at 0°C until required for analysis.

(e) *Eggs*.

1. Break and discard shells of each egg. Blend whole egg together, or separate yolk and white as required.

2. Blend a subsample (100 gm) with 150 ml acetonitrile for 10 minutes, finally decanting the supernatant acetonitrile into a concentration flask. Add 150 ml acetonitrile to the egg residue in the blender jar and repeat the extraction. Again decant the acetonitrile and concentrate the combined acetonitrile extracts down to 60 ± 5 ml on a rotary evaporator, taking care to keep the water bath temperature below 30°C.

3. Transfer the aqueous acetonitrile concentrate to a separating funnel (250 ml) using 10 ml water to wash out the concentration flask.

4. Add 100 ml 5% aqueous sodium sulfate solution to the separating funnel together with 100 ml *n*-hexane. Shake vigorously for 1 minute and allow the two layers to separate.

5. Drain off the lower aqueous acetonitrile layer, including any interfacial emulsion, into another separating funnel, and repeat the extraction with a further 100 ml *n*-hexane.

6. Combine the *n*-hexane layers and dry by passing through a short column of anhydrous granular sodium sulfate. Wash the column with 10 ml hexane and combine washings with *n*-hexane extracts.

7. Allow the *n*-hexane extract to percolate down a column of 25 gm 2% deactivated Florisil (prepared as described below), not allowing the liquid level to fall below the level of the column packing.

8. Collect the *n*-hexane eluent, and then elute the column successively with hexane–diethyl ether (7:3).

9. Follow by elution with diethyl ether–acetone (9:1), collecting 100 ml of each fraction.

10. Concentrate each fraction separately to the required volume (nominally 50 ml), using a rotary evaporator or Kuderna–Danish concentrator at 30°C, for GLC determination.

Note 8: Preparation of 2% deactivated Florisil columns. Heat the Florisil overnight at 650°C and allow to cool in a vacuum desiccator.

Store permanently in an oven at 130°C until required for deactivation, which is done by adding 2 ml water to 98 gm Florisil and mixing on a roller mill for 1 hour. Each batch is prepared as required. The column for cleanup is prepared by adding 20 gm deactivated Florisil to a glass column (40 cm × 2.5 cm i.d.) and topping with 15 gm anhydrous granular sodium sulfate. The characteristics of the Florisil column may vary with different types of sample extract and should be checked accordingly by eluting a fortified extract with the appropriate solvents and analyzing 10 ml fractions for compound recovery.

Note 9: With some samples of eggs and some oily samples, the diethyl ether–acetone eluate is not sufficiently clean for GLC analysis, injections of the concentrated solution showing large negative peaks having similar retention times to pirimiphos-methyl and its metabolites. In such samples, the following method is used:

(a) Following stage (vi) above, divide the *n*-hexane layer into two equal portions and carry out the Florisil cleanup as described, eluting the column with hexane–diethyl ether (7 : 3) for the determination of the parent compound.

(b) Extract the remaining *n*-hexane twice with 50 ml 0.1 *N* HCl by shaking for 1 minute in a separating funnel.

(c) Wash the combined acid extract with 25 ml *n*-hexane, bring the acid layer to neutral with 1 *N* NaOH and phosphate buffer solution (pH 6.7).

(d) Extract the neutralized solution three times with 50 ml ethyl acetate, and dry the extract by passing it down a short column of granular anhydrous sodium sulfate. Wash the column with 15 ml of ethyl acetate.

(e) Concentrate the ethyl acetate down to a suitable volume for GLC determination of any metabolites present. The analysis should be completed as rapidly as possible to prevent hydrolysis of the pesticides.

(f) *Animal Tissue.*

1. Allow the sample to partially thaw without loss of fluid (i.e., blood).

2. Finely mince or chop the sample, adding dry ice if necessary during this process to avoid thawing.

3. By quartering, take a 50-gm subsample and grind in a large pestle and mortar with 50 gm sharp quartz sand and sufficient anhydrous granular sodium sulfate to give a uniform dry mixture.

4. Transfer the mixture quantitatively to a large beaker (500 ml), add 200 ml *n*-hexane, and warm gently with stirring on a water bath until the solvent boils gently. Allow to simmer gently for 1 minute and then

decant the solution through a Whatman no. 1 filter paper into a measuring cylinder (250 ml). Extract the residue in the beaker again by washing with 50 ml *n*-hexane, finally transferring the tissue residue to the filter paper and washing with sufficient *n*-hexane to make the volume of *n*-hexane extract in the measuring cylinder up to 250 ml (when cool).

5. Transfer a 50-ml portion of the *n*-hexane extract to a separating funnel (250 ml) together with 100 ml acetonitrile, and shake vigorously for 1 minute. Allow the layers to separate, and run the bottom acetonitrile layer into another separating funnel (250 ml). Repeat the extraction with a further 50 ml acetonitrile. Wash the combined acetonitrile solution by shaking for 1 minute with 50 ml *n*-hexane. Again allow the layers to separate, and run the lower acetonitrile layer into a round-bottom flask.

Note 10: The sample size is reduced to 10 gm by taking a 50-ml portion for analysis.

Note 11: Hexane extracts of samples such as mutton fat, beef suet, etc., may be cleaned up before acetonitrile partition by allowing the *n*-hexane extract to stand overnight under refrigeration at 0°C. This causes the bulk of the fat to precipitate, and the 50-ml portion for analysis is removed by pipette while the solution is still cold.

6. Extract the combined *n*-hexane washings by shaking for 1 minute with 50 ml acetonitrile in a separating funnel. Combine all the acetonitrile extracts.

7. Concentrate the combined acetonitrile extracts carefully and slowly down to approximately 7 ml on a rotary evaporator, keeping the temperature below 30°C.

8. Transfer the acetonitrile concentrate quantitatively to a 10-ml volumetric flask, and make up to the mark with acetonitrile washings of the concentration flask. Store this solution under refrigeration until required for analysis. It should be clear and pale yellow in color.

Note 12: Any concentration of solutions containing pirimiphos-methyl and its metabolites should be checked for losses by volatilization and/or hydrolysis. A Kuderna–Danish concentrator with a 3 ball Snyder column may be necessary for concentration down to low volumes.

(g) *Soil.*

1. Thoroughly mix and randomize the undried sample of soil and remove any stones, roots, or vegetable matter. Weigh a representative subsample (100 gm) into a blender jar and add 200 ml acetone – 1 *M* aqueous ammonium chloride (1 : 1). Blend for 1 hour.

2. Filter the extract through a moistened pad of Celite 545 on a Whatman no. 1 filter paper (11 cm) supported on a Büchner funnel, and suck dry. Transfer quantitatively the Celite pad and soil residuum back to the blender jar, and repeat the extraction with another 200 ml mixed solvent. Repeat the filtration, this time washing the soil residuum with 50 ml mixed solvent.

3. Combine the filtrates and mix thoroughly, adding further 1 : 1 acetone–1 *M* aqueous ammonium chloride to make the final volume up to 500 ml in a measuring cylinder.

4. Transfer a 50-ml portion of the sample extract (equivalent to 10 gm soil sample) to a separating funnel (250 ml) and add 50 ml phosphate buffer (pH 6.7) together with 100 ml *n*-hexane. Shake thoroughly for 1 minute. Allow the layers to separate, drain off the lower aqueous layer into another separating funnel (250 ml), and repeat the extraction with 50 ml *n*-hexane.

5. Combine the *n*-hexane layers and dry by passing through a short column of anhydrous sodium sulfate. Wash the column with 10 ml *n*-hexane and concentrate the combined *n*-hexane extracts and washings to 8 ± 1 ml on a rotary evaporator, keeping the temperature below 30°C.

6. Transfer the *n*-hexane concentrate quantitatively to a 10-ml volumetric flask, and make up to the mark, using *n*-hexane washings of the concentration flask for this purpose. Store this solution under refrigeration at 0°C until required for analysis.

Note 13: The concentration of the *n*-hexane extract should be carried out with great care to avoid losses of pirimiphos-methyl. If losses are found to occur, a Kuderna–Danish type concentrator equipped with a three-ball Snyder column should be used.

Note 14: Residues of pirimiphos-methyl oxon (VII) are not found in soil.

iii. *Determination*

1. Prepare a series of standard solutions of pirimiphos-methyl and metabolites in *n*-hexane to cover the residue range expected. For crops a range of concentrations from 5 μg to 500 μg in 320 ml *n*-hexane is normally sufficient.

2. Make a number of 5-μl injections of the strongest standard into the gas chromatograph until a constant response is obtained. This priming of the column may need to be repeated throughout the lifetime of the column.

Note 15: A 10-μl syringe is recommended for injections of solutions into the gas chromatograph. To obtain satisfactory results using peak height measurements, the method of injection used by each operator should be standardized. Five-microliter injections are recommended, but this may need to be reduced to 2–3 μl for thermionic detectors. After filling the syringe, it should be free from air bubbles, and the tip should be wiped clean with tissue paper. The plunger should be depressed rapidly and the needle allowed to remain in the injection port for 3 seconds before being withdrawn.

3. Make a number (up to six) repetitive injections of the standard solution whose concentration approximates to that of the sample extract and check that good replication of peak height and retention time is obtained. Variations of more than 5% in peak height and more than 0.2 minute in retention time make the injection suspect. Attention should then be given to operating parameters such as gas leaks, syringe faults, or contaminated column packings. Gas chromatographs equipped with thermionic detectors (alkali flame) are more prone to variations than those equipped with flame photometric detectors. Where a thermionic detector is being used, it is recommended that a standard solution be injected after every third sample injection to check the detector response.

4. Prepare a calibration graph from the standard solutions showing micrograms per liter of pirimiphos-methyl and metabolites versus detector response expressed as peak height multiplied by attenuation of the gas chromatograph amplifier. Alternatively bracket the response given by an injection of the sample extract by injecting two standards of the appropriate strengths.

5. Inject 5 μl of the extract prepared from a control untreated or check sample. Examine the chromatogram carefully to check that the control extract is free from any extraneous peaks occurring at the retention time of pirimiphos-methyl and metabolites. The attenuation of the amplifier should be set during this process at the sensitivity likely to be used in the analysis. Check also that injection of control extracts do not give increased noise level, rising base lines, excessive solvent responses, negative responses, or peaks of long retention time that would interfere with subsequent analysis. Observation of any response of this nature usually means heavy and unacceptable contamination with sample extracts, lipids, sterols, or plasticizers, such as dibutyl phthalate. The samples for analysis must then be subjected to further clean up before GLC determination.

6. Inject a portion (usually 5 μl) of the sample extract into the GLC using the same technique as described for the injection of the standards. If a peak occurs at the retention time of pirimiphos-methyl or its metabolites, repeat the injection and carefully compare the retention time of

the peaks to that of the standard solutions. If the identification is doubtful, change the column packing to one of those described under Section II,B,2,d,i and repeat the injection. Retention times should not differ by more than 0.2 minute. If the repeatability of peak heights, retention times, and identification is satisfactory, use the measurement of peak height of sample compared with standards to calculate the residue present using the simple formula

$$R(\text{ppm residue}) = W/S$$

where W = amount of pirimiphos-methyl and metabolites in micrograms (μg) found in the portion of sample injected (from standards); S = sample weight (gm) represented by the portion injected.

7. The recovery factor to be applied in calculation of the actual residue present in any particular sample should be obtained by each operator by fortifying untreated control samples at that residue level likely to be found in the actual field samples. Where residues of pirimiphos-methyl and metabolites are very low or undetectable, recovery experiments should be carried out by fortifying control untreated samples at, or near, the claimed residue detection limit.

8. Recoveries in crop samples at the 0.1–5 ppm level are usually in the range $90 \pm 10\%$. Recoveries in samples requiring cleanup by acetonitrile–*n*-hexane partition, such as milk and animal tissues, are usually of the order $80 \pm 10\%$. Higher recoveries in these samples can be obtained if desired at the expense of a longer method by increasing the number of countercurrent extractions in the *n*-hexane–acetonitrile cleanup stage.

iv. *Limits of Detection*

The limits of detection obtained in the residue method depend obviously on the sensitivity of the detector and concentration of the sample extract. Defining the limit of detection as four times the noise level of amplifier at the maximum attenuation (sensitivity) used, limits of detection of 0.01 ppm for crops and 0.005 ppm for milk can usually be obtained without difficulty.

v. *General Notes*

1. The top 7 cm of the column packing, which is the injection zone for on-column injection chromatographs, may need replacing regularly. This is especially the case for uncleaned-up crop extracts. Preconditioned column packing or silanized glass wool may be used for this purpose.

2. Injections of extracts of samples containing a high residue should

be followed by injections of pure solvents (acetone or *n*-hexane) to avoid spurious residues due to ghosting.

3. Fortification of crop samples for recovery experiments should be carried out by spiking macerated crop samples with solutions of pirimiphos-methyl and metabolites in acetone.

4. Residues of pirimiphos-methyl and metabolites on crops are stable under the conditions of storage normally employed (deep freeze at −14°C).

5. Analysis of water samples can be accomplished by extraction of the sample with *n*-hexane. Pirimiphos-methyl and its metabolites are readily hydrolyzed under aqueous conditions (half-life of approximately 1 day in sunlight under neutral conditions). They are also readily absorbed from aqueous solution by plastic containers. Water samples should therefore be stored in aluminum containers under refrigeration.

6. While the residue method has been developed for pirimiphos-methyl and its two potential organophosphorus-containing metabolites, (VII) and (VIII), a very large number of residue analyses on a wide variety of samples has indicated that significant residues of (VII) and (VIII) are rarely found. For many samples, therefore, the residue method is simplified to being determination of pirimiphos-methyl only. Pirimiphos-methyl oxon is very unstable, being easily hydrolyzed by atmospheric moisture. New samples of this compound may need to be prepared at regular intervals during experiments with it.

III. DISCUSSION OF METHOD

The gas chromatographic method using a flame photometric detector has been applied successfully to a wide variety of crops, stored products, and animal tissues. Recoveries at over 90% are obtainable, as might be expected, with direct injection of uncleaned extracts. Use of *n*-hexane–acetonitrile partitions or liquid–solid column cleanup reduces the recovery to the 75–85% range, although higher recoveries can be obtained at the expense of a longer method by using more stages in the *n*-hexane–acetonitrile countercurrent cleanup. Losses of residue are more likely to occur during concentration of sample extracts and the appropriate precautions as described in the text should be taken.

The detector response has been measured by peak height rather than peak area, and this has been found to be quite satisfactory with a Melpar flame photometric detector, remaining quite reproducible over long periods of time. The identity of each peak emerging from the chromatograph has been confirmed by double-beam mass spectrophotometry.

TABLE II
CHROMATOGRAPHIC DATA FOR PIRIMIPHOS-METHYL AND METABOLITES

Solvent system	R_f		
	Pirimiphos-methyl	VII	VIII
Diethyl ether–*n*-hexane–ethyl alcohol (66:33:10)	0.89	0.67	0.76
Hexane–acetone (4:1)	0.50	0.27	0.27
Chloroform–methyl alcohol (85:15)	0.84	0.80	0.79

Confirmation of residue identity may be obtained by using another column packing for GLC, or by thin-layer chromatography as described in the text. Table II gives chromatographic data for pirimiphos-methyl and metabolites on preprepared silica gel plates (0.25 mm) using three different solvent systems.

The extractability of aged or weathered residues of pirimiphos-methyl and its metabolites from crop samples by the solvents most usually employed, 20% acetone in *n*-hexane and methanol has been proved satisfactory by comparison of results with residue levels obtained by exhaustive extraction methods. Water should be added to dry crops before extraction. The use of chloroform as a crop extractant should be avoided, since under some conditions, artifacts of pirimiphos-methyl chlorinated in the 5 position of the ring can be formed.

REFERENCES

Bagness, J. E., and Sharples, W. G. (1974). *Analyst* **99**, 225.
Imperial Chemical Industries Limited. (1966). British Patent 1,019,227.
Imperial Chemical Industries Limited. (1970). British Patent 1,204,552.

10

Promacyl

D. S. GUNEW

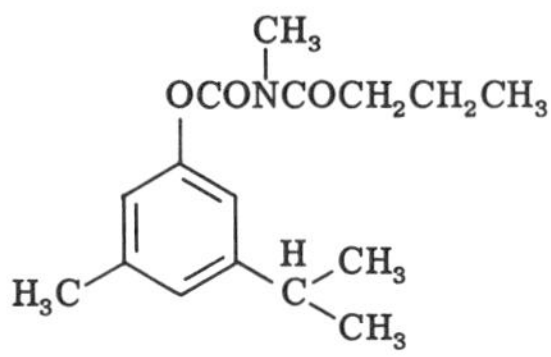

3-Isopropyl-5-methylphenyl *N*-butyryl-*N*-methylcarbamate

I. GENERAL

A. Empirical Formula

$C_{16}H_{23}NO_3$ (Mol. wt. 227.3).

B. Alternative Names

None. Promacyl is the recommended common name. Promicide® is the registered tradename of ICI Australia Limited for the formulated product.

C. Source of Analytical Standard

ICI Australia Limited, Central Research Laboratories.

D. Biological Properties

1. PESTICIDAL ACTIVITY

Promacyl is an acaricide and has shown excellent activity against susceptible and resistant strains of *Boophilus microplus*.

2. TOXICITY

The acute oral toxicity of technical promacyl is: LD_{50} and 95% confidence interval, for mouse 2000–4000 mg/kg, for rat 1220 mg/kg, for guinea pig 250 mg/kg, for rabbit 8000 mg/kg. The antidote is atropine sulfate.

E. History

Promacyl was developed by ICI Australia Limited in 1969 for the control of cattle tick by dipping or spraying.

F. Physical Properties

3-Isopropyl-5-methylphenyl *N*-butyryl-*N*-methylcarbamate is a colorless liquid with boiling point 158°C at 5 Torr; molar heat of vaporization 20 kcal/deg/mole; odor, faintly sweet; refractive index of technical material n_D^{24} 1.5052; density, 0.996 gm/ml at 20°C; solubility, miscible in all proportions with organic solvents, insoluble in water; stability, stable indefinitely at ambient temperatures.

G. Chemical Properties

1. Method of Synthesis

The manufacture of 3-isopropyl-5-methylphenyl *N*-butyryl-*N*-methylcarbamate is a two-step process.

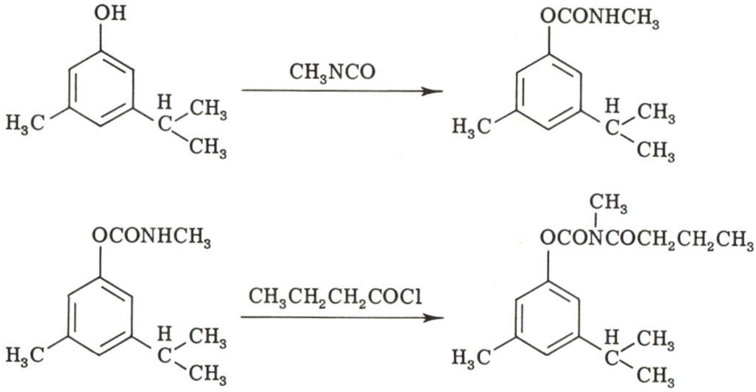

2. Chemical Reactions

3-Isopropyl-5-methylphenyl *N*-butyryl-*N*-methylcarbamate is hydrolyzed in the presence of strongly acidic or alkaline materials.

H. Formulations

Promacyl is available as Promicide®, an emulsifiable concentrate of 60% w/v active ingredient.

II. ANALYSIS

A. Formulation Analysis

1. REVIEW OF METHODS

There is no method available for the analysis of Promicide® other than that recommended. Methods for the analysis of carbamates that make use of the phenol moiety after hydrolysis without prior separation of related materials produce high results and are unsatisfactory for the purpose (Burnett and Lanham, 1970; Cohen *et al.*, 1970; Spengler and Hamroll, 1970).

2. RECOMMENDED METHOD

a. *Principle*

Technical promacyl and Promicide® emulsifiable concentrate are analyzed for purity by gas chromatography. Use is made of the ratio of peak area of 3-isopropyl-5-methylphenyl *N*-butyryl-*N*-methylcarbamate to that of an internal standard (phenanthrene).

b. *Reagents*

Analytical standard, obtainable from ICI Australia Limited, Central Research Laboratories.

Methylene chloride, analytical reagent.

Acetone, analytical reagent.

Ethanol, laboratory reagent, redistilled.

Phenanthrene, laboratory reagent, recrystallized from ethanol.

Chromosorb W, acid washed DMCS treated 60–70 mesh (Johns Manville).

DC QF-1 (Applied Science Laboratories, Inc.).

c. *Apparatus*

Gas chromatograph, ICI Australia Limited Mark 10 (or equivalent).

FID detector.

Recorder, Hitachi QPD53 (or equivalent), operated on 1 mV range, chart speed 24 cm/hour, pen travel less than 1 second f.s.d.

Integrator, ICI Australia Limited Pico-Ammeter (or equivalent).

Syringe, S.G.E. model 5A-RN-GP.

Flasks, 25-ml volumetric.

Flasks, 10-ml volumetric.

Bulb pipette, 5 ml.

d. *Experimental Procedure*

i. *Operating Conditions*

Column: 6 feet × $\frac{1}{8}$ inch o.d., 27 gauge, stainless steel.

Packing: 2 gm QF-1 on 10 gm Chromosorb W, DMCS treated, 60–70 mesh.

Carrier gas: nitrogen.

Retention times: methane 30.5 seconds, isothymyl butyrate 4.5 minutes, phenanthrene 11.0 minutes, promacyl 15.8 minutes.

Temperature

Injector: 240°C.
Column: 200°C.
Detector: 250°C.

ii. *Preparation of Calibration Graph for Promacyl*

Prepare the following standards in methylene chloride: 0.5, 1, 1.5, 2, 2.5, and 3% w/v, where allowance has already been made for the purity of the promacyl. **Do not use acetone** as solvent. Prepare a 2.5% w/v solution of phenanthrene (which has been previously recrystallized from ethanol) in acetone as internal standard solution. Mix in turn 5 ml of every promacyl standard solution with 5 ml of internal standard solution. Inject 1 μl of every mixture and determine ratio of peak areas of promacyl–internal standard. Plot these ratios versus the corresponding promacyl concentration.

iii. *Procedure*

Promacyl technical, about 1 gm; Promicide® 60% w/v emulsifiable concentrate (E.C.), about 1 gm.

Weigh about 1 gm of technical promacyl (or 60% emulsifiable concentrate) into a 25-ml volumetric flask and make up to volume with methylene chloride. Pipette 5 ml from the flask into a 10-ml volumetric flask and make up to volume with phenanthrene internal standard solution. Mix well and inject 1 μl into the gas chromatograph.

e. *Calculations*

Calculate the ratio (promacyl integral–phenanthrene integral). From the calibration graph find the concentration of promacyl in the methylene chloride solution described above.

$$\%\ \text{w/w promacyl} = \frac{\text{conc. of promacyl found from graph} \times 100}{4 \times \text{weight of original sample taken.}}$$

The concentration of a.i. in Promicide® E.C.'s is expressed as percent w/v. To convert from w/w, determine the density and multiply by it.

$$\% \text{ w/w} \times \text{density} = \% \text{ w/v}$$

f. *Discussion*

In production samples, complete resolution is obtained of the promacyl and the phenanthrene peaks from by-products in the technical material.

B. Cattle Dip Analysis

1. REVIEW OF METHODS

There is no method available for the analysis of Promicide® in cattle dip other than that recommended. However, a method by Clegg and Martin (1973) allows for the determination of promacyl and two metabolites in the tissues and milk of cattle.

2. RECOMMENDED METHOD

a. *Principle*

Use is made of tungstophosphoric acid, salting out, and pH adjustment to release promacyl for extraction from cattle dip samples. The extracts are analyzed for promacyl by gas chromatography.

b. *Reagents*

Analytical standard, obtainable from ICI Australia Limited, Central Research Laboratories.

Methylene chloride, analytical reagent.

Acetone, analytical reagent.

Diethyl ether, B.P. grade.

Ethanol, laboratory reagent, redistilled.

Dodecatungstophosphoric acid, analytical reagent.

Sodium chloride, laboratory reagent.

Phenanthrene, laboratory reagent, recrystallized from ethanol.

Chromosorb W, acid-washed, DMCS treated, 60–70 mesh (Johns Manville).

DC QF-1 (Applied Science Laboratories, Inc.).

c. *Apparatus*

See Section A,c.

d. *Operating Conditions and Preparation of Calibration Graph*

See Sections A,d,i and A,d,ii, respectively.

e. *Procedure*

Homogenize the sample (cattle dip, about 200 ml) thoroughly, take a 200 ml portion, measure the exact volume, and record. Note the pH. Add 20 ml of a 10% aqueous tungstophosphoric acid solution. Adjust the pH to 2.5 with phosphoric acid. Add 19 gm sodium chloride. Extract with 100 ml diethyl ether. After the two phases have separated, draw off the ether layer. Repeat the extraction twice with 100 ml and twice with 50 ml ether. Evaporate the combined ether layers to dryness. Transfer the residue into a 25-ml volumetric flask and make up to volume using methylene chloride. Pipette a 5-ml portion into a 10-ml volumetric flask and make up to volume with phenanthrene internal standard solution. Mix well and inject 1 μl into the gas chromatograph.

f. *Calculations*

Calculate the ratio promacyl integral/phenanthrene integral. From the calibration graph, find the concentration of promacyl in the dip sample.

$$\%\ \text{w/v promacyl} = \frac{\text{conc. of promacyl found from graph} \times 100}{4 \times \text{volume of dip sample taken}}$$

g. *Discussion*

Promacyl is strongly adsorbed onto solids, mainly clay, present in cattle dips. The addition of tungstophosphoric acid counteracts the effect of nonionic surfactants on the extraction, whereas pH adjustment and salting out are designed to help during the desorption of promacyl. Recoveries of promacyl from dips at the 0.1–0.2% w/v level have been 97–100%, even when very hard water was used for the preparation of dips.

Acknowledgment

The author wishes to thank M. Tursky and Mrs. M. D. McManus for technical support given in establishing these methods.

References

Burnett, E. M., and Lanham, A. E. I. (1970). *Pestic. Sci.* **1,** 197–199.

Clegg, D. E., and Martin, P. R. (1973). *Pestic. Sci.* **4,** 447–457.

Cohen, I. C., Norcup, J., Ruzicka, J. H. A., and Wheals, B. B. (1970). *J. Chromatogr.* **49,** 215–221.

ICI Australia Limited (1969). British Patent 1,252,105.

Spengler, D., and Hamroll, B. (1970). *J. Chromatogr.* **49,** 205–214.

11

Prothoate

B. Bazzi

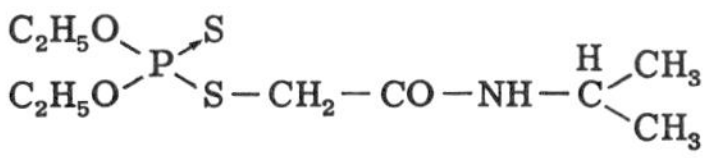

O,O-Diethyl *S*-(*N*-isopropylcarbamoylmethyl) phosphorodithioate

I. GENERAL

A. Empirical Formula

$C_9H_{20}NO_3PS_2$ (Mol. wt. 285.3).

B. Alternative Name

Fac®, trademark of Montedison Company.

C. Sourcc of Analytical Standard

Agricultural Division, Montedison Company, Milan, Italy.

D. Biological Properties

Prothoate is a contact and systemic compound that shows a high degree of acaricidal and insecticidal activity. It is used at rates of 0.02–0.03% a.i. (300–500 gm a.i./ha) for the protection of fruit, citrus, and vegetable crops from tetranychid and some eryophyid mites and some insects, notably aphids, tingidae, thysanoptera, and psyllidae. The acute oral LD_{50} of the technical product for male rats is 8 mg/kg, for female rats 8.9 mg/kg, for male mice 19.8 mg/kg, for female mice 20.3 mg/kg. In 90-day tests on rats, the highest dose producing no toxic effect was 0.5 mg/kg/day, on mice 1 mg/kg/day.

In tests on fish the LC_0 (10-day exposure) for *Carassius auratus* was 6–8 ppm, the LC_{100} (4-day exposure) was 50–70 ppm; on *Gambusia affinis,* the corresponding figures were 2–3 ppm and 7–8 ppm.

TABLE I
PHYSICAL PROPERTIES OF PROTHOATE

Property	Pure prothoate	Technical prothoate
Physical state	Crystal	Partly crystals at 20–25°C
Color	Colorless	Amber yellow
Specific gravity at 32°C	1.151	1.147
Melting point	28.5°C	21–24°C
Refractive index, n_D^{32}	1.5128	1.513
Flash point (Cleveland)	–	Over 160°C
Vapor pressure	1×10^{-4} mm Hg at 40°C	–
Solubility at 20°C	0.25% in water	Completely miscible with alcohols, aromatic hydrocarbon solvents, chlorinated hydrocarbon solvents, acetone, cyclohexanone, acetonitrile, methylcellosolve, butylcellosolve, carbon disulfide. Slightly soluble in water (below 0.5%), glycerol (below 1%), *n*-hexane, and light petroleum (below 2%).

E. History

Prothoate was introduced as an insecticide in 1956 by Montedison Company under the code L 343 and the protection of BP 791,824 and others.

F. Physical Properties

Important physical properties of prothoate are given in Table I.

G. Chemical Properties

1. METHOD SYNTHESIS

Prothoate is prepared by interaction of sodium *O,O*-diethyl phosphorodithioate with *N*-isopropyl-2-chloroacetamide. Alternatively, it may be synthesized by treating the reaction product of sodium *O,O*-diethyl phosphorodithioate with phenyl chloroacetate, with isopropylamine.

2. CHEMICAL REACTIONS

The compound decomposes slowly at high temperatures. It is stable in neutral, moderately acid, or slightly alkaline conditions, but at pH 9.2 and 50°C it is hydrolyzed in about 48 hours, to form *O,O*-diethyl phosphorothioate salts. Prothoate can be converted by oxidizing agents to its oxygen analogue.

H. Formulations

Prothoate is available as emulsifiable concentrates.

II. ANALYSIS

A. Formulation Analysis

1. REVIEW OF METHODS

Prothoate can be hydrolyzed with acid and, in a procedure resembling a Kjeldahl determination, can be determined titrimetrically from the evolved isopropylamine (Anonymous, 1957). However, this method is subject to interference from any compound that upon hydrolysis will generate ammonia or volatile bases; better results have been obtained using thin-layer chromatographic or gas chromatographic methods. The GLC procedure is normally preferred, but in some cases (e.g., where analytical standard is not available), the TLC method could be the procedure of choice. Both the methods are described below (Bazzi *et al.*, 1971, 1972).

2. RECOMMENDED METHOD—GLC PROCEDURE

a. *Principle*

Prothoate, or a convenient amount of an acaricidal preparation, is diluted to appropriate volume with xylene, together with malathion as an internal standard, and analyzed by gas chromatography using a thermal conductivity detector.

b. *Reagents*

Insecticide standard, prothoate Analytical Grade of known purity (Montedison Company, Milan, Italy).

Internal standard, malathion, pretested (American Cyanamid Company) (see Note 1 below).

GLC stationary phase, SE-30 silicone gum rubber (Cat. no. 50350, Applied Science Laboratories).

GLC solid support, Gas Chrom Z 60–80 mesh (Cat. no. 50024, Applied Science Laboratories).

Solvents, Reagent Grade.

c. *Apparatus*

Gas chromatograph, Fractovap model C, type ATC/f (Carlo Erba S.p.A., Milan, Italy), with a thermal conductivity detector or equivalent equipment.

Gas chromatographic column, 100 × 0.4 cm i.d. stainless steel, packed with 5% SE-30 on 60–80 mesh Gas Chrom Z.

Millimetric lens, 8×, 10-mm scale width, and 0.1-mm subdivisions (see Note 2 below).

d. *Experimental Procedure*

Accurately weigh about 1.2 gm malathion and about 1 gm pure reference prothoate into a 50-ml volumetric flask, dissolve in xylene and dilute the solution to volume (solution 1).

In another 50-ml volumetric flask, accurately weigh about 1.2 gm malathion and an amount of sample equivalent to approximately 1 gm of prothoate, and dilute to volume with xylene (solution 2). Analyze solutions 1 and 2 by gas chromatography with the following operating parameters:

Column temperature:	170°C
Injection port temperature:	240°C
Helium flow rate:	70 ml/minute (hot column)
Recorder sensitivity:	1 mV full scale
Recorder chart speed:	1 cm/minute

Adjust the gas chromatographic conditions so that the injections of 2–3 μl will give the peak height between one-third and two-thirds full scale height and half height base from 0.4 to 1 cm (Fig. 1). Inject 6 portions of solution 1 and 6 portions of solution 2, alternating 3 injections of solution 1 with 3 injections of solution 2.

e. *Calculation*

Calculate the individual correction factor F for each chromatogram from solution 1 as below and determine the mean $\bar{F}$ (for 6 values):

$$F = \frac{A_s}{A_f} \times \frac{W_f}{W_s}$$

where A_s = peak area of internal standard, A_f = peak area of prothoate, W_s = weight of internal standard, W_f = weight of prothoate reference standard.

From the chromatograms of solution 2, calculate the active ingredient content of the samples as mean of 6 individual values. Calculate each individual value as follows:

$$\%\ \text{prothoate} = \frac{A_f^1}{A_s^1} \times \frac{W_s^1}{W_f^1} \times \bar{F} \times t$$

where A_f^1 = peak area of prothoate, A_s^1 = peak area of internal standard, W_f^1 = weight of prothoate sample, W_s^1 = weight of internal standard,

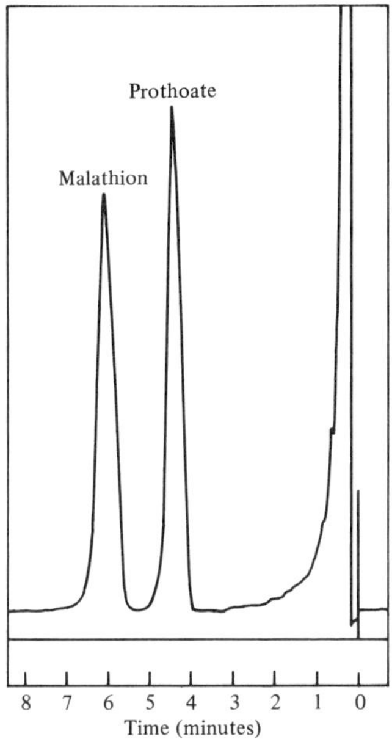

FIG. 1. Gas chromatogram of technical prothoate with malathion internal standard (Bazzi *et al.*, 1972).

$\bar{F}$ = mean of 6 correction factor values, t = purity of prothoate analytical standard (%).

f. *Notes on the Procedure*

Note 1: Samples of the internal standard from other sources may be used provided they are shown by gas chromatography to contain no impurities that appear at or near the prothoate peak.

Note 2: Measurement of the peak area by means of an automatic electronic integrator connected to the gas chromatograph is highly desirable and should result in better precision than specified. In the event that such an instrument is available, the micrometric lens is not required.

3. ALTERNATE METHOD—TLC PROCEDURE

a. *Principle*

Prothoate, or a convenient amount of an insecticidal preparation, is dissolved in benzene and the solution is spotted on a TLC plate. After the plate is developed and sprayed with the chromogenic reagent, the area containing prothoate is scraped off and the pesticide is eluted with methanol. The active ingredient content in the eluate is obtained by determining phosphorus as quinoline phosphomolybdate.

b. *Reagents*

Silica Gel G, TLC grade according to Stahl (Cat. no. 7731 Merck).

TLC developing solvent, n-hexane, 70 ml; acetone, 30 ml.

Chromogenic reagent. Mix 0.25 gm palladium(ous) chloride with 1 ml concentrated HCl and 5 ml water. Dilute to 100 ml with 95% ethanol.

Citric–molybdic acid reagent. Dissolve 54 gm 100% molybdic anhydride (MoO_3) and 12 gm NaOH, with stirring, in 400 ml hot water, and cool. Dissolve 60 gm citric acid in a mixture of 140 ml concentrated HCl and 200 ml water, and cool. Gradually add the molybdic solution to the citric acid solution with stirring. Cool, filter, and dilute to 1 liter with water (the solution may be green or blue; the color depends on exposure to light). If necessary, add a 0.5% $KBrO_3$ solution, drop by drop, until the green color becomes pale. Store in the dark in a polyethylene bottle.

Quinoline solution. Dissolve 50 ml synthetic quinoline, with stirring, in a mixture of 60 ml concentrated HCl and 300 ml of water. Cool, dilute to 1 liter with water and filter. Store in the dark in a polyethylene bottle.

Quimociac reagent. Dissolve 70 gm sodium molybdate dihydrate in 150 ml water. Dissolve 60 gm citric acid in mixture of 85 ml HNO_3 and 150 ml water, and cool. Gradually add the molybdate solution to the citric acid–HNO_3 mixture, with stirring. Dissolve 5 ml synthetic quinoline in mixture of 35 ml HNO_3 and 100 ml water. Gradually add this solution to the molybdate–citric acid–HNO_3 solution, mix and let stand 24 hours. Filter, add 280 ml acetone, dilute to 1 liter and mix. Store in the dark in a polyethylene bottle.

Sulfuric acid, Reagent Grade (sp. gr. 1.84).

Nitric acid, Reagent Grade (sp. gr. 1.40).

Perchloric acid, Reagent Grade (sp. gr. 1.073).

c. *Apparatus*

Chromatographic tank, 22 × 10 × 22 cm.

Glass plates, 20 × 20 cm.

Glass crucible, 10 G4, with fritted disk, Gooch-type, low-form.

Filtering crucible, porcelain, 25 ml with medium porosity disk (9-μm pores) fused in bottom.

Rotating vacuum evaporator, Ascenso, Milan, Italy, or equivalent.

Filtration vacuum apparatus, Fisher filtrator, Cat. no. 9-788, Fisher Scientific Company, or equivalent.

d. *Preparation of the TLC Plates*

Carefully mix 50 gm silica gel G with 85–90 ml of water for 3 minutes in an Erlenmeyer flask; immediately apply 1 mm thick layer to the glass plates and dry them 90 minutes in an oven at 110°C. With a needle, mark a thin line on the plate at 3 cm from the upper edge and remove 5 mm of silica gel from both sides of the plate.

e. *Procedure*

In a 50-ml volumetric flask, weigh an amount of thoroughly homogenized sample corresponding to 1 gm of prothoate, dissolve in benzene, and dilute to volume. With a controlled pipette apply 1 ml of the pesticide solution in uniform and thin streak, 14 cm long and 3 cm from the bottom edge of the plate. After the solvent has completely evaporated, place the TLC plate in the chromatographic tank containing the solvent system, and let the plate develop until the solvent front has reached the upper scored line. Remove the plate and air dry at least 30 minutes; then repeat the TLC operation three times, each time restoring the initial volume of solvent in the tank. When the plate is completely dried, spray with the chromogenic reagent and let stand at least 2 hours.

Prothoate appears as a bright yellow band at R_f 0.55 (Fig. 2). Mark the prothoate area, allowing a safety margin depending on the presence of other compounds, then remove the silica gel around this zone with a microscope slide and thoroughly clean the glass with a filter paper wetted with methanol.

Scrape off the prothoate area into a glass crucible mounted on a vacuum assembly and elute with methanol (30–40 ml total) into a 50-ml Kjeldahl flask. After each washing with methanol, release the vacuum, and with a glass rod, carefully stir the silica gel in the crucible. Using the rotating evaporator, cautiously evaporate the filtrate at 35°C (maximum) just to dryness; add 2.5 ml H_2SO_4, followed by 1.5 ml HNO_3, heat gently to white fumes, and cool. Add 1 ml $HClO_4$ and repeat the digestion to white fumes. After cooling, transfer the solution to a 600 ml beaker, carefully rinse the flask with water and dilute to 150 ml. Precipitate the phosphorus by (1) or (2) below:

1. Add 30 ml citric–molybdic acid reagent and gently boil 3 minutes; remove from the heat and swirl carefully. Immediately add 10 ml of quinoline solution from a burette with continuous swirling (add the first 3–4 ml dropwise and the remainder in a steady stream).
2. Add 50 ml of quimociac reagent, cover with a watch glass, place on a hot plate and boil for 1 minute.

After the treatment by (1) or (2), cool the solution to room tempera-

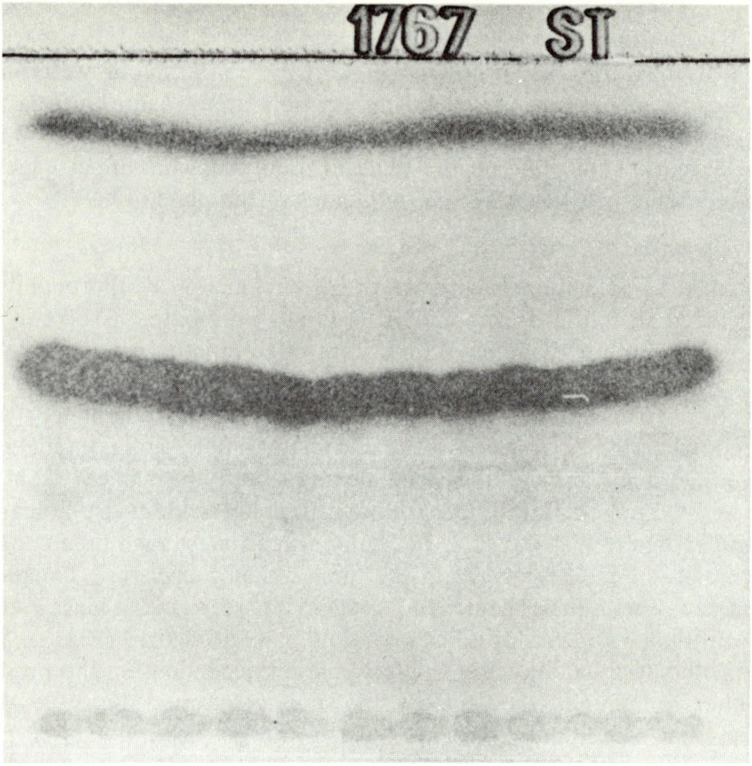

FIG. 2. Thin-layer chromatogram of a technical prothoate sample (20 mg) (Bazzi *et al.*, 1971).

ture, swirling carefully 3–4 times during the cooling. Filter the solution into a porcelain filtering crucible, previously dried at 250°C and weighed, and wash the filter five times with 25-ml portions of water. Dry crucible and contents 30 minutes at 250°C, cool in a desiccator to constant weight, and weigh as $(C_9H_7N)_3H_3(PO_4 \cdot 12MoO_3)$.

f. *Calculation*

$$\% \text{ prothoate} = \frac{A \times 0.0140 \times 9.21 \times 100}{W}$$

where A = weight of quinoline phosphomolybdate (mg)
0.0140 = factor to convert quinoline phosphomolybdate to phosphorus
9.21 = factor to convert phosphorus to prothoate
W = weight of sample applied to TLC plate (mg).

4. DISCUSSION OF THE METHODS

Both of the methods have been applied on technical products and liquid formulations and have given excellent results. The various impurities in the technical prothoate, as well as the solvents, emulsifiers, wetting agents, etc. present in the formulations do not interfere. The precision of the GLC method described is estimated to be ±1.2%, while precision of the TLC procedure can be better than ±1%.

B. Residue Analysis

1. REVIEW OF METHODS

In the past, the only procedure that has been used for analyzing prothoate residues in olive oil is based on the final colorimetric determination of phosphorus (Bazzi, 1957–1958). In the author's laboratory, this method worked very satisfactorily also for the determination of prothoate residues in peaches, apples, and pears. The preferred method, given below, uses gas–liquid chromatography with alkali flame ionization detection to analyze prothoate and its oxygen analogue residues in several vegetables (Bazzi *et al.*, 1974).

2. RECOMMENDED METHOD

a. *Principle*

An analytical procedure is described for the determination of trace amounts of prothoate and its oxygen analogue in fruit down to levels of 0.005 ppm for both compounds. The method is based on extraction of the compounds with acetone from the vegetable being considered, purification of the extract by partitions with couples of immiscible liquid phases, and final gas chromatographic determination of the active ingredient and its oxygen analogue using an alkali flame ionization detector.

b. *Reagents*

Acetone, Reagent Grade, redistilled.

Benzene, Reagent Grade, redistilled.

Methylene chloride, Reagent Grade, redistilled.

Nitrogen, highly purified.

Anhydrous sodium sulfate, Reagent Grade.

Prothoate and prothoate oxygen analogue, Analytical Grade, known purity (Montedison Company, Milan, Italy).

GLC stationary phase, UCCW 982, methylvinyl silicone polymer (Cat. no. 8501-3840 Hewlett-Packard).

GLC solid support, Gas Chrom Q, 80–100 mesh (Applied Science Laboratories, Inc., State College, Pennsylvania).

Check all reagents for impurities using the thermionic detector gas chromatograph before they are used.

c. *Apparatus*

Homogenizer, Omni-Mixer, Sorvall Inc., or equivalent.
Rotating vacuum evaporator, Ascenso, Milan, Italy, or equivalent.
Filtration vacuum apparatus, equipped with Büchner funnel.
Filter paper, glass fiber, Whatman GF/C.
Gas chromatograph, Hewlett-Packard model 5755 with KCl thermionic flame ionization detector or equivalent equipment.
Recorder, Hewlett-Packard model 7128 or equivalent.
Gas chromatographic column, 180 × 0.4 cm i.d. Pyrex glass, packed with 3.8% UCCW 982 on 80–100 mesh Gas Chrom Q. The column is silylated in the vapor phase at 250°C by injecting 5 × 20 μl of hexamethyldisilazane.

d. *Experimental Procedure*

i. *Extraction and Cleanup*

Homogenize 50 gm representative sample with 125 ml acetone in the Omni-Mixer immersed in an icewater bath for 15 minutes. Filter the mixture under vacuum through filter paper in a Büchner funnel into a 500-ml separatory funnel fitted with a Teflon stopcock and containing 265 ml benzene. Wash the pulp three times with 25 ml acetone, filter into the separatory funnel, and then shake for 3 minutes. Let the mixture stand about 20 minutes and discard the separated aqueous phase. Filter the organic layer through 30 gm anhydrous sodium sulfate into a 500-ml volumetric flask and dilute to volume with acetone. Transfer 100 ml of this solution, equivalent to 10 gm of sample, into a 250 ml pear-shaped flask, evaporate on the rotating evaporator until a wet residue is obtained, and then dry it with a filtered jet of nitrogen. Dissolve the residue with 10 ml acetone, transfer into a 250-ml separatory funnel, dilute to 100 ml with water and extract twice with dichloromethane (100 and 50 ml). Evaporate the extract containing prothoate and its oxygen analogue as above, and transfer the residue in a graduated test tube with 10 ml of acetone. Concentrate the solution to a final volume of 1 ml.

ii. *Gas Chromatography*

Column temperature: 210°C.
Injection port temperature: 250°C.
Detector temperature: 380°C.
Carrier gas *Helium flow rate:* 60 ml/minute.
Hydrogen flow rate: 43 ml/minute.
Air flow rate: 400 ml/minute.

Recorder sensitivity: 1 mV full scale.
Recorder chart speed: 75 cm/hour.

Inject 4–6 μl of the solution into the gas chromatograph and measure the height of the peaks corresponding to prothoate and its oxygen analogue. Relative retention times with respect to parathion are 0.71 and 0.53, respectively. By the same procedure, inject 2, 4, 6, 8, and 10 ng of prothoate and its oxygen analogue into the gas chromatograph and prepare a suitable calibration curve. Calculate the concentration of both compounds in the crops as follows:

$$\text{ppm prothoate (or oxygen analogue)} = W/S$$

were W = weight of prothoate or its oxygen analogue in nanograms found in the portion of sample injected (from calibration curve); S = weight of the sample in milligrams represented by the portion injected.

iii. *Determination Limit*

Considering the numerous blanks examined, the detection limit is about 0.005 ppm for each compound.

iv. *Recoveries*

Recovery of prothoate and its oxygen analogue from the crops examined was checked by adding known amounts of standard at the blending step. All recoveries were run at the 0.05 ppm level and were generally in the 80–100% range for both prothoate and its oxygen analogue.

v. *Interferences*

An interference study was carried out to determine the specificity of the method. Since the mode of detection is specific for phosphorus containing compounds, only organophosphates could interfere. But it was found that the retention times of a lot of organophosphorus pesticides are quite different from those of prothoate and its oxygen analogue, except for dimethoate and diazinon. These are retention times that are close to

TABLE II
RETENTION DATA OF SOME PHOSPHORUS-CONTAINING COMPOUNDS

Compound	Retention time (minutes)	Relative retention time
Dimethoate	2.14	0.44
Diazinon	2.22	0.46
Prothoate P=O	2.54	0.53
Prothoate	3.41	0.71
Parathion	4.76	1.00

that of prothoate oxygen analogue, but they are sufficiently different in order to exclude any error of interpretation (see Table II).

3. APPLICABILITY OF RECOMMENDED METHOD

The method has been tested on apples, pears, peaches, oranges, and grapes at different degrees of ripening. It has been successfully applied also to some wine samples.

REFERENCES

Anonymous, (1957). Technical Bulletin, p. 7. Montecatini, Milan, Italy.

Bazzi, B. (1957–1958). *Contributi Istituto Ricerche Agrarie,* Vol. 2, pp. 27–37. Montedison Company, Milan, Italy.

Bazzi, B., Fabbrini, R., and Radice, M. (1971). *J. Ass. Offic. Anal. Chem.* **54,** 1313–1315.

Bazzi, B., Galluzzi, G., and Abbruzzese, L. (1972). *J. Ass. Offic. Anal. Chem.* **55,** 1336–1338.

Bazzi, B., Abbruzzese, L., and Galluzzi, G. (1974). *Pestic. Sci.* **5,** 511–515.

12

Pyrethrum*

JOSEPH SHERMA

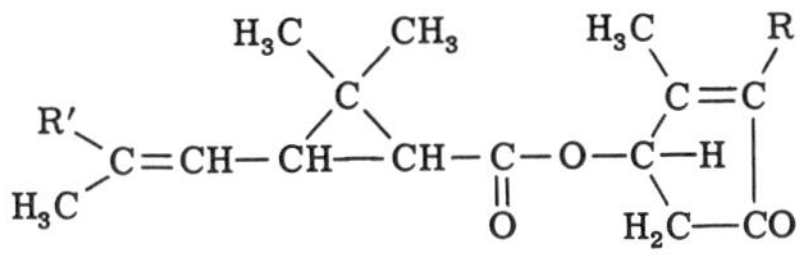

	R	R′	Empirical formula
Pyrethrin I	—CH_2—CH=CH—CH=CH_2	—CH_3	$C_{21}H_{28}O_3$
Pyrethrin II	—CH_2—CH=CH—CH=CH_2	—$COOCH_3$	$C_{22}H_{28}O_5$
Cinerin I	—CH_2—CH=CH—CH_3	—CH_3	$C_{20}H_{28}O_3$
Cinerin II	—CH_2—CH=CH—CH_3	—$COOCH_3$	$C_{21}H_{28}O_5$
Jasmolin I	—CH_2—CH=CH—CH_2—CH_3	—CH_3	$C_{21}H_{30}O_3$
Jasmolin II	—CH_2—CH=CH—CH_2—CH_3	—$COOCH_3$	$C_{22}H_{30}O_5$

I. GENERAL

Pyrethrum is a botanical insecticide whose active ingredients, pyrethrins I and II, cinerins I and II, and jasmolins I and II, are collectively known as pyrethrins. The pyrethrum plant is *Chrysanthemum cinerariaefolium* and the flower heads are the source of the six insecticidal principles. The flowers and flower extracts are mainly imported from Kenya and Tanzania. General information was presented in Volume II, pp. 399–402, and some updated information is included below. A book by Casida (1973) covers the history, chemistry, toxicology, pharmacology, use areas, tolerances, and advantages of pyrethrum.

A. Formulations

The pyrethrum flower is a perennial, and blossoms are hand picked at full maturity during a 10-month growing season. The active ingredients reach their highest concentration (by weight) in the flower when all but the last two to three rows of disc florets are open. The flowers are dried, ground, and extracted with organic solvents. The crude extract is further refined to remove resins, waxes, pigments, and allergens, is stan-

*See also Volume II, Chapter 35 and Volume VI, Chapter 47.

dardized, and is then ready for formulation. In practice, pyrethrins are combined in formulations with synergists, principally piperonyl butoxide (PBO) and MGK 264. Research has shown that this dual-synergist system has certain advantages over a single system for pyrethrum sprays and aerosols. Various commercial pyrethrin dusts, liquid intermediates, concentrates, sprays, and aerosols are available from McLaughlin Gormley King Co., Minneapolis, Minnesota.

B. Uses

Liquid, dust, and aerosol formulations of pyrethrins, usually containing synergists, find wide use as household and pet insecticides, in restaurants and all types of food processing plants, on agricultural crops before and after harvest, for insect proofing of paper and cotton food bags, for control of lice on humans, and in mosquito repellent coils.

C. Toxicology

Pyrethrins have low chronic and acute toxicities, are readily metabolized by mammals, and degrade rapidly in the environment (Camougis, 1971). The toxicology data shown in Tables I and II were supplied by D. C. Kassera (Chief Chemist, McLaughlin Gormley King Co., Minneapolis, Minnesota, personal communication, 1974).

D. Tolerances

Tolerances for residues of pyrethrins in or on raw agricultural commodities range from 0.05 ppm in or on potatoes from post harvest application and 0.1 ppm in eggs, meat, fat, and meat by-products to 3 ppm from post harvest application in or on barley, birdseed, buckwheat, corn, rice, rye, and wheat. Pyrethrum and pyrethrins are exempted from the requirement of a tolerance on growing crops (40 CRF 180.1001 Subpart D., as ammended, 37 F. R. 11725; Section 180.128; Section 121.1075; Section 121.1205; 37 F. R. 9464, 21 CFR).

E. Metabolism and Decomposition Studies

Yamamoto *et al.* (1969) studied the oxidative metabolism of pyrethroids in houseflies, and Chen and Casida (1969) studied the photodecomposition, especially modifications in the acid moiety, of pyrethrin I, allethrin, phthalthrin, and dimethrin.

F. Piperonyl Butoxide Technical Synergist

Piperonyl butoxide (3,4-methylenedioxy-6-propylbenzyl butyl diethyleneglycol ether) is registered with the Environmental Protection

TABLE I
ACUTE TOXICITY FOR HIGHER ANIMALS[a,b]

Animal	Route	Dose	Dosage (mg/kg)	Remarks
Mouse	IP	LD_{100}	200	Given in petroleum oil
Mouse	IP	LD_{50}	40	Pyrethrin II in lauryl glycol
Mouse	IP	LD_{100}	60	Pyrethrin II in lauryl glycol
Mouse	IP	LD_{66}(ca.)	240	Pyrethrin II in sesame oil; death in 21–141 minutes
Mouse	IP	LD_{100}	>480	Pyrethrin II in sesame oil; death in 21–141 minutes
Mouse	IP	LD_{83}(ca.)	480	Pyrethrins I, II in sesame oil
Rat	Oral	LD_{50}	820 (680–1000)	In a 20% oil base
Rat	Oral	LD_{50}	1870 (1340–2600)	In a 20% oil base
Rat	Oral	LD_{50}	ca. 1500	In a 20% oil base
Rat	IP	LD_{90}	200	In petroleum oil
Guinea pig	IP	LD_{90}	200	In petroleum oil
Guniea pig	IP	LD_{50}	100–150	In petroleum oil
Guinea pig	IP	LD_{100}	120	Pyrethrin I in lauryl glycol
Guinea pig	Oral	LD_{50}	1500	Death in 48 hours
Dog	IV	LD_{50}	6–8	
Fish	Medium	Toxic dose	20 ppm (0.2 ppm pyrethrins)	As pyrethrum flowers
Carp	Medium	Harmful	2 ppm	Affects movements; as pyrethrum flowers
Carp	Medium	LC	5–10 ppm	Paralysis before death; as pyrethrum flowers
Trout (fingerling)	Medium	Disabling dose	5–10 ppm (0.1 ppm pyrethrins)	As pyrethrum flowers
Acellus aquaticus (Crustacea)	Medium	LC	0.002 ppm	As pyrethrum flowers

[a] Average LD_{50} oral (acute) for all animals tested was 1500 mg/kg; the chronic MLC, 500 ppm.

[b] Data from D. C. Kassera (personal communication, 1974).

Agency as a synergist for pyrethrins, allethrin, *d-trans*-allethrin, and dimethrin. Due to the widespread use of piperonyl butoxide as a synergist, various tolerances and exemptions from tolerances have been established. In insecticidal formulations, the proportion of toxicant to piperonyl butoxide ranges from 1 : 1 to 1 : 10, depending on the desired biological effect. Piperonyl butoxide is also used extensively in combina-

TABLE II

FURTHER TOXICOLOGY INFORMATION FOR PYRETHRUM[a]

1. *Subacute and chronic toxicity for higher animals*
   a. *Rats:* Received 1000 ppm in diet for 2 years without tissue pathology
   b. *Rats:* Received 5000 ppm in diet; tissue damage and gross signs in one series, but not in another
   c. *Rabbits:* Dermal applications at 200–400 mg/kg gave toxic signs
   d. *Guinea pigs, rats:* Survived 480 mg/animal, with diarrhea as only symptom
   e. *Rabbits:* Survived, with permanent tremor and spastic incoordination, 240 mg/kg of pyrethrin II
2. *Nonallergenic in the refined extract.*
3. *Phytotoxicity:* Apparently nonphytotoxic; no reports of phytotoxicity with pyrethrins, as such, appear in the literature; solvents of pyrethrins, improperly applied, may present a hazard as is the case with many insecticide formulations
4. *Toxicity for insects:* Pyrethrolone and cinerolone (keto alcohol moieties) and the chrysanthemum mono- and dicarboxylic acid moieties of the pyrethrins in uncombined form are nontoxic to insects; no esters compare with the naturally occurring esters in general insecticidal activity

[a] Data from D. C. Kassera (personal communication, 1974).

tion with MGK 264 as a dual synergist system with various pyrethroids. Specifications of MGK Piperonyl Butoxide Technical are as follows:

Assay: 88.0% as determined by colorimetry or gas liquid chromatography.

Color (Gardner–Holt): 8 maximum (typical color = 6).

Specific gravity 20/20: 1.040–1.070 (typical weight/gallon = 8.79 lbs.).

Flash point: 340°F minimum.

100% soluble in: Petroleum distillate, Propellent 11, Propellent 12, methylene chloride, and most organic solvents.

Refractive index (25°C): 1.4850–1.5000.

Acid number: 0.30 maximum.

Chloromethyl dihydrosafrole: 0.10 maximum.

Dihydrosafrole: Nil.

Piperonyl butoxide can be stored indefinitely at ambient temperatures with no detectable deterioration in quality.

The purification and characterization of PBO was reported by Albro *et al.* (1972). PBO in greater than 99% purity was obtained from technical material by column chromatography on Florisil, and numerous impurities were identified by preparative thin-layer chromatography (TLC), gas chromatography (GC), and spectroscopic analysis.

G. MGK 264 Synergist

MGK 264 is practically pure *N*-octylbicycloheptene dicarboximide. For formulating purposes it is considered 100% active.

1. PHYSICAL AND CHEMICAL CHARACTERISTICS

Nitrogen content: 5%.
Color (Gardner–Holt): Maximum = 4, typical = 1+.
Specific gravity: 1.050 ± 0.010 at 20°C.
Refractive index: 1.4990 ± 0.0030 at 25°C.
Boiling point: Above 150°C at 2 Torr.
Flash point: Above 350°F, open cup method.
Solidification: Below −20°C.
Miscible: With kerosenes and white oils, benzene hydrocarbons; aliphatic alcohols, esters, and ketones; and halogenated hydrocarbons such as methylene chloride, ethylene chlorides, and fluorinated hydrocarbon propellents.

Water emulsions of MGK 264 can be prepared with or without other solvents by using proper emulsifying agents at use concentrations. Solutions in a "white" solvent, such as deodorized kerosene, are practically colorless. The weight per U.S. gallon is 8.75 pounds. MGK 264 will dissolve approximately 35% of methoxychlor by weight.

2. STORAGE STABILITY

MGK 264 is stable for at least 2 years at room temperature.

3. BIOLOGICAL PROPERTIES

Biologically, MGK 264 acts as a synergist for pyrethrins and allethrins. With reduced amounts of pyrethrins and pyrethroids, MGK 264 is of value as a knockdown agent. It may be used in aerosols, pressurized sprays, oil-type sprays, and dust concentrates. The synergist value of MGK 264 is particularly evident when combined with the allethrins and pyrethrins for roach control in aerosols and pressurized and liquid sprays. Its high solvent power is helpful in aerosols containing materials difficult to get into solution such as methoxychlor. With more soluble ingredients, less of the odorous aromatic solvents are required.

MGK 264 is used extensively in combination with pyrethrins and piperonyl butoxide in liquid and pressurized sprays to control roaches and flies. Data show that pyrethrins and the dual synergists (piperonyl butoxide and MGK 264) are more effective against resistant flies and roaches than an equivalent amount of pyrethrins with either of the synergists alone.

MGK 264 is also used in combination with MGK's Repellents 11 and 326, because tests indicate that the effectiveness of the repellents are significantly enhanced by its presence. Generally, a 1:1 ratio of MGK 264 and repellent has proven to be the minimum amount of MGK 264 that should be included with the repellents.

MGK 264 extends and augments the repellency of MGK roach repellents 11 and 874. A 2:1 or higher ratio of MGK 264 to the repellents has proven helpful and economical in this use. MGK Repellent 11 cannot be used on dairy or meat animals.

4. TOXICOLOGY

The acute oral LD_{50} for MGK 264 to mice is 1.5 gm/kg and 3.64 gm/kg to rats. The acute dermal LD_{50} to rabbits applied as a 5% suspension in water was approximately 470 mg/kg (National Academy of Sciences, 1959).

Long-term toxicological studies on MGK 264 produced the following results:

a. *Two-Year Feeding Study on Rats for McLaughlin Gormley King Company by Wisconsin Alumni Research Foundation*

During a 2-year feeding study on rats, no effects with respect to weight gain, food consumption, clinical analysis, or pathological changes were noted as a result of administration of MGK 264 at levels of 62.5, 250, and 1000 ppm in the basal diet when compared to similar observation in rats receiving the unsupplemented basal diet. No correlation was noted between the occurrence of lesions and tumors and the administration of MGK 264. Under the conditions of this study, MGK 264 is nontoxic to the rat when included in the daily diet at levels 62.5, 250, and 1000 ppm for a period of 2 years.

b. *Two-Year Feeding Study on the Dwarf Pig for McLaughlin Gormley King Company by Harris Laboratories*

MGK 264 was fed for 2 years to dwarf pigs at the rate of 25, 100, and 300 ppm in their basal diet. No adverse effect was found from any feeding level. Periodic examination of urine and blood showed no effect. Weight gain was normal. At the end of the test, all pigs were active and appeared to be in good health. No gross pathology was noted in postmortem examination. Histological examinations were made on all animals, and all tissues examined were found to be normal.

c. *Three-Generation Teratogenic Study on Rats for McLaughlin Gormley King Company by Industrial Biology Laboratories, Inc.*

MGK 264 insecticide synergist, fed at a dosage level up to and including 36.4 mg/kg/day through three generations of rats, produced no systemic effects or gross teratologic effects, but at 364 mg/kg/day, some effect upon female fertility and the litter size and growth ratio was noted. It appears, however, that this dosage level is below that level that would produce a consistent dose effect response.

II. FORMULATION ANALYSIS

A. Review of Methods

A colorimetric method for analysis of pyrethrin formulations is described in Volume II of this treatise, and several GC procedures are reviewed in Volume VI. Details of colorimetric methods for PBO and MGK 264 in insecticide formulations are available from McLaughlin Gormley King Co., but the GC procedures developed by D. C. Kassera and V. J. Meinen of this company and presented below are recommended for increased accuracy and selectivity for these compounds compared to the colorimetric methods.

Kawano and Bevenue (1972) studied the analysis of water-based formulations and found that extraction of pyrethrins and synergists from aerosols was improved and simplified by the use of acetone. This procedure minimized interferences in subsequent analysis by GC on a 5% SE-30 column at 190°C with flame ionization detection after Florisil column cleanup (Bevenue and Kawano, 1970, described in Volume VI, pp. 462–64). Kawano *et al.* (1974) applied GC techniques to the six pyrethrum esters and compared refined and crude pyrethrin extracts from various world sources. Thirty different GC column materials and three sets of operating conditions were examined, and it was concluded that a 2 foot × $\frac{1}{4}$ inch column of 2.5% XE-60 on Chromosorb W AW-DMCS, 60–100 mesh and matrix programming gave the most satisfactory resolution of the six pyrethrum esters. Matrix programming involved an initial oven temperature of 155°C and programmed changes when various peaks appeared on the recorder chart, to a final temperature of 205°C. Coupled GC-MS was used to identify components present in the GC peaks. Murano (1972) described the GC determination of optical isomers of chrysanthemic acid obtained by hydrolysis of pyrethrins. The diastereoisomeric esters of the acid with *l*-menthol were separated on a column of 2% QF-1 coated on Chromosorb W.

Rickett (1972) separated the six pyrethrum esters on a preparative scale by pressurized liquid–liquid column partition chromatography using nitromethane supported on Celite as the stationary phase and hexane followed by CCl_4–hexane (1:3 v/v) as the mobile phases. Preliminary cleanup was on a silica gel column eluted with ether–hexane (1:3 v/v).

Olive (1973) described a color-specific thin-layer chromatography reagent for the identification and semiquantitation of pyrethrins and piperonyl butoxide in pesticide formulations. Approximately 1 μg of either pyrethrin or PBO was detected as pink and blue spots, respectively, in aerosol and solid formulations. Silica gel GHR/UV layers were developed with benzene–ethyl acetate (85:15 v/v) solvent; sprayed, after air drying, with the reagent, 85% H_3PO_4–0.33% (w/v) tannic acid in acetic acid–acetone (3:1:26 v/v, mixed just before use); and dried at 85°C for about 15 minutes to bring out the spots. Furmanec and Schilling (1969) separated pyrethrins I and II from inert products by development on silica gel thin layers using hexane–ethyl acetate (75:25 v/v) solvent in a saturated tank. The pesticide zones were scraped off, reacted with orthophosphoric acid, and determined colorimetrically at 550 nm. This assay method was applied to pyrethrin concentrations as low as 1%.

B. Recommended Method for Formulations Containing Pyrethrins, PBO, and MGK 264

1. PRINCIPLE

Dibutyl phthalate internal standard is used for the GC determination of formulations with flame ionization detection. The following procedure (Kassera and Meinen, 1974) is based on a formulation containing 9% pyrethrins, 18% PBO, and 30% MGK 264 (MGK Pyrocide Intermediate 5192) plus 43% petroleum distillate.

2. REAGENTS

Solvent, Reagent Grade acetone.

GC packing, 2% SE-30 on Chromosorb W(HP), 80–100 mesh.

Internal standard, Reagent Grade di-*n*-butyl phthalate (DBP).

Insecticide standard, formulation prepared in the laboratory, preferably from material used in manufacturing the formulation to be analyzed. In this way, one can eliminate some of the errors caused by MGK 264 impurities, which interfere somewhat at the high sensitivity needed for detection of the pyrethrins. Also, the possibility of different ratios of pyrethrin I to pyrethrin II can be accounted for. Since an impurity peak from MGK 264 interferes with cinerin II and an impurity peak from PBO interferes with pyrethrin II, only the pyrethrin I and cinerin I peaks can be used for calculating.

3. APPARATUS

Gas chromatograph, Nuclear-Chicago 5000 (or equivalent) with flame ionization detector. Use the following GC parameters.

Column: 6 feet × 4 mm i.d. glass.
Temperatures
Column: 150–200°C programmed at 7.5°C/minute.
Injection port: 230°C.
Flame ionization detector: 240°C.
Chart speed: $\frac{1}{3}$ inch/minute.
Gas flows
Nitrogen: 125 ml/minute.
Air: 350–400 ml/minute.
Hydrogen: 40–45 ml/minute.
Sensitivities (AFS = amperes full scale): Pyrethrins, 10^{-10} AFS, attenuation = 4; MGK 264, 10^{-10} AFS, attenuation = 32; PBO, 10^{-10} AFS, attenuation = 16.

4. PROCEDURE

Weigh approximately 1.25 gm of sample and standard into different 100-ml volumetric flasks. Amounts of sample and standard weighed out and GC sensitivities may have to be altered for formulations containing varying percentages from those stated above. Pipette 50 ml of a 4 mg/ml acetone solution of DBP into each flask and bring to volume with acetone.

Inject approximately 2 μl of the solutions into the chromatograph. GC conditions may have to be altered slightly to obtain a retention of PBO of approximately 8 minutes. Also vary either injection volume or detector sensitivity to attain peak heights of at least 100 mm, preferably nearer 150 mm. (The 100 mm peak height minimum does not apply to cinerin I.) Approximate amounts injected are pyrethrins, 1.25 μg, MGK 264, 7.5 μg, and PBO, 4.5 μg. Inject the standard twice and determine the ratios of the PBO, MGK 264, and pyrethrins by dividing the peak height of each by the peak height of the internal standard (which elutes approximately 3–4 minutes before MGK 264). Peak heights are determined by drawing a base line through the minima on either side of the peak of interest (Fig. 1.). The ratios of successive injections should agree within ±1.5%. If not, repeat the series of injections. Follow the same procedure for the sample solution. If more than three series of sample injections are made, the standard injections must be repeated. Calculate percentages of sample ingredients by comparing respective standard and sample ratios.

5. RESULTS

Approximate retention times in minutes from the solvent peak are:

MGK 264	4
Cinerin I	6

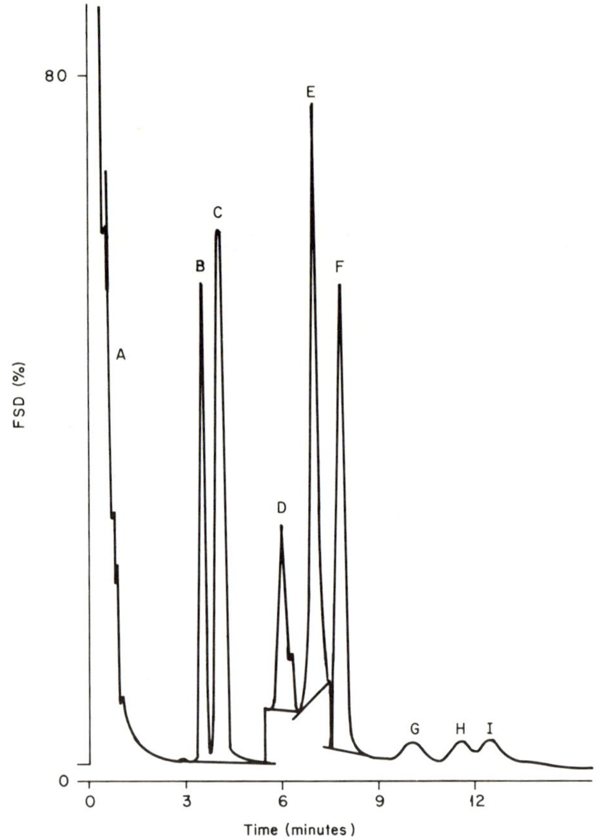

FIG. 1. Typical gas chromatogram of formulation containing pyrethrins, PBO, MGK 264, and DBP internal standard under conditions described in the text (1.25 gm sample, 100 ml final volume, 4 μl injected). A = solvent peak, B = DBP, C = MGK 264, D = cinerin I, E = pyrethrin I, F = PBO, G = cinerin II + MGK 264 impurity, H = PBO impurity, I = pyrethrin II. FSD% = percentage full scale deflection on recorder chart. (Kassera and Meinen, 1974.)

Pyrethrin I	7
Cinerin II	10 (also retention of small MGK 264 impurity peak)
Pyrethrin II	12
PBO	8 (PBO impurity at 11.4 minutes)
DBP	3.4

See Fig. 1. An OV-1 column can be substituted for SE-30 with similar results.

C. Recommended Method for MGK 264

1. PRINCIPLE

The sample is diluted in acetone containing dibutyl phthalate as an internal standard. The GC peak area ratios of MGK 264 to dibutyl phthalate are compared for quantitation. Peak areas must be used for quantitation because MGK 264 is a mixture of two isomers (which are partially resolved using this procedure), usually in an approximate 1:1 ratio. The ratio of the isomers is not constant from one lot to another. The method can be applied to many aerosol formulations.

2. REAGENTS AND SOLUTIONS

Solvent, Reagent Grade acetone.

Internal standard, Reagent Grade di-*n*-butyl phthalate.

Internal standard solution, 15 mg/ml DBP in acetone.

MGK 264 standard, obtainable from McLaughlin Gormley King Co., Minneapolis, Minnesota.

Standard solution, prepared by accurately weighing about 0.3 gm standard into a 50-ml volumetric flask, adding by pipette 10 ml internal standard solution, and diluting to volume with acetone.

3. SAMPLE PREPARATION

Technical material. Accurately weigh about 0.3 gm sample into a 50-ml volumetric flask. Pipette 10 ml internal standard solution into the flask and dilute to volume with acetone.

Formulations. Accurately weigh a sample equivalent to 300 mg MGK 264 into a 50-ml volumetric flask. Pipette 10 ml internal standard solution into the flask and dilute to volume with acetone.

4. GAS CHROMATOGRAPHY

Use an instrument equipped with a flame ionization detector and a 4 foot × 4 mm i.d. glass column packed with 5% OV-1 on 80–100 mesh Chromosorb W(HP). Condition the column for several hours at 275°C with N_2 at 50 ml/minute. Operate at the following approximate conditions:

Column: 150°C.

Injection port: 230°C.

Nitrogen carrier gas: 125 ml/minute.

Air: 350–400 ml/minute.

Hydrogen: 40–50 ml/minute.

Sensitivity: 10^{-9} A full scale attenuated at 2.

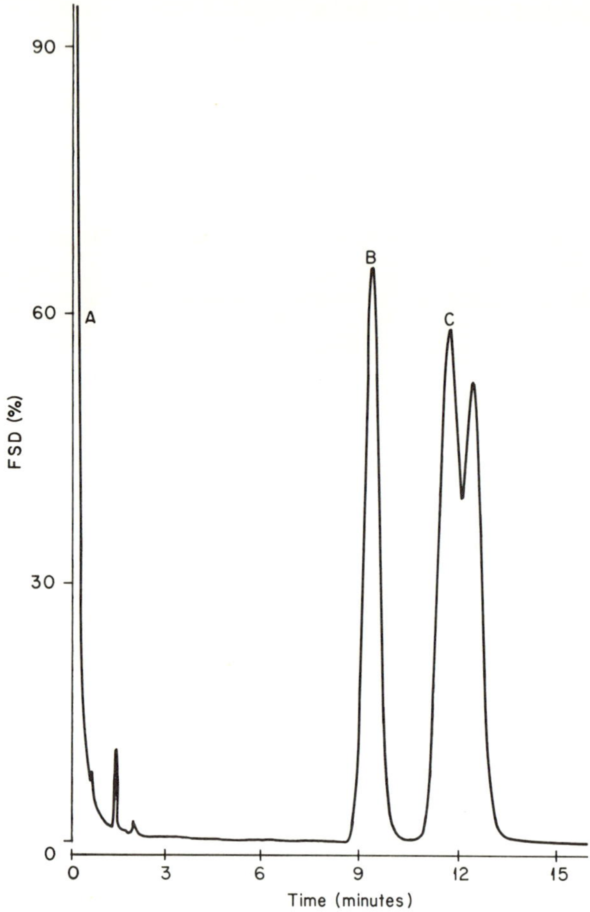

FIG. 2. Typical gas chromatogram of 24 μg MGK 264 and 12 μg DBP internal standard under conditions described in the text. A = solvent peak, B = DBP, C = MGK 264. FSD% = percentage full scale deflection on recorder chart. (Kassera and Meinen, 1974.)

Retention times: dibutyl phthalate approximately 9 minutes; MGK 264 approximately 11–12 minutes. If necessary, vary column temperature to attain stated retention times. Also vary detector sensitivity or injection volume to attain at least 75% of full-scale deflection of the internal standard peak (Fig. 2.).

5. DETERMINATION

Inject portions (about 4.5 μl) of standard solution until peak area (peak heights should not be used) ratios (MGK 264:dibutyl phthalate) vary less than 2% for successive injections. After following the same procedure for the sample solution, repeat duplicate injections of the standard solution. Peak area ratios should be within ±1.5% of the previous standard injections. If not, repeat the series of injections.

6. CALCULATIONS

Average the standard and sample ratios. The two values should be within ±1.5%. If not, repeat the analysis using a new sample portion. Use the equation

$$\%\ \text{MGK}\ 264 = \frac{W_s P_s R_x}{W_x R_s}$$

where W_s = grams standard, W_x = grams sample, P_s = purity of standard, R_x = ratio of sample, and R_s = ratio of standard.

7. DISCUSSION

This method, in which peak areas are employed, is more accurate for MGK 264 in technical material than the method described in Section II,B. Accuracy requirements are usually not as stringent when analyzing formulations, so the first method described is usually sufficient for MGK 264 in formulations. Application of this procedure to formulations results in a chromatogram that requires 2–3 hours for elution of all components.

III. RESIDUE ANALYSIS

See Volume II, pp. 408–412, and Volume VI, pp. 465–469, of this treatise for analytical methods applicable to pyrethrin residues in food products, animal tissues, and milk. The method in Volume VI for pyrethrin residues in milk and milk products is presented with some modifications in Chapter 16 of the book by Casida (Moore, 1973). This method involves extraction; cleanup of extracts by acetonitrile–petroleum ether partitioning, column chromatography, and planar chromatography on microfiber strips; and analysis by electron capture GC. Changes from the method in Volume VI include use of pentane rather than petroleum ether as the third extraction solvent, silica gel rather than Florisil for the cleanup column, and a longer GC column (91 cm compared to 15 inches) and Chromosorb W, 80–100 mesh, instead of Aeropak-30, as the support.

Moore (1973) also described a cleanup procedure for residues of piperonyl butoxide and MGK 264 from samples of eggs and tissue extracts. Samples are extracted with a mixture of ethanol, ether, and hexane, saponified with alcoholic KOH for removal of fat and removal or conversion of many chlorinated pesticides, and then cleaned up chromatographically as for the pyrethrins. The amount of synergist is determined down to 1 ppb levels in wheat flour by analysis with an especially sensitive electron capture detector fitted inside a Parr bomb using a 30.5 cm × 0.32 mm stainless steel column packed with Chromosorb W supporting 1.0% Epon 1001, 0.3% DEGS, 0.2% Viton A, and 0.1% DC-710 at 140–172°C.

Acknowledgment

The author is indebted to Mr. Dean C. Kassera of McLaughlin Gormley King Co. for supplying most of the material included in this chapter and for reviewing the manuscript.

References

Albro, P. W., Fishbein, L., and Fawkes, J. (1972). *J. Chromatogr.* **65,** 521.

Bevenue, A., and Kawano, Y. (1970). *J. Chromatogr.* **50,** 49.

Camougis, G. (1971). "An Evaluation of Potential Risks Resulting from the Use of Pyrethrins as Insecticides." New England Research, Inc. (report available from McLaughlin Gormley King Co., Minneapolis, Minnesota).

Casida, J. E., ed. (1973). "Pyrethrum: The Natural Insecticide." Academic Press, New York.

Chen, Y.-L., and Casida, J. E. (1969). *J. Agr. Food Chem.* **17,** 208.

Furmanec, D., and Schilling, F. A. E. (1969). *Pyrethrum Post* **10,** 21.

Kassera, D. C., and Meinen, V. J. (1974). Procedures supplied by D. C. Kassera, Chief Chemist, McLaughlin Gormley King Co., Minneapolis, Minnesota.

Kawano, Y., and Bevenue, A. (1972). *J. Chromatogr.* **72,** 51.

Kawano, Y., Yanagihara, K. H., and Bevenue, A. (1974). *J. Chromatogr.* **90,** 119.

Moore, J. B. (1973). *In* "Pyrethrum: The Natural Insecticide" (J. E. Casida, ed.), Chapter 16, pp. 293–306. Academic Press, New York.

Murano, A. (1972). *Agr. Biol. Chem.* **36,** 917.

National Academy of Sciences. (1959). "Handbook of Toxicology," Vol. III. Saunders, Philadelphia, Pennsylvania.

Olive, B. M. (1973). *J. Ass. Offic. Anal. Chem.* **56,** 915.

Rickett, F. E. (1972). *J. Chromatogr.* **66,** 356.

Yamamoto, I., Kimmel, E. C., and Casida, J. E. (1969). *J. Agr. Food Chem.* **17,** 1227.

13

Thiometon

M. WISSON, C. VAN HOEK AND H. H. SAUER

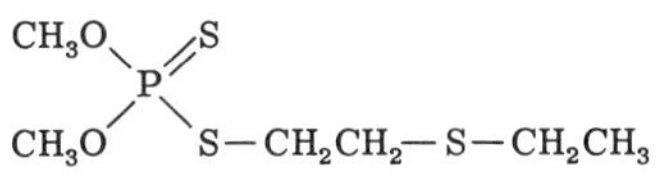

O,O-Dimethyl *S*-2-(ethylthio)ethyl phosphorodithioate

I. GENERAL

A. Empirical Formula

$C_6H_{15}O_2PS_3$ (Mol. wt. 246.35).

B. Alternative Names

Dithiometon; Ekatin® is the registered trademark of Sandoz Ltd., Basel, Switzerland.

C. Source of Analytical Standard

Sandoz Ltd., Agrochemical Research Division, Basel, Switzerland.

D. Biological Properties

1. INSECTICIDAL ACTIVITY

Thiometon is a selective systemic aphicide that is quickly absorbed and translocated in the sap stream of the plant. It possesses an outstanding efficacy with a moderate persistence, and can be used on all main crops. Sucking pests are controlled from within the plant, and thus thiometon presents no harmful effects to useful predators. This selectivity allows its use in integrated control schedules. The complete aphicidal activity is particularly valuable for the control of the vectors of virus diseases.

2. TOXICITY

Thiometon may be described as a systemic organophosphate with relatively low mammalian toxicity. The acute oral LD_{50} for rats is about

100 mg/kg, the acute dermal LD_{50} about 700 mg/kg. After inhalation of 60 mg thiometon/liter of air for 1 hour, there were no toxic symptoms or histological changes. As a result of a 90-day feeding trial conducted with doses of up to 15 ppm thiometon with rats, it is reasonable to recommend the rate of 5.0 ppm as no-effect level. Thiometon is relatively harmless to honey bees.

E. History

Thiometon was discovered and developed by the Farbenfabrik Bayer A.G. in 1952 and independently by Sandoz Ltd., Basel, in 1953. It is prepared by two different syntheses. Each company holds its own rights.

F. Physical Properties

1. Thiometon is a colorless oily liquid of characteristic odor. It has a boiling point of 110°C at 0.1 Torr and a vapor pressure of 1.7×10^{-4} Torr at 20°C.

2. The density d_4^{20} is 1.209; the refractive index n_D^{20} 1.5515.

3. It is miscible with most organic solvents, but only slightly soluble in petroleum ether. Its water solubility is 200 ppm at 25°C.

G. Chemical Properties

Thiometon shows a poor stability in the form of the pure active ingredient, whereas it is stable in nonpolar solvents. It is hydrolyzed easily under all conditions.

H. Formulations

Ekatin® insecticide is available as emulsifiable liquid concentrate containing 25% active ingredient, and as ultra low volume (ULV) formulation containing 15% active ingredient.

II. ANALYSIS

A. Formulations

1. REVIEW OF METHODS

Thiometon in technical products and liquid formulations can be determined by three different methods: (1) phosphorus combustion; (2) IR spectrophotometry; and (3) gas chromatography.

Methods (1) and (2) are approved and recommended by the Collaborative International Pesticides Analytical Council Ltd. (CIPAC) after extensive work of collaborative analytical laboratories. Method (3) was developed most recently. The first method is based on the separation of thiometon from by-products and formulation additives by paper chromatography, followed by combustion of the chromatogram spots according to Schöniger (1955, 1956). The content of thiometon is determined colorimetrically by the phosphorus molybdenum blue method described by Faderl (1962). This procedure has been shown to be an accurate and specific method for the assay of thiometon. Other materials present, such as solvents, emulsifiers, dyestuffs, or other additives, do not interfere. By-products containing phosphorus are separated by paper chromatography using appropriate stationary phases and solvent mixtures. Furthermore, this method does not need an analytical standard of thiometon.

Method (2) utilizes the strong characteristic IR absortion of thiometon solutions in well-defined solvent systems. Thiometon preparations are dissolved in carbon disulfide, and the analysis is accomplished by measuring the absorbance at 658 cm^{-1} corrected for the base line between 725 and 710 cm^{-1}. The contents are calculated from a calibration curve resulting from reference solutions of known amounts of thiometon. The procedure is outlined for thiometon determinations in technical and formulated preparations, assuming that no substances are present that will interfere in the final solution at the wavelengths for measurement. Interferences can usually be detected by anomalous absorptions in the scanned region.

Both methods yield identical and satisfactory results. They were tested by a CIPAC collaborative study and accepted as approved CIPAC methods, the combustion method being the referee method. While the IR method is a very quick procedure, the phosphorus combustion method is certainly the more complicated method, needing skilled manual performance. However, it is applicable to all organophosphorus compounds and is independent of analytical organophosphorus standards.

Method (3) is based on gas–liquid chromatography (GLC) as a rapid, accurate, and selective procedure for analyzing thiometon. Formulations are diluted with a hexane–acetone mixture and injected, preferably by an automatic injector, into the gas chromatograph equipped with a flame ionization detector. Dimethoate is used as internal standard. Quantitation is achieved by comparison of the peak area ratios for the unknown sample with the corresponding peak area ratios of a reference solution containing known amounts of thiometon and internal standard.

2. RECOMMENDED METHOD—PHOSPHORUS COMBUSTION METHOD

See "CIPAC Handbook," *Analysis of Technical and Formulated Pesticides* (Raw, 1975).

3. RECOMMENDED METHOD—IR SPECTROPHOTOMETRY

See "CIPAC Handbook" as under Section 2 above.

4. SUPPLEMENTARY METHOD—GLC METHOD

a. *Reagents*

Toluene, Reagent Grade.
Hexane, Reagent Grade.
Acetone, Reagent Grade.
Reference sample of purified thiometon with a known content.
Reference sample of dimethoate, analytical standard.

The standard should be at least 95% pure and should not contain impurities interfering with the retention time of thiometon.

b. *Apparatus*

Gas chromatograph, Varian Aerograph 1520 or equivalent, equipped with flame ionization detector (FID).

Automatic injector, Hewlett-Packard 7670 A or equivalent.

Integrator, Infotronics CRS 104 with Olivetti Terminal TE 300 or equivalent.

Balance, Mettler HE 10 or equivalent.

Injection syringe, Hamilton 10 μl type 701 N or equivalent.

Bottles, 5-ml sizes, with a Mininert valve or equivalent closure.

c. *Experimental Procedure*

i. *Preparation of Samples*

Weigh accurately an amount of the sample corresponding to 50 mg of thiometon into a 5-ml bottle and add 1 ml of the standard solution containing 100 mg dimethoate in toluene. Dilute to a volume of 5 ml with a hexane–acetone mixture (2:1), close tightly, and shake. Inject 1.0 μl into the GLC column.

ii. *GLC Operating Conditions*

Column: 1.0 m × 3.6 mm i.d., glass spiral column.

Stationary phase: 3% OV 225 on Chromosorb W hp, 80–100 mesh.

Temperature

Column: 195°C.
Injector: 220°C.
Detector: 220°C.

Carrier gas, flow rate: 60 ml N_2/minute.
Theoretical plate: about 2300 for dimethoate.

iii. *Qualitative Identification*

Qualitative identification of thiometon is performed by relating the net retention times to the internal standard–dimethoate: 10.9 minutes = 1.00; thiometon: 3.0 minutes = 0.27.

iv. *Quantitative Determination*

The thiometon content is assayed from the respective integrated peak by comparison to the integrated internal standard peak, and calculated from the equation below.

d. *Calculation*

$$\% \text{ Thiometon} = \frac{Z_{STD} \times W \times f \times P}{E \times W_{STD}}$$

where

Z_{STD} = weight of internal standard in milligrams in the sample solution
W = peak area of thiometon in the sample solution
f = correction factor
P = percent of thiometon in the reference solution
E = weight of sample in milligrams
W_{STD} = peak area of internal standard in the sample solution

The correction factor is determined as follows: Prepare the reference solution with thiometon reference sample as described under Section II,A,1,c,i and evaluate gas chromatographically. Calculate the correction factor from this reference sample as

$$f = \frac{a \times w_{STD}}{a_{STD} \times w}$$

where

a = weight of thiometon in milligrams in the reference solution
w_{STD} = peak area of internal standard in the reference solution
a_{STD} = weight of internal standard in milligrams in the reference solution
w = peak area of thiometon in the reference solution

e. *Discussion*

Using the GLC method, consideration must be given to column aging and detector contamination. It is therefore necessary to examine the conditions of the column and detector by repeated injections of the reference solution until the peak area ratios (thiometon to dimethoate) are almost constant for successive injections (deviation less than 1%).

For the final determination, the sample must be preceded and followed by reference samples. Additionally, every sample has to be determined at least three times, and all peak area ratios of the reference solutions must be within ±1% of the first accepted standard value.

For percent calculations the average *f* value of the preceding and following reference solution is taken. Considering the points discussed above, the reproducibility is very satisfactory with a relative standard deviation of <1%.

B. Residues

1. INTRODUCTION

According to early investigations on the fate of thiometon in plants (Jucker, 1958), the molecule is very rapidly metabolized. Hydrolysis is a major breakdown mechanism yielding nontoxic breakdown products. By oxidation, metabolites are formed that are of similar toxicity as the parent compound. In particular, the thiometon sulfoxide and thiometon sulfone are shown to occur in significant amounts and are to be determined in residue investigations. The corresponding P=O analogs were found in lower quantities. The method described below was developed most recently and includes all potential oxidation metabolites.

2. RECOMMENDED METHOD

a. *Principle*

The method for determining residues in plant materials and soil is based on extraction with acetonitrile containing acetic anhydride, followed by oxidation with potassium permanganate in aqueous buffer solution of pH 7. By this treatment, thiometon and its sulfoxide are converted to thiometon sulfone, whereas the P=O metabolite (phosphorothiolate) and its sulfoxide are oxidized to the corresponding sulfone. Additionally the extract is cleaned up utilizing partition between water and methylene chloride–petroleum ether.

Quantitative determination of the two sulfone derivatives is achieved by GLC using the phosphorus sensitive alkali flame ionization detector (AFID).

b. *Reagents*

Celite 545 (see FDA, 1971).

Magnesium sulfate, anhydrous, Analytical Grade.

Acetonitrile, technical, redistilled from P_2O_5/H_3PO_4 (see FDA, 1971).

Acetic anhydride, Analytical Grade.

Buffer Titrisol for buffer solution pH 7.00, phosphate, dissolved in 100 ml redistilled water (E. Merck, Darmstadt, Germany) or equivalent.

Potassium permanganate, Analytical Grade, 0.1 *N* in redistilled water.

Formic acid, Analytical Grade, 20% in redistilled water (v/v).

Methylene chloride, redistilled and purified over alumina basic, activity grade I (M. Woelm, Eschwege, Germany) or equivalent.

Petroleum ether, as described for methylene chloride.

Sodium sulfate, anhydrous, Analytical Grade.

Toluene, Analytical Grade.

Thiometon sulfone, analytical standard of known purity.

O,O-Dimethyl S-2-(ethylsulfonyl)ethyl phosphorothiolate, analytical standard of known purity.

c. *Apparatus*

Blender, Waring or equivalent, with 1-liter jars.

Shaking machine.

Rotary evaporator equipped with a water bath equilibrated at a maximum temperature of 50°C.

Gas chromatograph, Varian Aerograph 1200 or equivalent equipped with an alkali flame ionization detector.

Automatic injector, Hewlett Packard 7670 A or equivalent.

Integrator, Infotronics CRS 208 or equivalent.

Syringe for injection, Hamilton, 10 μl size.

d. *Experimental Procedure*

i. *Extraction Procedure*

Reduce the entire sample to small pieces and mix thoroughly. Place 100–200 gm of material in a blender jar, add 20 gm Celite, 50 gm anhydrous magnesium sulfate, 200 ml acetonitrile, and 0.5 ml acetic anhydride. Blend for 3 minutes.

Transfer the homogenate into a 1 liter screw-cap bottle. Rinse the jar with 50 ml acetonitrile and add it to the bottle. Shake the slurry mechanically for 30 minutes. Filter with suction through a filter paper placed in a suction funnel, and collect the filtrate. Return the filter cake into the bottle using 200 ml acetonitrile. Shake again for 30 minutes, and filter into the same filtrate. Evaporate in a rotary evaporator to a final volume of about 40 ml.

ii. *Oxidation and Cleanup*

Add to the solution successively 30 ml redistilled water, 5 ml buffer solution pH 7, and 50–100 ml 0.1 *N* potassium permanganate previously warmed to 40°C. The purple color of the potassium permanganate mixture should remain for at least 5 minutes.* Shake and place the flask into a water bath of 40°C. Allow to stand for 15 minutes.

Remove the excess of permanganate by dropwise addition of a 20% formic acid solution. Filter with suction through a filter paper placed in a suction funnel, to remove solid materials. Wash the filter cake twice with 10 ml each redistilled water and acetonitrile. Evaporate the eluates until all of the acetonitrile has been removed.

Pour the remaining solution into a 250-ml separatory funnel, rinse the flask successively with 30 ml redistilled water and 80 ml of a mixture of methylene chloride–petroleum ether (9 : 1), and add both rinsings to the separatory funnel. Shake vigorously, allow to separate, and drain the lower phase through anhydrous sodium sulfate into a round-bottom flask. Extract the aqueous phase again 3 times with 80 ml of the methylene chloride–petroleum ether mixture. Evaporate the combined organic phases in a rotary evaporator. Transfer quantitatively to a 10-ml volumetric flask with the aid of toluene, and dilute to the volume. Quantitate the content by GLC using the external calibration method.

iii. *GLC Determination*

Column: glass spiral (100 cm × $\frac{1}{8}$ inch).

Stationary phase: 10% Dexsil 300 on Gas Chrom Q (60–80 mesh).

Temperature

Column oven: 180°C.

Injection port: 200°C.

Detector: 250°C.

Gas flow rates

Nitrogen: ca. 20 ml/minutes.

Hydrogen: ca. 40 ml/minutes.

Air: ca. 200 ml/minutes.

Retention times

Thiometon sulfone: 8.6 minutes.

PO analogue sulfone: 5.7 minutes.

* In some cases, e.g., sugarbeet leaves, it is necessary to add a slight excess of 1 *N* potassium permanganate solution to the reaction mixture.

Inject 1.0 μl of the solution containing up to 3.0 ng of each compound.

iv. *Calibration*

Weigh accurately 200 mg each of thiometon sulfone and its PO analogue into 20-ml flasks and dissolve in toluene. Pipette 1 ml of each solution into a 100-ml volumetric flask and fill up to the mark with toluene. Dilute this stock solution with toluene to yield standard solutions of 0.2, 0.5, 1.0, 2.0, and 3.0 ng/μl of each compound.

Equilibrate the gas chromatograph, inject each solution, and evaluate by plotting the peak areas versus nanograms injected for each compound. Ascertain that the linearity of the calibration curve ranges from 0.2 to 3.0 ng.

v. *Calculation*

Read the amount of thiometon sulfone and its PO analogue from the calibration curve. Calculate the residues by

$$\text{ppm} = \frac{\text{ng read} \times \text{ml final extract}}{\mu\text{l injected} \times \text{gm sample}} \times \text{dilution factor}$$

e. *Applicability of the Method*

This procedure has been applied successfully to apples, potatoes, sugarbeets, and wheat.

f. *Modification of the Method for Residues in Soil*

Mix the entire sample thoroughly and eliminate rocks larger than 1-cm diameter. Take subsamples from these undried samples for analysis. To determine the water content of the soil samples, dry a 100–200-gm part of the undried sample in the open air overnight and calculate the factor f_w

$$f_w = \frac{\text{gm soil undried}}{\text{gm soil airdried}}$$

Place 100 gm of the undried sample in a 1 liter screw-cap bottle, add 20 gm Celite, 50 gm anhydrous magnesium sulfate, 200 ml acetonitrile, and 0.5 ml acetic anhydride. Shake mechanically for 30 minutes, then filter with suction through a filter paper and collect the filtrate. Return the filter cake into the bottle with the aid of 200 ml acetonitrile. Shake the slurry again for 30 minutes and filter into the same flask. Then treat the crude extract as described in Section II,B,2,d. Calculate the final residues as shown in Section II,B,2,d,v, and correct them with the factor f_W. These results will give the residue values in ppm related to airdried soil.

g. *Discussion*

As already established for analyses of disulfotone residues (Thornton and Anderson, 1968), the method described was found to be an excellent procedure for the analysis of total residues of thiometon. The mild oxidation with permanganate in neutral solution converts the active ingredient and its metabolites quantitatively to their respective sulfones, which are more stable to the subsequent GLC determination than the corresponding sulfoxides. Many tests have shown that slight oxidation of the P=S group takes place under the conditions tested; furthermore, only minor degradation, amounting to less than 10%, was observed. The oxidation procedure also converts plant extracts to a water-soluble form that can be easily removed by the subsequent cleanup step.

Recovery experiments were conducted by adding known amounts of thiometon and its metabolites to the samples at the extraction step. In general, the recoveries at 0.1–1.0 ppm levels ranged between 70 and 90% for both thiometon sulfone and the oxygen analogue sulfone. The limit of detection was in the range of 0.1 ppm.

REFERENCES

Faderl, N. (1962). *Mitt. Geb. Lebensmittelunters. Hyg.* **53,** 154–175.

Farbenfabrik Bayer A. G. (1952). D.B.P. 917, 668; A.P. 1,759,010.

Food and Drug Administration. (1971). "Pesticide Analytical Manual," Vol. 1, 121. U.S. Department HEW, Washington, D.C.

Jucker, O. (1958). *Mitt. Geb. Lebensmittelunters. Hyg.* **5,** 1–24.

Raw, G. R. (1975). "CIPAC Handbook," Vol. IA (in press).

Sandoz Ltd. (1953). Swiss Patent 319, 579.

Schöniger, W. (1955). *Mikrochim. Acta* **1,** 123–129.

Schöniger, W. (1956). *Mikrochim. Acta* **2,** 869–876.

Thornton, J. S., and Anderson, C. A. (1968). *J. Agr. Food Chem.* **16,** 895–898.

Part III
Fungicides

14

2-Aminobutane

E. W. DAY, JR., AND J. R. KOONS

$$\begin{array}{c} \quad NH_2 \\ \quad | \\ CH_3CHCH_2CH_3 \end{array}$$

I. GENERAL

A. Empirical Formula

$C_4H_{11}N$ (Mol. wt. 73.14).

B. Alternative Name

sec-Butylamine

C. Source of Analytical Standard

Suitable *sec*-butylamine is available from a number of chemical supply companies. Reference standard 2-aminobutane hydrochloride is readily prepared by reacting anhydrous HCl with the amine in hexane solution.

D. Biological Properties

1. BIOLOGICAL ACTION

2-Aminobutane has been found to be effective for the control of fungal pathogens, especially *Penicillium* species, causing postharvest decay in citrus, apples, and peach fruit. It is also effective against organisms causing decay of commercial cut flowers during storage and transit, including a number of *Botrytis* species.

2. TOXICOLOGY

Hazards relating to the application of 2-aminobutane are due primarily to its alkalinity. These effects are minimized both by dilution and by neutralization. The single dose oral LD_{50} for 2-aminobutane is 380 mg/kg for rats, 250 mg/kg for hens, and 250 mg/kg for dogs. The dermal LD_{50} for rabbits is >2500 mg/kg, and the LC_{50} for bluegill fingerlings is between 50 and 100 ppm. In extended feeding and reproduction studies, 2500 ppm was safe to rats and 5000 ppm was safe to dogs.

E. History

The decay-controlling properties of 2-aminobutane were discovered by Eckert and Kolbezen (1962, 1963, 1964) at the University of California Agricultural Experiment Station at Riverside, California. Its use as a fungicide is covered by U.S. Patent 3,522,030 issued to the University of California.

F. Physical Properties

Pure 2-aminobutane is a colorless liquid with a density of 0.724 at 20°C. It has a boiling point of 63°C and the melting point is −104.5°C. The chemical is miscible with water, alcohols, ethers, and most other organic solvents.

G. Chemical Properties

2-Aminobutane is a primary aliphatic amine that undergoes all of the reactions of this class of compounds. Its basic dissociation constant (K_b) is 4.4×10^{-4} at 25°C, which makes it a slightly stronger base than ammonia.

H. Formulations

2-Aminobutane is formulated as an aqueous solution containing 26% of the carbonate salt and is equivalent to 15.5% of the free base. The free base is a liquid that usually contains 98% 2-aminobutane. Treatment solutions are prepared by diluting the free base or the formulated material to 0.5–1% with water.

II. ANALYSIS

A. Formulation Analysis

1. Review of Methods

Direct titrimetric methods are generally not applicable to formulated 2-aminobutane or to treatment solutions containing 2-aminobutane because of the presence of interfering chemicals. The active ingredient must be separated from these solutions prior to measurement. Two methods of separation and measurement are suggested. In the first, 2-aminobutane is distilled from an alkaline solution into aqueous boric acid and is subsequently titrated with standardized hydrochloric acid. The second method involves reacting the amine with 2,4-dinitrofluorobenzene (DNFB) and extracting the resulting aniline into cyclohexane for subsequent spectrophotometric measurement. The procedure described here is adapted from the methods of McIntire *et al.* (1953) and of Kolbezen *et al.* (1962). Both methods are suitable for a wide range of concentrations, and the choice of method is dictated by the availability of reagents and equipment. The spectrophotometric method is the more specific of the two.

2. DISTILLATION—TITRIMETRIC METHOD

a. *Principle*

The test solution containing 2-aminobutane is made alkaline with sodium hydroxide and subjected to conventional distillation. The distillate is collected under a solution of boric acid. The weak basic salt is then titrated directly with standardized hydrochloric acid.

b. *Reagents*

Sodium hydroxide, 1 *N*, 40 gm/liter.

Boric acid solution, 4%, 40 gm/liter.

Hydrochloric acid, 0.1 *N* standardized solution (Fisher Scientific No. SO-A-54).

Litmus or Hydrion paper.

Methyl purple indicator solution (Fisher Scientific No. SO-I-9).

c. *Apparatus*

Boiling flask, round bottom, short neck, 24/40 standard taper joint, 500 ml (Pyrex No. 4321).

Heating mantle, hemispherical for 500-ml round-bottom flask (Glas-Col, No. M-106).

Autotransformer-Variable Powerstat (E. H. Sargent and Co., No. 5-30940).

Distilling head, 24/40 standard taper (Ace Glass, Inc., No. 6608-06).

Stopper, glass, 10/30 standard taper (Pyrex No. 7570) or *thermometer* 3 inches immersion (Ace Glass Inc. No. 8315).

Adapter, angle type, 24/40 standard taper (Ace Glass Inc., No. 5080), lengthen by connecting glass or Tygon tubing to end such that it reaches to within 1 inch of the bottom of a 250-ml graduated cylinder.

Graduated cylinder, 250 ml.

Erlenmeyer flask, 500 ml.

Burette, 50 ml.

Boiling chips.

d. *Experimental Procedure*

i. *Distillation*

Transfer a 50.0-ml portion of treatment solution or 10.0 ml formulated product to a 500-ml round-bottom boiling flask. Add 50 ml 1 *N* sodium hydroxide and 100 ml deionized water. Add boiling chips and connect the flask for distillation. Measure 100 ml boric acid solution into the 250-ml graduated cylinder. Connect the adapter to the end of the condenser such that it dips well into the boric acid solution.

Heat the solution in the flask to an even boil. Use cold running water to cool the condenser. Collect distillate until the volume of liquid in the graduated cylinder reaches the 200-ml mark. Disconnect the adapter and test the distillate from the end of the condenser with pH paper. If the distillate is still basic, reconnect the adapter and distill an additional 10–20 ml. Cease the distillation when the condensate is neutral.

ii. *Titration*

Quantitatively transfer the contents of the graduated cylinder to a 500-ml conical flask. Add 5 drops of methyl purple indicator solution and titrate with 0.1 *N* hydrochloric acid. The end point is detected by a color change from green to blue or violet.

e. *Calculations*

The percent of 2-aminobutane in the sample is calculated directly from the following equation:

$$\%\ \text{2-aminobutane} = (N_a)(V_a)(7.314)/V_{sa}$$

where N_a = normality of HCl; V_a = volume of HCl required; V_{sa} = volume of sample subjected to distillation.

f. *Discussion*

This distillation–titrimetric method is relatively nonspecific in that any volatile, strongly basic amine will be measured. The calculations assume that all of the titrant used is consumed by 2-aminobutane. 2-Aminobutane is a very volatile material, and precautions should be taken during the distillation to insure against losses. All connections should be sealed with a silicone stopcock grease.

3. SPECTROPHOTOMETRIC METHOD

a. *Principle*

A dilute aqueous solution of 2-aminobutane is reacted with 2,4-dinitrofluorobenzene (DNFB) in borate buffer to form *N-sec*-butyl-2,4-dinitroaniline (BDNA). After hydrolysis of the excess reagent, the BDNA is extracted into cyclohexane and is measured spectrophotometrically at 334 nm.

b. *Reagents*

Standard 2-aminobutane hydrochloride. 2-Aminobutane hydrochloride is easily prepared by bubbling anhydrous hydrogen chloride through a hexane solution of the free base, filtering, and drying the solid at 110°C. Prepare aqueous solutions of the hydrochloride at concentrations equivalent to 100 and 1.0 μg/ml of free base (1.5 gm hydrochloride is equivalent to 1.0 gm free base).

Borate buffer solution, 2.5 gm sodium borate powder ($Na_2B_4O_7 \cdot 10\ H_2O$) in 100 ml water.

Dinitrofluorobenzene reagent, 0.75 ml of DNFB in 50 ml spectroquality or redistilled dioxane. This reagent should be prepared fresh daily.

Sodium hydroxide, 2 *N*, 8 gm of sodium hydroxide pellets in 100 ml water.

Cyclohexane, analytical reagent, further purified by passing 500 ml through a 40-gm column of silica gel, 0.2–0.5 mm. Discard the first 25 ml through the column.

Sodium carbonate, 0.1 *N*, 5.3 gm anhydrous sodium carbonate per liter of water.

Sodium sulfate, anhydrous, analytical reagent.

c. *Apparatus*

Water bath, capable of maintaining a temperature of 60 ± 2°C.

Spectrophotometer, Beckman DU or equivalent capable of making measurements at 334 nm with 1-cm cells.

d. *Experimental Procedure*

Choose a representative sample size such that it is convenient to make dilutions to a final concentration of 1 μg/ml of 2-aminobutane (2-AB). For formulations, weigh a sample equivalent to 1 gm of 2-AB into a tared 100-ml volumetric flask and dilute to volume with water. Make two successive 1:100 dilutions with water to obtain a 1 μg/ml solution. For treatment solutions, simply dilute to approximately 1 μg/ml and calculate the results on a w/v percent basis.

For the dinitrophenylation, pipette duplicate 10.0-ml portions of (a) working standard solution, (b) water, and (c) sample solutions into separate 25-ml conical flasks. Add 2 ml borate buffer solution and 2 ml DNFB reagent to each flask. Swirl the flasks briefly and place them in the water bath at 60°C for 20 minutes. Then add 2 ml 2 *N* NaOH to each flask and heat for 30 minutes to hydrolyze the excess reagent.

Remove the flasks from the bath, cool to room temperature, and quantitatively transfer the reactant solution to 125 ml separatory funnels. Pipette exactly 10.0 ml of purified cyclohexane into each funnel and shake vigorously. After the layers have separated, discard the lower aqueous layer. Wash the cyclohexane extract with three 15-ml portions of 1 *N* sodium carbonate.

Pour the cyclohexane solutions into clean, dry 25-ml conical flasks or 20-ml vials containing a few grams of anhydrous sodium sulfate. Swirl briefly to clarify the solutions. Measure the absorbance of each solution in 1-cm silica cells at 334 nm with the spectrophotometer. Use purified cyclohexane as the reference.

e. *Calculations*

i. *Formulations*

$$\%\ \text{2-Aminobutane} = \frac{A_{sa} - A_{bl}}{A_{st} - A_{bl}} \times \frac{W_{st}}{W_{sa}} \times 0.6673 \times 100$$

where A_{sa} = absorbance of sample solution
A_{st} = absorbance of standard solution
A_{bl} = average absorbance of reagent blanks
W_{st} = weight of standard 2-aminobutane hydrochloride in 100 ml of stock solution (grams)
W_{sa} = weight of sample (grams).

ii. *Treatment solutions*

$$\%\ \text{2-Aminobutane (w/v)} = \frac{A_{sa} - A_{bl}}{A_{st} - A_{bl}} \times \frac{W_{st}}{10^6} \times \frac{D}{V_{sa}} \times 0.6673 \times 100$$

where A_{sa}, A_{st}, A_{bl}, and W_{st} are as above and D is the dilution factor for the sample. For example, if 1.0 ml is diluted 1:100 twice, the dilution factor is 10^4. V_{sa} is the volume of sample solution in milliliters.

f. *Discussion*

The use of a high-quality dioxane as the solvent for the DNFB is a prerequisite for obtaining low blanks and good reproducibility. If spectroquality dioxane is not available, it may be necessary to purify the dioxane by distilling in the presence of sodium metal. The pH of the reactant solution is fairly critical, and must be buffered with the borate solutions. At high pH, the reagent hydrolyzes and the 2-aminobutane may be lost by volatilization. The reaction does not proceed reproducibly at low pH values. Spectroquality cyclohexane may be used if purification of reagent grade cyclohexane is undesirable.

The purpose of the sodium carbonate wash is to remove the dinitroaniline, which would have formed through reaction with any ammonia that may have been present in the sample. The dinitroaniline derivative of 2-aminobutane is not extracted from cyclohexane with 0.1 *N* sodium carbonate.

B. Residue Analysis

1. REVIEW OF METHODS

Two methods are available to determine residue levels of 2-aminobutane. The spectrophotometric procedure of Kolbezen *et al.* (1962) reacts dinitrofluorobenzene (DNFB) with 2-aminobutane. However, this method lacks specificity, since all primary and secondary amines react with DNFB to form colored products with essentially the same absorption maxima.

The recommended method, described by Day *et al.* (1968), also depends on the DNFB reaction to form the dinitrophenylamines, and selectivity is accomplished by a combined thin-layer and gas chromatographic procedure utilizing an electron affinity (electron capture) detector.

2. RECOMMENDED METHOD

a. *Principle*

The 2-aminobutane and other volatile amines are separated from the tissue by Kjeldahl distillation, followed by a carbon tetrachloride wash to remove some codistilled interfering compounds. After dinitrophenylation, the determination of 2-aminobutane is made by electron capture gas chromatography. In certain instances, the DNP derivative must be purified by thin-layer chromatography (TLC) prior to gas chromatographic measurement (see Section e, below).

b. *Reagents*

Benzene, analytical reagent.

Carbon tetrachloride, analytical reagent.

Ether, analytical reagent.

Hexane, analytical reagent.

Calcium chloride, 1 *M*. Dissolve 147 gm dihydrate salt in water and dilute to 1 liter.

Magnesium oxide, USP heavy powder (Mallinckrodt Chemical Works).

Sodium carbonate, 0.1 *N*.

Sodium hydroxide, 0.3 *N*.

Sodium hydroxide, 2 *N*.

Sodium sulfate, anhydrous.

Sulfuric acid, 0.3 *N*.

Antifoam agent, Antifoam 60 silicone emulsion (General Electric Silicone Products Dept., Waterford, New York).

Borate buffer. Dissolve 2.5 gm sodium borate powder ($Na_2B_4O_7 \cdot 10H_2O$) in 100 ml water.

DNFB reagent. Dissolve 0.75 ml 2,4-dinitrofluorobenzene (Distillation Products, Inc.) in 50 ml spectroquality *p*-dioxane (Matheson, Coleman, Bell). Prepare fresh daily.

Cyclohexane, purified. Pass 500 ml Reagent Grade cyclohexane through 125-mm column (25 mm diameter) of 0.2–0.5 mm silica gel (Brinkmann Instruments, Inc., Westbury, New York). Discard first 25 ml of eluate and collect remainder in a clean, dry reagent bottle.

Standard solutions. 2-Aminobutane hydrochloride is easily prepared by bubbling anhydrous hydrogen chloride through a hexane solution of the free base, filtering, and drying the solid at 110°C. Prepare aqueous solutions of the hydrochloride at concentrations equivalent to 100 and 1.0 μg/ml of free base (1.5 gm hydrochloride is equivalent to 1.0 gm free base).

c. *Apparatus*

Toledo chopper, Model 5125 (Toledo Scale Corp., Toledo, Ohio).

Kjeldahl unit, manufactured by Laboratory Construction Co., Kansas City, MO, or conventional distillation apparatus capable of handling about 200 ml of sample mixture.

Water bath, constant temperature 60°C, Model 1120A (Blue M Electric Co., Blue Island, Illinois).

Gas chromatograph, equipped with an electron affinity detector.

TLC plates, precoated, Sil G 0.25 mm (Brinkmann Instruments, Inc., Westbury, New York).

Pasteur capillary pipette (Matheson Scientific Co., no. 55698-00).

i. *Gas Chromatographic Conditions*

Column: 6 foot $\times \frac{1}{4}$ inch o.d. borosilicate glass.

Packing: diethyleneglycol succinate (DEGS) 2% w/w on Gas Chrom Q 80–100 mesh.

Column temperature: 188°C.

Injector temperature: 250°C.

Electron affinity detector temperature: 210°C.

Detector potential: 26 V.

Electrometer: 1×10^{-9} A full scale.

Nitrogen gas: 80 ml/minute.

Retention time N-sec-butyl-2,4-dinitroaniline: approximately 8 minutes.

d. *Experimental Procedure*

i. *Preparation of Sample*

Cut a sample of whole fruit, peel, or pulp into small pieces and grind to a fine pulp in a Toledo chopper. Weigh 50 gm ground tissue into an 800-ml Kjeldahl flask. Fortify recovery samples with 100 μg of 2-aminobutane (equivalent to 2.0 ppm) by adding 1.0 ml of the 100-μg/ml standard solution to a 50-gm sample (preferably an untreated control) in the flask. Swirl the flask to mix the standard and sample and allow it to set for a few minutes before adding reagents.

Add to the flask, in order, 75 ml 1 *M* calcium chloride, 50 ml water, 5–10 drops antifoaming agent, and a slurry of 10 gm magnesium oxide in 50 ml water. Rinse the sides of the flask with 50 ml water. Connect the flask to the Kjeldahl or other distillation apparatus and apply sufficient heat to effect distillation. Collect about 80 ml of distillate in a 100-ml graduated cylinder containing 10 ml 0.3 *N* sulfuric acid. The acidic distillate may be held in a closed vessel for several days, if necessary, without affecting the determination.

ii. *Dinitrophenylation*

Adjust the distillate to pH 7 (use pH meter) by adding 0.3 *N* NaOH. Add enough deionized water to bring the total volume to 100 ml. Pipette a 10.0-ml portion of the neutralized distillate into a small separatory funnel. Use a 10.0-ml portion of the 1.0-μg/ml solution of standard 2-aminobutane as the direct standard and a 10.0-ml portion of deionized water as the reagent blank. Wash the solutions with three 15-ml portions of carbon tetrachloride and transfer to 50-ml conical flasks. Add 5 ml borate buffer and 2 ml DNFB reagent. Place the flasks in the 60°C water bath for 20 minutes, add 2 ml 2 *N* NaOH and heat for 30 minutes. Cool and transfer the yellow solutions to 125-ml separatory funnels. Extract with 10.0 ml purified cyclohexane, discard the aqueous layer, and wash the cyclohexane solutions with three 15-ml portions of 0.1 *N* sodium carbonate. Pour the cyclohexane solutions into 20-ml conical flasks containing about 2 gm anhydrous sodium sulfate. Analyze these solutions directly by gas chromatography unless TLC is required (see Section e below).

iii. *Thin-Layer Chromatography*

If TLC is required, pipette 5.0 ml of cyclohexane solution into a small pear-shaped flask or a conical centrifuge tube. Evaporate the solvent carefully with a stream of air. Add 0.5 ml benzene to the residue in the vessel and swirl to dissolve the sample. Quickly spot 100 μl of this solution on a TLC plate. Also spot a portion of dinitrophenylated direct standard on the plate to aid in locating the desired zone. Chromatograph the plate in the usual fashion, using hexane–ether (70:30) as the developing solvent. Do not presaturate the chamber.

Mark the zone on the plate to encompass the spot arising from the DNP derivative of 2-aminobutane. Place a small glass wool plug at the constriction of a disposable Pasteur capillary pipette. Rinse the pipette plug with 10 ml chloroform and dry by suction. Loosen the adsorbent in the desired zone on the plate and remove by suction into the pipette. Elute the adsorbent in the pipette with about 5 ml chloroform, collect the eluate in a small vial, and evaporate the solvent with a stream of air. Dissolve the residue in 1.0 ml benzene and analyze by gas chromatography.

iv. *Gas Chromatography*

Perform the analysis, using normal gas chromatographic techniques, with instrument settings described previously. The retention time of the 2-aminobutane DNP derivative is approximately 8 minutes. Measure the appropriate peak responses in the reagent blank, control, control plus standard, direct standard, and treated samples. The reagent blank should not have a peak at the retention time of the dinitrophenylated direct standard. Calculate the recovery from Eq. (1) and the residue of 2-aminobutane in the treated samples in ppm from Eq. (2).

$$\%\ \text{recovery} = \frac{P_r - P_c}{P_{st}} \times 100 \tag{1}$$

$$\text{ppm 2-aminobutane} = \frac{P_{sa}}{P_r - P_c} \times 2.0 \tag{2}$$

where P_r = peak response of the fortified recovery
P_c = peak response of the control sample used for the recovery
P_{st} = peak response of the 1.0 μg/ml dinitrophenylated standard
P_{sa} = peak response for the sample.

e. *Discussion of Method*

The pH adjustment prior to the carbon tetrachloride washing step is critical. If the pH is greater than 8, some 2-aminobutane is extracted by the carbon tetrachloride and low recoveries result. The carbon tetrachloride washing step is not required for some crops. However, it does remove some volatile components of certain varieties of apples and citrus fruit yielding less complex gas chromatograms. Hence, for uniformity, the step is performed on all samples.

The TLC step is only rarely required to complete the analysis of 2-aminobutane. Dimethylamine is the only low molecular weight amine that interferes with the gas chromatographic measurement. This amine is possibly endogenous to some fruit, but it is of no consequence unless low residue levels of 2-aminobutane are sought. Thus, if the control sample exhibits little or no response at the retention time of DNP-2-aminobutane, TLC is not performed. However, three fungicides—ferbam, thiram, and ziram—hydrolyze under the conditions described to yield dimethylamine. These chemicals have established tolerances on a variety of fruit. Thus, if fruit are known to have been treated with these fungicides or if no control tissue is available, the TLC step should be performed to eliminate any ambiguities that might arise. The effect of sixty-five other pesticides were tested at their maximum permissible tolerance level (as listed in the *Federal Register*) on fruit. None of these interfered with the determination of 2-aminobutane. The recoveries from 50 gm of ground sample tissue at the 2 ppm level fell within the 80–100% range.

The limit of detection of 2-aminobutane under the conditions described is about 0.1 ppm. The sensitivity can be increased by using the total distillate instead of a portion, by concentrating the cyclohexane extract before gas chromatographic analysis, or by distilling from a larger sample. Samples of ground tissue up to 200 gm have been conveniently and successfully analyzed for amine content. Reproducibility and linearity studies on 2-aminobutane have been reported previously by Day *et al.* (1966).

REFERENCES

Day, E. W., Jr., Golab, T., and Koons, J. R. (1966). *Anal. Chem.* **38,** 1053–1058.

Day, E. W., Jr., Holzer, F. J., Tepe, J. B., Eckert, J. W., and Kolbezen, M. J. (1968). *J. Ass. Offic. Anal. Chem.* **51,** No. 1, 39–44.

Eckert, J. W., and Kolbezen, M. J. (1962). *Nature (London)* **194,** 888.

Eckert, J. W., and Kolbezen, M. J. (1963). *Phytopathology* **53,** 1053.

Eckert, J. W., and Kolbezen, M. J. (1964). *Phytopathology* **54,** 978.

Kolbezen, M. J., Eckert, J. W., and Bretschneider, B. F. (1962). *Anal. Chem.* **34,** 583.

McIntire, F. C., Clements, L. M., and Sproull, M. (1953). *Anal. Chem.* **25,** 1757.

15

Chlorothalonil

D. L. BALLEE, W. C. DUANE, D. E. STALLARD,
AND A. L. WOLFE

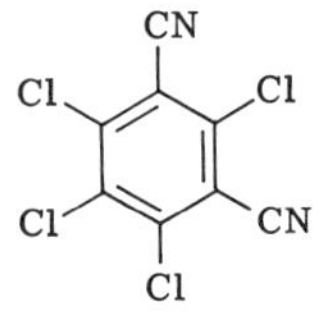

Tetrachloroisophthalonitrile

I. GENERAL

A. Empirical Formula

$C_8Cl_4N_2$ (Mol. wt. 265.9).

B. Alternative Names

Chlorothalonil is the common name for tetrachloroisophthalonitrile. Daconil 2787®, Bravo®, Termil®, and Exotherm Termil® are registered trademarks of Diamond Shamrock Corporation. Wiswesser line notation: NCR BG CG DG FG ECN.

C. Source of Analytical Standard

Agricultural Chemicals Division, Diamond Shamrock Corporation, Cleveland, Ohio.

D. Biological Properties

Chlorothalonil is a broad-spectrum fungicide effective against many economically important plant pathogens affecting vegetables; agronomic, tree and small fruit crops; turf; and ornamentals. The diverse fungicidal activity of chlorothalonil is characteristic of its mode of action, the alkylation of cellular thiols–glutathione and thiol dependent glycolytic and respiratory enzymes (Vincent and Sisler, 1968; Tillman *et al.*, 1973). The selective activity of chlorothalonil against fungi is correlated with selective uptake by fungal cells commensurate with the oil–water partition value (Turner and Battershell, 1969, 1970). In the presence of moisture, chlorothalonil undergoes a breakdown in soil. High moisture and temperature accelerate the rate of breakdown. In temperate regions with moderate levels of moisture, chlorothalonil has a half-life of $1\frac{1}{2}$–3 months, depending upon type of soil. Chemical analysis of soils treated for years at recommended rates showed that chlorothalonil does not accumulate in soil.

E. Toxicity

The acute oral toxicity (LD_{50}) of chlorothalonil to male and female albino rats is greater than 10,000 mg/kg of body weight. The acute oral toxicity for female dogs is greater than 5000 mg/kg. No estimate of the quantity necessary to produce toxic signs could be determined because of emesis.

The acute oral LC_{50} for 7-day-old mallard ducklings is greater than 21,500 ppm and for 7-day-old Bobwhite quail is estimated at 5200 ppm. The LC_{50} values for fish were 250 ppb for rainbow trout, 386 ppb for bluegill sunfish, and 432 ppb for channel catfish.

The acute dermal toxicity (LD_{50}) of chlorothalonil for albino rabbits is greater than 10,000 mg/kg of body weight. Repeated application of Bravo W-75 to shaved skin of male and female albino rabbits at a concentration of 500 mg of active material per kilogram of body weight per application for a 21-day period produced mild to moderate dermal irritation. A single application of 3.0 mg of chlorothalonil to the eyes of rabbits produced mild irritation consisting of conjunctivitis, which either completely or largely subsided within 7 days or less following application.

In two-year chronic toxicity tests, the no-effect level for rats was greater than 60 ppm and less than 120 ppm in the daily diet. For dogs, the no-effect level was greater than 120 ppm.

F. History

Chlorothalonil was first synthesized at the T. R. Evans Research Center of the Diamond Shamrock Corporation in 1962. The composition of matter patent U.S. 3,290,353 and the fungicidal use patent U.S. 3,331,735 were issued on December 6, 1966, and July 18, 1967, respectively. Chlorothalonil was first registered in the United States for use on potatoes for the control of early blight and late blight in 1969. Registration for uses on seventeen additional economically important vegetable and agronomic crops was accepted in 1971.

G. Physical Properties

Odorless, white, crystalline, solid (Technical Grade, slight pungent odor).

Melting point: 250–251°C.
Boiling point: 350°C at 760 mm Hg.
Specific gravity: 1.73.

Temperature (°C)	*Vapor pressure* (mm Hg)
40	0.01
170.4	9.2
190.8	17.4
211.5	27.3
229.5	43.3

Solubility in (at 25°C)	*Percent by weight*
Xylene	8%
Dimethylformamide	3%
Acetone	2%
Dimethylsulfoxide	2%
Water	0.6 ppm

H. Chemical Properties

1. METHOD OF SYNTHESIS

The synthesis of chlorothalonil proceeds via the ammoxidation of *m*-xylene to produce isophthalonitrile, which is in turn chlorinated to produce the technical material.

2. STABILITY

Chlorothalonil is thermally stable under normal temperature of storage. Above 200°C, it is readily sublimed without decomposition. This property has been employed to apply fungicidally active "thermal dusts" of chlorothalonil (Turner and Lamont, 1965). In organic solvent-based solutions, chlorothalonil may undergo a slow photodegradation to unknown product(s) at normal temperatures. Chlorothalonil does not hydrolyze in either moderate alkaline ($pH < 9$) or acidic aqueous media. At pH 9 or greater, chlorothalonil hydrolyzes to give 3-cyano-2,4,5,6-tetrachlorobenzamide and 4-hydroxy-2,5,6-trichloroisophthalonitrile.

Chlorothalonil is highly reactive toward thiol-containing compounds. The $t_{1/2}$ for reaction with 4-nitrothiophenoxide is about 0.56 minutes at 20°C in millimolar aqueous solutions (Turner and Battershell, 1969).

I. Formulations

Chlorothalonil is commercially available as:

1. *Daconil 2787® W-75 and Bravo® W-75,* wettable powder formulations containing 75% active ingredient by weight.
2. *Bravo® 6F,* a water dispersible concentrate containing 6 pounds of active ingredient per gallon.
3. *Exotherm Termil®*, a powder formulation containing 20% active ingredient by weight.
4. *Termil®*, compacted tablets containing 90% chlorothalonil.

Termil® and Exotherm Termil® are special formulations for applying chlorothalonil as a thermal dust in greenhouses.

II. ANALYSIS

A. Formulation Analysis

1. INTRODUCTION

The appropriate weight of the formulated product is dissolved in a suitable volume of the specified extracting solvent. When solution of the chlorothalonil is complete, portions from 2 to 6 μl are injected into a gas chromatograph equipped with either flame ionization or thermal conductivity detectors. The concentration of chlorothalonil is determined by the ratio of the peak height obtained for the sample to peak height obtained for a standard of known concentration. The inherent differences in sensitivity of the two detectors to chlorothalonil requires a higher initial concentration for assay using the TC detector. The recommended sample and standard weights for the analyses of formulations using either detector are given in Table I.

2. SAMPLE PREPARATION

a. *Bravo® W-75 and Daconil 2787® W-75*

The recommended weight of Bravo® W-75 or Daconil 2787® W-75 is quantitatively transferred to a 25 ml volumetric flask using 20 ml 1,4-dioxane. The flask is stoppered and shaken for 15 minutes using a mechanical shaker. The sample is then diluted to volume using 1,4-dioxane, mixed thoroughly, and allowed to stand or is centrifuged to remove insolubles. The sample and an appropriate standard prepared

TABLE I
SAMPLE AND STANDARD WEIGHTS FOR ANALYSIS OF CHLOROTHALONIL FORMULATIONS

Formulation	Solvent[a]	Sample weights (gm)		Weight of standard (gm)	
		TC[b]	FID[c]	TC	FID
Daconil 2787® W-75	1,4-Dioxane	2.00	0.25	1.50	0.1875
Bravo® W-75	1,4-Dioxane	2.00	0.25	1.50	0.1875
Bravo® 6F	Mixed solvent[d]	2 00	0.50	1.10	0.275
Termil®	1,4-Dioxane	1.50	0.25	1.35	0.220
Exotherm Termil®	1,4-Dioxane	2.50	0.75	0.50	0.150

[a] All samples and standards are dissolved in 25 ml solvent.
[b] TC = thermal conductivity detector.
[c] FID = flame ionization detector.
[d] Methyl ethyl ketone–carbon disulfide–1,2-dimethoxyethane (6:3:1) by volume.

using the same procedure are reserved for analysis by gas chromatography.

b. *Bravo® 6F*

The sample is vigorously shaken to insure a homogeneous suspension, and the appropriate amount weighed directly into a tared 25-ml volumetric flask. Twenty milliliters of the mixed solvent, methyl ethyl ketone–carbon disulfide–1,2-dimethoxyethane (6:3:1 v/v) is added, and the flask is stoppered and shaken for 15 minutes using a mechanical shaker. The sample is then diluted to volume using the mixed solvent and is mixed thoroughly. The sample and an appropriate standard prepared using the same procedure are reserved for analysis by gas chromatography.

c. *Termil®*

A minimum of two Termil® tablets are broken, and approximately 5 gm are ground to a fine powder using a mortar and pestle. The recommended weight of the finely ground tablet is quantitatively transferred to a 25-ml volumetric flask. Twenty milliliters of 1,4-dioxane is added and the flask is stoppered and shaken for 15 minutes using a mechanical shaker. The sample is then diluted using 1,4-dioxane and mixed thoroughly. The sample and an appropriate standard prepared using the same procedure are reserved for analysis by gas chromatography.

d. *Exotherm Termil®*

The recommended weight of Exotherm Termil® is quantitatively transferred to a 25-ml volumetric flask, 20 ml 1,4-dioxane is added, and the flask is stoppered and shaken for 15 minutes. The sample is then diluted to volume using 1,4-dioxane, mixed thoroughly, and allowed to stand or centrifuged to remove insolubles. The sample and an appropriate standard prepared using the same procedure is reserved for analysis by gas chromatography.

3. ANALYSIS BY GAS CHROMATOGRAPHY

a. *Apparatus*

Gas chromatograph equipped with either a TC or flame detector. The flame detector is preferred because of its superior response to organic compounds and its insensitivity to water. The instrument should have a heated injection port to enhance component resolution and elution.

Recorder, 1 mV full scale, 1 second or better response time, capable of 0.25 inches/minute chart speed.

b. *Instrument Conditions*

Column: 6 feet $\times \frac{1}{8}$ inches stainless steel packed with 5% XE-60 on 60–80 or 80–100 mesh Gas Chrom® Z (Applied Science Labs, Inc., P. O. Box 440, State College, Pennsylvania).

Column temperature: 210°C.

Detector temperature: Flame, 300°C; TC, 275°C.

Injection port temperature: 250°C.

Carrier gas: Nitrogen for flame and helium for TC; helium flow = 40 ml/minute; nitrogen flow = 40 ml/minute; air flow = 250–300 ml/minute; hydrogen flow = 30 ml/minute.

Recorder chart speed: 0.25 inches/minute.

c. *Quantization*

Portions from 2 to 6 μl are injected until a peak response of 75–85% of full scale deflection is obtained. Failure to get the desired response requires an adjustment of the attenuator or TC power supply. Injections of the selected sample portion are repeated until responses as measured by peak height or total area agree within 1%. A portion of the appropriate standard, equal in volume to that taken for the sample, is injected under identical conditions until a precision of 1% is obtained. It is recommended that samples be analyzed in triplicate and averaged.

d. *Calculations*

$$\% \text{ chlorothalonil} = \frac{\text{peak height or area of sample}}{\text{peak height or area of standard}} \times \frac{\text{weight of standard}}{\text{weight of sample}} \times 100$$

B. Residue Analysis

1. PRINCIPLE

Chlorothalonil and its metabolite, 4-hydroxy-2,5,6-trichloroisophthalonitrile (DAC-3701), as the *n*-propyl ether derivative, are readily determined by gas chromatography coupled with electron capture detection. To obtain the residue, chlorothalonil is surface stripped from crops by methylene chloride extraction. Alternately, chlorothalonil and DAC-3701 are extracted using an acidic acetone extraction medium. Subsequent separation by selective partitioning generally provides interference-free extracts of chlorothalonil and its metabolite for gas chromatographic analysis.

2. REAGENTS

Chlorothalonil standard, serial dilutions in benzene.
DAC-3701 standard, serial dilutions in acetone.

Do not expose standard solutions to sunlight or strong artificial light.

Acetone, practical grade.
Acetonitrile, practical grade.
Alumina, Woelm W200 acid (available from ICN Pharmaceuticals, Inc., Cleveland, Ohio).
Benzene, certified ACS.
Ethyl ether, anhydrous, certified ACS.
Isopropyl ether, certified ACS.
n-Hexane, certified ACS.
Methylene chloride, practical grade.
1:1 concentrated H_2SO_4*–water.*
1:3 HCl–n-propyl alcohol, Analytical Reagent Grade.
2% paraffin oil in benzene.
0.4 M $NaHCO_3$, aqueous solution.
1-n-Propyl-3-tolyltriazene (available from Willow Brook Laboratories, Inc., P. O. Box 526, Waukesha, Wisconsin 53186).

3. APPARATUS

Gas chromatography system. The instrument is equipped with glass columns for on-column injection and an electron capture detector (preferably ^{63}Ni operating at 285–325°C). The column packing can be coated with any of several liquid phases, as the compounds to be analyzed elute well over a range of column polarities. Some examples are: 6% DC-200, 3% Silar 5-C-P, or a mixture of 1.5% OV-17 and 1.95% SP-2401 liquid phase coated on 80–100 mesh Gas Chrom Q (Applied Science Laboratories, Inc., P. O. Box 440, State College, Pennsylvania).

Waring Blendor® with stainless steel container.

Wrist-action shaker.

Controlled-temperature water bath.

Dry air stream source.

pH meter.

Tumbling apparatus, 2-quart jars.

Normal laboratory glassware, i.e., flask, separatory funnels, volumetric flasks, etc.

4. EXPERIMENTAL PROCEDURE

a. *Surface Stripping*

This technique is applicable to rapid analysis of chlorothalonil from crop surfaces. A wide variety of crops, e.g., leafy vegetables, fruits, can be processed this way. A portion (100–200 gm) of the crop is placed in a 2-quart wide-mouth glass jar. A 5 : 1 ratio of methylene chloride to the weight of crop is added. The jar is sealed with an aluminum foil-lined cap and the contents tumbled for 2 hours. The jars are allowed to stand until phase separation is complete (approximately 1 hour). A portion of the extraction solvent, equivalent to 10 gm of crop, is transferred to a 125-ml Erlenmeyer flask, and 0.2 ml 2% paraffin oil in benzene is added. The flask is placed in a water bath maintained at 35°C, and a gentle stream of dry air is directed toward the liquid surface to aid solvent removal. This technique of sample concentration is used throughout all methods presented. After solvent removal, the resulting residue is dissolved in a suitable volume of benzene and analyzed for chlorothalonil by electron capture gas chromatography.

b. *Extraction of Chlorothalonil and Its Metabolite, DAC-3701*

This technique is applicable to the complete extraction of products resulting from the application of chlorothalonil. Modifications of the basic procedure allow determination of a wide array of materials.

The basic procedure involves the extraction into an acidic acetone medium (5 ml of 1 : 1 concentrated H_2SO_4–water per 95 ml acetone) followed by selective partitioning to separate chlorothalonil and DAC-3701. All glassware must be exceptionally clean to prevent loss of the very polar DAC-3701 by random binding. It is recommended that after normal laboratory washing, glassware should be rinsed further in the following order: acetone, hot tap water, acidic water (pH 1), distilled water, and acetone.

c. *Sample Preparation*

i. *Soil*

Soil samples are air-dried and thoroughly mixed by dry grinding in a Waring Blendor®. A 50-gm sample is placed in a 500-ml glass-stoppered Erlenmeyer flask, and 250 ml of the acidic acetone extraction medium is added. The flask is sealed and shaken on a wrist-action shaker for 2 hours. After soil solids have settled, a 50-ml portion of the supernate (10 gm equivalent) is placed in a 250-ml beaker and concentrated as previously described to remove acetone. 100 ml 0.4 *M* $NaHCO_3$ in water is added and the resulting solution adjusted to pH 4.5 with the aid of a pH meter and 1:1 concentrated H_2SO_4–water.

ii. *Water*

A 100-ml sample of water is placed in a 250-ml beaker and adjusted to pH 4.5 with the aid of a pH meter and 1:1 concentrated H_2SO_4–water.

iii. *Green Leafy Vegetables (Spinach, Lettuce, Kale, Collards, and Turnip Greens)*

A 50-gm sample of the leafy vegetable is homogenized with 200 ml of the acidic acetone extraction medium for 1 minute in a Waring Blendor®. The homogenate is quantitatively transferred to a glass-stoppered 500-ml Erlenmeyer flask using an additional 50 ml acetone to complete the transfer. The flask is shaken for 2 hours on a wrist-action shaker. The flask contents are vacuum filtered through Whatman no. 42 filter paper supported on a Büchner funnel. The resulting filtrate is brought to a total volume of 400 ml by the addition of acetone. An 80 ml portion (10 gm equivalent) of the filtrate is transferred to a 250-ml beaker, concentrated, and adjusted to pH 4.5 as described for soil.

iv. *Tomatoes*

Whole tomatoes (350–450 gm) are blended in a Waring Blendor® with 500 ml of the acidic acetone extraction medium and quantitatively transferred to a 2-quart wide-mouth glass jar. The jar is sealed with an aluminum foil-lined cap and tumbled for 2 hours. The jar contents are filtered as described in the procedure for green leafy vegetables. The resulting filtrate is brought to a total volume of 700 ml by again adding acetone. A 100-ml portion (50–65 gm equivalent, depending upon tomato size) is transferred to a 250-ml beaker, concentrated, and adjusted to pH 4.5 as described for soil.

v. *Dry Oily Crops, Such as Soybeans, Dry Beans, and Peanuts*

These samples are thoroughly mixed by dry grinding in a Waring Blendor®. A 20-gm ground sample is placed in a glass-stoppered 500-ml Erlenmeyer flask and a mixture of 5 ml water, 10 ml 1:1 concentrated H_2SO_4–water, and 185 ml acetone is added and the flask contents are shaken on a wrist-action shaker for 2 hours. A 100-ml portion (10 gm equivalent) is transferred to a 250-ml beaker and concentrated. After the acetone is evaporated, the beaker contents are transferred to a 125-ml separatory funnel with the aid of 50 ml acetonitrile and 50 ml hexane. The contents are shaken, the upper hexane layer discarded, and the extraction repeated with another 50 ml hexane. The lower layer is added to 100 ml 0.4 *M* $NaHCO_3$ in water, an additional 50 ml water is added and the mixture adjusted to pH 4.5 using either 0.4 *M* $NaHCO_3$ or dilute H_2SO_4.

d. *Selective Partitioning*

The aqueous solution at pH 4.5 is quantitatively transferred to a 250-ml separatory funnel and extracted two times with 50 ml isopropyl ether. These essentially neutral ether extracts, containing chlorothalonil, are combined in a 125-ml Erlenmeyer flask and concentrated as described. The concentrated ether fraction is adjusted to an appropriate volume with benzene and analyzed for chlorothalonil by electron capture gas chromatography.

The remaining aqueous solution is acidified to pH <2 with 1:1 concentrated H_2SO_4–water and extracted twice with two 50-ml portions of isopropyl ether. This combined acidic ether extract, containing DAC-3701, is concentrated in a 125-ml Erlenmeyer flask as per the chlorothalonil extract treatment and is then reserved for derivatization of the DAC-3701 to a volatile ether.

e. *Alkylation of DAC-3701*

The metabolite DAC-3701 may be determined as either the methyl, ethyl, or *n*-propyl ether derivative using electron capture gas chromatography. The alkylating agent used for derivatization is 1-alkyl-3-*p*-tolyltriazene, with alkyl being either methyl, ethyl, or *n*-propyl (Willow Brook Laboratories, Inc., 1971). The *n*-propyl ether is routinely prepared because of its greater gas chromatographic retention time relative to interferences often found in crop and soil samples. To the concentrated DAC-3701 residue is added 15 μl of a 1:3 mixture of concentrated hydrochloric acid–*n*-propyl alcohol followed by 5 ml of the alkylating agent (0.2 gm 1-*n*-propyl-3-*p*-tolyltriazene in 110 ml ethyl ether), and the reaction mixture is allowed to stand at room temperature for 30 minutes. The solvent is then removed by a stream of dry air. To ensure complete propylation, an additional 3 ml of alkylating reagent is added, and the mixture is allowed to stand for 10 minutes, after which the solvent is again evaporated by air stream. The sample is brought to an appropriate volume in benzene and analyzed by electron capture gas chromatography.

f. *Removal of Interferences*

Interferences are occasionally encountered in the gas chromatographic analyses of chlorothalonil and DAC-3701. The analysis of chlorothalonil in green leafy crops employing the acidic acetone extraction procedure is the most common example. When interferences are encountered, liquid chromatography on alumina is used to clean up the residue before gas chromatography. The cleanup procedure may be adapted wherever necessary:

Cleanup procedure. A glass column, 40 cm × 2 cm, fitted with a Teflon stopcock, is filled to a height of 10 cm with alumina (Woelm W200 acid, heat activated for 24 hours). Unless the alumina is sufficiently activated, chlorothalonil will elute in the methylene chloride fraction. Therefore, it is critical that this separation be checked in each laboratory. The alumina is supported on a 2-cm bed of glass wool. The column is prewashed with 50 ml methylene chloride, and the sample added in either methylene chloride or benzene. The column is eluted with 200 ml methylene chloride, which is discarded. Chlorothalonil is recovered in excellent yield by elution with 100 ml of 5% acetone in methylene chloride.

Interferences in the analysis of DAC-3701 have generally been traced to solvent constituents and can be reduced by obtaining interference-free solvents (by distillation, etc.) or by resolution on an alternative liquid phase in gas chromatography.

g. *Calculations and Recoveries*

The linear response range for chlorothalonil and the *n*-propyl ether of DAC-3701 on the gas chromatographic system of choice is determined using benzene solutions of the standards and plotting peak height versus concentration. The samples are then diluted until a portion injection is within the linear range defined by the standard. The *n*-propyl ether of DAC-3701 can be synthesized and used as a standard; in this case, sample values are multiplied by 0.855 to correct for molecular weight differences between DAC-3701 and its propyl ether. Alternatively, small, exact amounts (10–50 μg) of DAC-3701 may be alkylated at the same time and in the same manner as the samples to generate the standard for gas chromatographic analysis. Experience has shown either procedure equally useful, as the propylation of DAC-3701 is essentially quantitative. Recoveries of fortified residues over a wide range of samples have shown consistent values of 85–100% for both chlorothalonil and DAC-3701.

h. *Sensitivity*

Most electron capture gas chromatographic systems are capable of the detection of 0.1 ng of each compound; therefore, a 10-gm sample can easily provide the 0.01 ppm level of sensitivity. If the given instrumentation does not have this capability, it is a simple matter to increase sample size to reach the desired level of sensitivity.

References

Tillman, R. W., Siegel, M. R., and Long, J. W. (1973). *Pestic. Biochem. Physiol.* **3,** 160.
Turner, N. J., and Battershell, R. D. (1969). *Contrib. Boyce Thompson Inst.* **24,** 139.
Turner, N. J., and Battershell, R. D. (1970). *Contrib. Boyce Thompson Inst.* **24,** 203.
Turner, N. J., and Lamont, D. (1965). *Contrib. Boyce Thompson Inst.* **23,** 51.
Vincent, P. G., and Sisler, H. D. (1968). *Physiol. Plant.* **21,** 1249.
Willow Brook Laboratories, Inc. "1-Alkyl-3-*p*-Tolyltriazenes," Technical Data Sheet. (March, 1971). Waukesha, Wisconsin.

16

Dinobuton

V. P. LYNCH

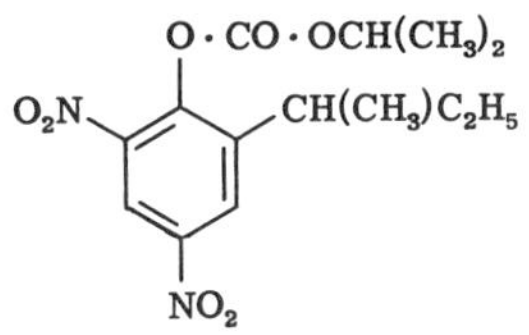

2-*s*-Butyl-4,6-dinitrophenyl isopropyl carbonate

I. GENERAL

A. Empirical Formula

$C_{14}H_{18}N_2O_7$ (Mol. wt. 326).

B. Alternative Name

Acrex® is the registered trade mark of Murphy Chemical Ltd.

C. Source of Analytical Standard

Murphy Chemical Ltd., Wheathampstead, St. Albans, Hertfordshire, United Kingdom.

D. Biological Properties

1. ACTIVITY

Dinobuton is a nonsystemic acaricide having particular activity against red spider mites (*Tetranychus* spp.), including races that are resistant to other acaricides. Dinobuton is also a fungicide and is active against members of the order Erysiphales (powdery mildews). It has both protective and some curative effect.

2. TOXICITY

The acute oral LD_{50} for mice is 2540 mg/kg, for rats 140 mg/kg, for hens 150 mg/kg. The acute dermal LD_{50} for rats is greater than 5000 mg/kg and for rabbits is 2500 mg/kg. The maximum no-effect level for dogs is 4.4 mg/kg/day, and for rats the value was between 3 and 6.2 mg/kg/day. Dinobuton acts as a metabolic stimulant, and at high intake levels causes a loss of body weight.

E. History

Dinobuton was first described by Pianka and Smith (1965). The structure and acaricidal activity of it and related compounds was later described by Pianka (1966).

F. Physical Properties

Dinobuton is a faintly yellow crystalline solid of melting point 61–62°C. It is practically insoluble in water. It is soluble in aliphatic hydrocarbons, highly soluble in acetone and aromatic hydrocarbons. At room temperatures the vapor pressure is neglible. The technical material is about 97% pure and melts at 57–60°C.

G. Chemical Properties

1. METHOD OF SYNTHESIS

Dinobuton is prepared by the condensation of an alkali salt of dinoseb with isopropyl chloroformate.

2. CHEMICAL REACTIONS

Dinobuton is stable to acid but hydrolyzed by alkali to give salts of dinoseb. In homogeneous nonaqueous solutions, e.g., alcohol or acetonitrile, it is rapidly hydrolyzed after the addition of a trace of butylamine.

H. Formulations

Dinobuton is available as a 50% wettable powder, 30% emulsifiable concentrate, 50% aqueous suspension, and a 20% atomizing concentrate.

II. ANALYSIS

A. Analysis of Technical Material and of Formulations

1. RECOMMENDED METHOD

a. *Principle*

The nitro groups in dinobuton are reduced to the corresponding amino groups when a sample is heated under reflux with excess titanium

(III) sulfate in glacial acetic acid. The excess titanium (III) sulfate is titrated with a standard ammonium ferric sulfate solution.

b. *Reagents*

Ammonium ferric sulfate solution $[NH_4Fe(SO_4)_2 \cdot 12H_2O]$, 0.2 M, transfer 96.44 gm ammonium ferric sulfate to an 800-ml beaker. Add 500 ml distilled water and 30 ml concentrated sulfuric acid and warm to dissolve. Cool the solution and transfer to a 1-liter volumetric flask and make up to volume with water.

Titanium (III) sulfate, approximately 0.2 M, transfer 400 ml 15% titanium (III) sulfate solution to a 1500-ml beaker. Add 300 ml water containing 50 ml concentrated sulfuric acid and then add 300 ml glacial acetic acid. Mix and transfer this solution to a stock bottle and keep the solution under an atmosphere of carbon dioxide.

Ammonium thiocyanate indicator, 20% aqueous solution.

Sulfuric acid, specific gravity about 1.84.

Glacial acetic acid.

c. *Apparatus*

Round-bottom flask with two necks, 250 ml, one fitted with a Liebig condenser and the other carrying a lead-in tube reaching just above the liquid level in the flask and connected to a carbon dioxide source.

Automatic pipette, 25 ml, connected to the reservoir of the titanium (III) sulfate solution.

Burette, 50 ml.

d. *Procedure*

i. *Blank*

Place glacial acetic acid (30 ml) and sulfuric acid (5 ml) in the flask and connect to the carbon dioxide supply. Allow an interval of a few minutes to displace the air, add titanous sulfate solution (25 ml), fit the flask to the condenser, and heat the contents under gentle reflux for 15 minutes. Rinse down the condenser, and cool the flask to room temperature. Detach the flask and titrate with the ammonium ferric sulfate solution. Add the ammonium thiocyanate indicator (5 ml) at the first sign of fading of the titanium (III) color. The end point is reached when a faint red-brown color persists (*b* ml). Maintain a current of carbon dioxide gas through the apparatus during the entire procedure.

ii. *Analysis of Sample*

Accurately weigh sufficient sample to contain about 0.1 gm of the active ingredient. Transfer to the flask as described above. Add acetic acid (30 ml) and sulfuric acid (5 ml) and allow the sample to dissolve (warming if necessary). Connect to the carbon dioxide source, add 25 ml of the titanous sulfate solution, and proceed as described above (t ml).

iii. *Calculation*

$$\% \text{ dinobuton} = \frac{(b - t) \times 0.2 \times 326}{\text{weight taken} \times 12 \times 10}$$

2. ALTERNATIVE METHOD FOR ASSAYING TECHNICAL MATERIAL

The following method is a modification of the one described by Crossley and Lynch (1968) for determination of dinobuton residues.

a. *Principle*

Dinobuton undergoes deesterification to a salt of dinoseb on neutral alumina. The dinoseb is eluted from the column as its butylammonium salt and determined spectrophotometrically. Any free dinoseb or its salts present in dinobuton are removed prior to the chromatographic stage using an aqueous alkali extraction.

b. *Reagents*

n-Butylamine solution, 1% in 1:1 acetone–water (v/v).
Xylene–hexane 2:3 v/v, solvent mixture I.
Acetonitrile–propan-2-ol 3:2 v/v, solvent mixture II.
Chloroform.
Sodium hydroxide, 1% aqueous solution.
Alumina, Woelm neutral grade I.

c. *Apparatus*

Volumetric flasks, 10 and 100 ml.
Pear-shaped separating funnels, 100 ml.
Pipettes.
Chromatographic column, a glass tube 10 cm long, 1.4 cm i.d. fitted with a tap.
Spectrophotometer, Pye Unicam SP 1800 or similar instrument (1 cm cells).

d. *Procedure*

i. *Preparation of Column*

Weigh 8 gm neutral grade 1 alumina, place a cotton wool support in the column, add the alumina and tamp it to a depth of about 5 cm. Place a cotton wool pad on top of the alumina and tamp it firmly with a glass rod.

ii. *Preparation of Standard Curve*

1. Prepare xylene–hexane solution of pure dinobuton to contain 10, 20, 30, 40, and 50 μg/ml.

2. Transfer 5-ml portions of these solutions to the chromatographic columns previously wetted with solvent mixture I.

3. Rinse the walls of the columns with 3 × 1 ml of solvent I; elute to waste.

4. Add 20 ml chloroform; allow it to wash through and then add 20 ml solvent II and allow to wash through. Discard these eluates.

5. Elute the colored band with the butylamine solution and start collecting the eluate when the colored band reaches a point 1 cm from the cotton wool support. Collect a total volume of 10 ml.

6. Read the absorbance of the solution against a reagent blank, similarly prepared, at 420 nm.

7. Plot absorbance versus concentration.

iii. *Assay*

1. *Solution A:* Weigh accurately about 0.5 gm of technical dinobuton into a 100-ml volumetric flask. Dissolve in solvent I and make to volume.

2. Pipette a 20-ml portion of solution A into a 100-ml pear-shaped separating funnel and extract this with 2 × 10 ml sodium hydroxide solution. Bulk and retain these aqueous extracts for the free dinoseb determination.

3. *Solution B:* Pipette 10 ml of the organic phase in the separating funnel (step 2) and transfer to 100-ml volumetric flask. Dilute to volume with solvent I.

4. *Solution C:* Withdraw a 10-ml portion of solution B and dilute to 100 ml volume with solvent I.

5. For analysis, transfer 3 ml solution C to the chromatographic column and proceed from step 3 in the preparation of the standard curve. Calculate the purity of the sample from the standard curve.

iv. *Determination of Free Dinoseb in Technical Dinobuton*

(a) *Outline of Method.* The absorbance of the alkaline solution of the dinoseb is measured at 420 nm. The dinoseb content is obtained from a calibration curve prepared from pure dinoseb.

(b) *Reagents.*

Dinoseb, pure.

Dinoseb, standard solutions. Prepare solutions in 1% aqueous sodium hydroxide to contain 1, 5, 10, and 15 μg/ml.

(c) *Procedure.*

Preparation of calibration graph: Measure the absorbance of each of the standard dinoseb solutions at 420 nm against 1% sodium hydroxide.

Determination of free dinoseb in the sample: Filter the bulked alkaline extracts retained from the method above, through a Whatman no. 2 filter paper and read the absorbance at 420 nm against 1% sodium hydroxide. Determine the dinoseb content from the calibration curve and calculate the dinoseb content in the sample.

B. Residue Analysis

The following method has been previously described by Crossley and Lynch (1968).

a. *Principle*

Dinobuton is extracted from crops by hexane or by a mixture of hexane and acetone. After removing the extracting solvent, the residue is dissolved in benzene, and this solution is subjected to a cleanup procedure involving acid potassium permanganate solution and chromatography on neutral alumina. Dinobuton undergoes deesterification on the alumina to dinoseb, and the latter is eluted as its butylammonium salt and determined spectrophotometrically.

b. *Apparatus*

Macerator.

Büchner flask, 1 liter.

Büchner sintered glass funnel.

Chromatographic column comprising a glass tube 10 cm long, 1.4 cm i.d., fitted with a tap at the lower end and a ground joint (B24) at the upper end to which is attached a 100-ml vessel serving as a reservoir.

c. *Reagents*

Benzene, Reagent Grade.

Hexane, b.p. 67–69°C.

Chloroform, Reagent Grade.

Acetonitrile–propan-2-ol 3:2 v/v.

Neutral alumina, grade 1 Woelm.

Acetone–water 1:1 (v/v), containing 1% *n*-butylamine

Potassium permanganate, 0.1 *M* in 1 *N* sulfuric acid.

Hydrogen peroxide, 20 vol.

Sodium sulfate, anhydrous, neutral.

d. *Procedure*

i. *Preparation of Column*

Weigh 8 gm neutral grade 1 alumina. Place a cotton wool support in the column, add the alumina, and tamp it to a depth of about 5 cm. Place a cotton wool pad on top of the alumina and tamp it firmly with a glass rod.

ii. *Preparation of Standard Curve*

Prepare a benzene solution of dinobuton to contain 10 μg/ml. Transfer portions containing solute in the 0–200 μg range to 500-ml pear-shaped separating funnels. Adjust the volume in each funnel to 200 ml with benzene. Add 50 ml 0.1 *M* potassium permanganate solution and shake the funnels vigorously for 5 minutes. Add sufficient 20 vol. hydrogen peroxide to destroy the excess potassium permanganate. Allow the phases to separate and run off the lower, aqueous layer into a 250-ml separating funnel. Extract this aqueous solution with 50 ml benzene and bulk the benzene extracts. Wash the combined benzene extract with 25 ml water and discard the washing. Dry the benzene solution over sodium sulfate, filter, and concentrate it to about 20 ml on a rotary-film evaporator. Transfer the concentrate to the chromatographic column, wash the flask three times with 5 ml benzene and add each wash to the column. Allow the benzene to drain through the column and then rinse the walls of the reservoir with about 5 ml benzene and allow the rinsing to drain through. Add 50 ml chloroform, allow it to wash through, and then add 25 ml acetonitrile–propanol mixture and allow it to wash through. Elute the colored band with the aqueous acetone containing 1% *n*-butylamine and start collecting the eluates when the colored band reaches a point 1 cm from the cotton wool support. Collect a total volume of 10 ml. Read the absorbance of the solutions against a reagent blank, similiarly prepared, at 378 nm. Plot absorbance versus concentration.

iii. *Extraction Procedure for Apples and Cucumbers*

Weigh 100 gm finely shredded sample into a macerator jar and add 250 ml hexane. Macerate the sample thoroughly for 2–3 minutes and then add sufficient anhydrous sodium sulfate to break any emulsion formed. Decant the supernatant liquid onto a sintered glass funnel containing a 1 cm layer of sodium sulfate. Repeat the extraction with a further 250 ml hexane. Finally, transfer the residue from the macerator jar onto the filter funnel and wash it with about 50 ml hexane. Evaporate the extract to incipient dryness on a rotary-film evaporator and remove final traces of solvent under a gentle stream of nitrogen. Dissolve the residue in 200 ml benzene and then proceed as for the standard curve, starting at the addition of permanganate stage.

iv. *Extraction Procedure for Black Currants*

The extraction procedure used for apples and cucumbers is inapplicable to black currants. Instead of hexane as extracting solvent, a 1:4 acetone–hexane mixture is used.

Weigh 100 gm fruit into the macerator jar, add 200 ml acetone–hexane, and macerate the mixture thoroughly for 2–3 minutes. Add 200 gm anhydrous sodium sulfate and mix it well in. Filter the supernatant liquid through a sintered glass funnel containing a 1-cm layer of sodium sulfate. Repeat the extraction with a further 200 ml acetone–hexane. Finally, transfer the residue to the filter funnel and wash it with about 50 ml solvent mixture. Evaporate the extract to incipient dryness on a rotary-film evaporator and remove the last traces of solvent under a gentle stream of nitrogen. Dissolve the residue in 200 ml benzene and proceed as for the standard curve, starting at the addition of permanganate stage.

v. *Additional Cleanup Technique*

Crop blanks tend to vary with variety and maturity of the crop and, where the control values are too high, the following additional cleanup may be employed.

Collect the acetone–water eluate in a 50-ml separating funnel. Acidify with hydrochloric acid, add 20 ml hexane, and shake it for 1 minute. Allow the phases to separate and discard the aqueous layer. Extract the hexane with 2 ml and then 1 ml 2% *n*-butylamine in acetonitrile. Collect each extract (lower layer) in a 5 ml volumetric flask and make it up to volume with acetone. Read the absorbance of the solution against a reagent blank at 378 nm. Refer to a standard calibration curve prepared by this method.

vi. *Recovery Experiments*

To determine the efficiency of extraction and overall recovery, the finely shredded crop samples were treated with hexane solution of dinobuton at suitable levels. The samples were then extracted as described above. Examples of recovery values obtained are given in Table I.

e. *Discussion*

Hexane was chosen as extracting solvent since it removed a minimum of interfering substances from crops and yielded satisfactory

TABLE I

RECOVERY OF DINOBUTON FROM APPLES, BLACK CURRANTS, AND CUCUMBERS[a]

Amount of dinobuton added (ppm) in recovery experiments	Recovery of dinobuton (%)		
	Apples	Black currants	Cucumbers
0.20	109, 100	93	83
0.30	–	–	98, 95
0.40	94, 100	–	–
0.60	–	97	98, 86
0.80	96	–	–
0.90	–	–	93
1.00	–	85, 85, 84	–
2.00	86, 90	–	–

[a] Control values gave < 0.05 ppm (6 determinations), except for black currants for which they gave 0.06 ppm.

recovery values. Emulsion problems normally encountered with solvents such as benzene rarely occurred with hexane. With black currants, the acetone–hexane mixture proved to be more satisfactory than hexane alone.

The cleanup step using potassium permanganate was first tried on the hexane extract, but this led to low recoveries since some degradation of dinobuton occurred. Benzene was chosen instead because, under the prescribed conditions, dinobuton is unaffected by potassium permanganate. In certain cases, e.g., surface deposits, the hexane extract or washing may be chromatographed directly on the alumina column without the preliminary potassium permanganate cleanup.

REFERENCES

Crossley, H., and Lynch, V. P. (1968). *J. Sci. Food Agr.* **19**, 57–59.

Pianka, M. (1966). *J. Sci. Food Agr.* **17**, 47–56.

Pianka, M., and Smith, C. B. F. (1965). *Chem. Ind.* (*London*) p. 1216.

17

Ethirimol

M. J. EDWARDS

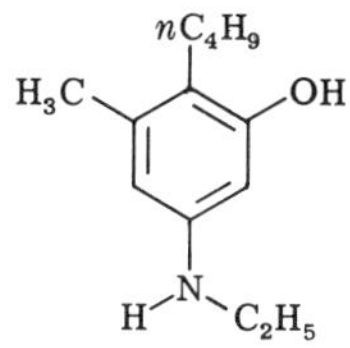

5-Butyl-2-ethylamino-4-hydroxy-6-methylpyrimidine

I. GENERAL

A. Empirical Formula

$C_{11}H_{19}N_3O$ (Mol. wt. 209.3).

B. Alternative Name

PP 149.

C. Source of Analytical Standard

ICI Plant Protection Division, Yalding, Kent, England.

D. Biological Properties

Ethirimol is a fungicide that is effective against powdery mildews of cereals and grasses, particularly *Erysiphe graminis hordei* of barley. It is also active against powdery mildew of certain dicotyledonous plants such as melons and cucumbers. It is systemic, being translocated in the xylem of plants, and is particularly effective when applied in the root zone as a seed dressing. It is also active as a foliar spray. It has no significant insecticidal properties and little effect upon soil fauna.

Ethirimol is a compound of low toxicity. The percutaneous LD_{50} for the rat is greater than 1000 mg/kg. In 90-day feeding studies with rats, a no-effect level of 200 ppm in the diet, equivalent to 12 mg/kg body weight, was found. In 2-year feeding studies with dogs, the no-effect level was 30 mg/kg body weight. It is safe to bees and birds, and the

TABLE I
ACUTE ORAL LD_{50} VALUES FOR ETHIRIMOL

Species	LD_{50}(mg/kg)
Rat (female)	4000
Mouse (female)	4000
Rabbit (male)	1000–2000
Guinea pig (female)	500–1000
Cat (female)	>1000
Hen (adult)	>4000

mean toxic level for brown trout fingerlings (96-hour exposure) is 20 ppm. The acute oral LD_{50} values for ethirimol are shown in Table I.

E. History

Ethirimol was introduced in 1968 by ICI Plant Protection Ltd. as a selective fungicide under the code number PP 149, trademark Milstem, and the protection of British patent 1,182,584. Its properties were described by Bebbington *et al.* (1969).

F. Physical Properties

Pure ethirimol is a white crystalline solid with melting point 159–160°C (phase change at 140°C); approximate solubility at room temperature: water 0.02%, acetone 0.5%, ethanol 2.4%, chloroform 15%; it is soluble in aqueous solutions of strong acids and bases. It has a vapor pressure of 2.3×10^{-6} mm Hg at 25°C and a density of 1.21 g/cm^3. The ultraviolet absorption spectrum is dependent on pH; in 0.1 *M* aqueous hydrochloric acid it has an absorption maximum at 270 nm. In methanol, the absorption maximum occurs at 297 nm ($\epsilon = 7820$).

G. Chemical Properties

1. METHOD OF SYNTHESIS

Ethirimol is prepared by condensation of ethyl-α-n-butyl acetoacetate with ethylguanidine.

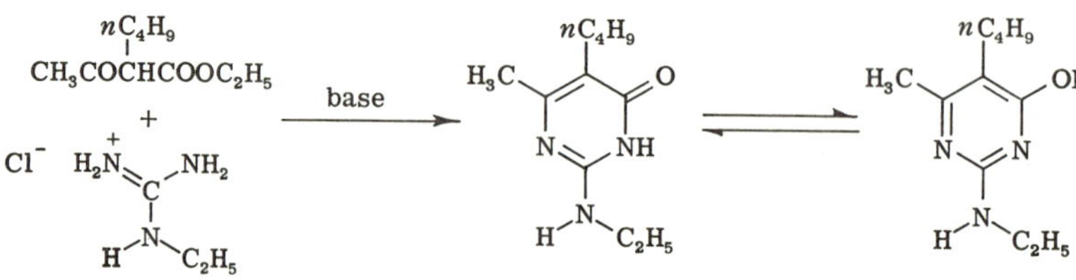

2. CHEMICAL STABILITY

Ethirimol is stable to heat, alkali, and acid. It is soluble in aqueous solutions of strong acids and bases. It is classed as a hydroxypyrimidine, but it exists predominantly in solution and as a solid in the tautomeric oxoform (see Note 5).

H. Formulations

Milstem Seed Dressing, an aqueous suspension of 500 gm/liter or 580 gm/liter used as a seed dressing on cereals.

Milgo, an aqueous suspension of 250 gm/liter used as a spray on cereals.

Milgo E, an aqueous suspension of 280 gm/liter used as a spray on cereals.

Milcurb Super, an aqueous suspension of 250 gm/liter used as a spray on cucurbits.

II. ANALYSIS

A. Formulation Analysis

1. REVIEW OF METHODS

Ethirimol may be determined by measurement in the ultraviolet at 297 nm, and by gas–liquid chromatography (GLC).

For ultraviolet determination the technical or formulated material is dissolved or extracted with hydrochloric acid. Ethirimol is then partitioned from the buffered (pH 6.5) solution into chloroform. The chloroform is evaporated and replaced with methanol, in which the ultraviolet absorption of ethirimol is measured at 297 nm ($\epsilon = 7820$).

A quantitative thin-layer chromatographic procedure has also been developed. After elution on Kieselgel GF254 plates, the band containing the pesticide is removed and extracted with methanol. The ultraviolet absorbance of the methanolic extract is measured at 297 nm. A chloroform–acetone–glacial acetic acid mixture (75:10:15 v/v) is used as the developing solvent. The thin-layer method will separate ethirimol from impurities, and its specificity should be comparable with that of gas–liquid chromatography. The ultraviolet absorption methods were developed by S.H. Yuen at The Jealott's Hill Research Station of ICI Plant Protection Ltd., Berkshire, England.

Ethirimol is determined by gas–liquid chromatography as its trimethylsilyl ether (Bagness and Sharples, 1974). Because of its specificity and ease of application, it is the preferred method for analyzing technical and formulated ethirimol products.

2. RECOMMENDED METHOD

a. *Principle*

Ethirimol, dissolved in a solution containing nonadecane as an internal standard is silylated with BSA [*N,O*-bis(trimethylsilyl)acetamide]. The solution is injected into a column packed with 100–120 mesh Celite 545 coated with 10% E301 silicone elastomer. The quantity of ethirimol present in the sample is obtained by comparing the ethirimol silyl ether and nonadecane peak areas with peak areas obtained from a calibration solution containing known amounts of pure ethirimol and nonadecane.

b. *Reagents*

Analytical standard, obtainable from ICI Plant Protection Division, Yalding, Kent, England.

Chloroform, analytical reagent.

Hydrochloric acid, analytical reagent, concentrated (11 *M*).

Silicone elastomer E301, available from gas chromatography suppliers.

BSA [*N,O-Bis(trimethyl silyl)acetamide*]. Because this material is decomposed by moisture, it is conveniently purchased and stored in Hypo-vials fitted with a septum cap. Available from Phase Separations Ltd., U.K., or Pierce Chemical Co., Rockford, Illinois.

Silyl 8, a silanizing agent for column treatment. Available from Phase Separations Ltd., U.K., or Pierce Chemical Co., Rockford, Illinois.

Nonadecane, pure.

Celite 545, 100–200 mesh, acid-treated. Commercially available grades of acid-treated Celite may be suitable but should be checked for residual acidity. By using column packing prepared as below, chromatograms should be reproducible and give peaks with negligible tailing. Distorted peaks and excessive retention times may indicate that the Celite was insufficiently washed following the acid treatment.

> Unwashed grades are treated as follows: Weigh 250 gm 100–120 mesh Celite 545 into a 2-liter beaker and add concentrated hydrochloric acid until the Celite is covered with a 2-inch layer of acid. Heat the mixture on a boiling water bath in a fume hood for 2–3 hours, stirring occasionally. Decant the acid and repeat the treatment with fresh acid. Filter off the Celite into a sintered-glass funnel, wash it several times with distilled water, transfer it into a clean 2-liter beaker, and stir it for 15 minutes with about 1 liter of distilled water. Filter off the Celite and repeat the procedure with fresh water until the filtrate remains neutral after stirring the mixture for 15 minutes.

Filter off the Celite and dry it overnight in an oven at 130°C. Sieve the dried Celite through 100 and 120 mesh sieves, and use the portion retained on the 120 mesh sieve to prepare the column packing.

Nitrogen, hydrogen, and air for gas chromatography.

c. *Apparatus*

Gas chromatograph fitted with a flame ionization detector. A suitable instrument is the Pye series 104, model 4 gas chromatograph equipped for flame-ionization detection and single-column operation with on-column injection. A Hewlett-Packard model 3370A integrator was used for peak area evaluation and the chromatograms were monitored with a 10 mV FSD potentiometric recorder.

Column, 150 cm × 4 mm i.d. glass.

Column packing. 10% silicone elastomer E301 coated on 100–120 mesh acid-treated Celite 545 prepared by dissolving 5 gm silicone elastomer E301 in 200 ml chloroform, which is most conveniently achieved by allowing the mixture to stand overnight in a stoppered flask. Add 45 gm of acid-treated Celite to the solution of stationary phase and distill off the chloroform under vacuum on a water bath by using a rotary evaporator. Remove the last traces of solvent by drying the packing in an open dish at 100°C. Sieve the dried material, retaining the 100–120 mesh fraction.

Syringe, 10 μl Hamilton or Terumo.

d. *Experimental Procedure*

i. *Gas–Liquid Chromatographic (GLC) Conditions*

The exact conditions for the analysis by GLC will depend upon the equipment available. The following operating conditions have been found to be satisfactory using a Pye series 104, model 4, gas chromatograph.

Column oven temperature, 220°C.

Carrier gas, nitrogen at 50 ml/minute measured at the detector outlet.

Detector gases, hydrogen at 50 ml/minute, air at 750 ml/minute, both measured at the detector outlet.

Column pretreatment. Before use, condition a freshly-packed column by heating overnight at 300°C with carrier gas flowing. Do not connect the column to the detector during this procedure. With the column temperature at 220°C inject three successive 10 μl portions of Silyl 8 at 1 minute intervals. The detector flame should not be alight during this treatment.

Attenuation and integrator settings. Select an appropriate setting to give for the standard injection volume about 80% full-scale deflection on a 10 mV recorder for peak area evaluation or an integrator count of about 1000×10^2 μV-seconds for each major peak. Settings for the Hewlett-Packard 3370A integrator were: slope sensitivity (up) 0.3 mV/minute, slope sensitivity (down) 0.1 mV/minute, baseline reset delay zero.

ii. *Preparation of Working Solutions*

Internal standard solution. Weigh accurately 1.00 gm nonadecane and dissolve it in 100 ml chloroform.

Ethirimol standard solution. Weigh accurately about 0.15 gm pure ethirimol into a 50-ml glass-stoppered test tube. Add 20.0 ml of the internal standard solution from a burette and shake to dissolve the ethirimol.

Sample solution. Weigh accurately sufficient sample to contain 0.15 gm ethirimol into a 50-ml glass-stoppered test tube. Add 20.0 ml internal standard solution from a burette and shake for 1 minute to dissolve the ethirimol. Allow sufficient time for any solid matter to settle, or in the case of aqueous suspensions, for the aqueous and chloroform layers to separate.

iii. *Determination of Ethirimol*

Transfer 1 ml of the sample and ethirimol standard solutions to separate small glass-stoppered tubes. Add 0.1 ml of BSA reagent to each tube, stopper, shake vigorously and allow to stand for 10 minutes. Inject three 1 μl portions of the ethirimol standard solution into the gas chromatograph to equilibrate the system. Inject 1 μl portions of the ethirimol standard and sample solutions into the gas chromatograph in turn. Record the integrator counts for the ethirimol and internal standard peaks for each solution. Alternatively, the peak areas from the chart-recorder traces can be calculated from the product of peak height and peak width at half-height.

iv. *Calculations*

$$\%\ (\text{w/w})\ \text{ethirimol} = \frac{A}{B} \times \frac{C}{D} \times \frac{E}{W} \times 100$$

where A = mean integrator count for sample ethirimol
B = mean integrator count for sample nonadecane
C = mean integrator count for standard nonadecane
D = mean integrator count for standard ethirimol
E = weight in grams of ethirimol in standard solution
W = weight in grams of sample.

If a liquid formulation is analyzed, the ethirimol content in grams per liter is

$$\frac{A \times C \times E \times F \times 1000}{B \times D \times W}$$

where F = density at 20°C in grams per milliliter.

e. *Discussion*

Replicate analyses of manufactured technical ethirimol showed that the standard deviation of the GLC procedure was less than ±2%. Analysis of an accurately prepared standard formulation of ethirimol (about 40.6% w/w aqueous suspension) gave a satisfactory recovery (mean of 99.2%) of ethirimol. The specificity of the GLC procedure can be utilized for the analysis of low-purity samples or materials that have been degraded, for example, as a result of storage at elevated temperatures.

B. Residue Analysis

1. RECOMMENDED METHOD

a. *Principle*

The analysis of ethirimol residues involves the following stages: (a) Crop samples are extracted with methanol; soils require a mixed solvent of acetone and aqueous ammonium chloride. (b) Interfering materials are removed by solvent partition using hexane and diethyl ether; column chromatography on silica gel may also be applied if necessary. (c) Ethirimol is converted to a more volatile methyl derivative by reaction with diazomethane. (d) The methyl derivative is determined by GLC using a nitrogen-selective thermionic detector.

b. *Reagents*

Analytical standard, obtainable from ICI Plant Protection Division, Yalding, Kent, England.

Acetone, redistilled.

Chloroform, redistilled.

Hexane, redistilled.

Methanol, redistilled.

Diethyl ether, analytical reagent. Check that this solvent does not produce interfering peaks on the gas chromatograph at the retention time of the methyl derivative of ethirimol. Concentrate 10 ml solvent to 0.1 ml in a centrifuge tube using a stream of clean dry air and inject 5 μl of the solution into the gas chromatograph. Serious contamination of this solvent can sometimes be caused by the plastic bottle tops and/or seals.

Hydrochloric acid, analytical reagent, concentrated (11 *M*).

Celite 545 filter aid, available from Johns Mansville Corp.

Sodium bicarbonate, solid.

Sodium sulfate, anhydrous granular.

Ammonium chloride, solid.

Silicone gum rubber, UCW-982 (methyl vinyl) available from gas chromatography suppliers.

Chromosorb W(AWDCMS), 80–100 mesh, available from gas chromatography suppliers.

Helium, hydrogen, and air for gas chromatography.

Silica gel, a suitable grade is Kieselgel 0.05–0.2 mm (for column chromatography) available from E. Merck, Darmstadt, West Germany. The adsorbent is used as received and usually contains 3.5% (w/w) moisture as determined by drying at 110°C overnight.

Diazomethane. Prepare an ethereal solution of diazomethane using a standard method of preparation (DeBoer and Backer, 1956). Store the diazomethane solution in a closed glass vessel (use a tight cotton-wool plug—**not** a glass stopper) at −10°C or below, and use only behind a safety screen in a well ventilated fume hood.

> **CAUTION:** *Treat diazomethane solutions with great care as the chemical is carcinogenic and potentially explosive.*

Standard solutions of ethirimol, 5 μg/ml, 10 μg/ml, 50 μg/ml in redistilled methanol.

c. *Apparatus*

Crop preparation equipment suitable for crushing or grinding grain and straw to a fine powder.

Soil extraction equipment, e.g., end-over-end tumbler, roller, or high-speed shaker.

Macerator or blender, e.g., Sorvall Omni-Mixer Homogenizer.

Glass bottles for soil extraction, fitted with screw-caps sealed with aluminum disks.

Graduated tubes, 10 ml capacity accurately calibrated to 1.0 ml in 0.1 ml units.

Chromatography columns for cleanup, glass, 1 cm i.d. and 20-cm long fitted with glass or Teflon taps.

Syringe, 10 μl Hamilton fixed-needle type.

Gas chromatograph fitted with a nitrogen-selective thermionic detector. A suitable instrument is the Hewlett-Packard 5750 series chromatograph fitted with a rubidium bromide nitrogen detector model no. 15161A.

d. *Experimental Procedure*

i. *Gas–Liquid Chromatographic (GLC) Conditions*

The exact conditions for the analysis by GLC will depend upon the equipment available. The following operating conditions have been found to be satisfactory using a Hewlett-Packard Series 5755 gas chromatograph.

Column: 180 cm × 2 mm i.d. stainless steel.

Packing: 10% (w/w) silicone gum rubber UCW-982 on 80–100 mesh Chromosorb W(AWDCMS). Other nonpolar or low-polarity columns may also be used, e.g., 5% (w/w) DC200 silicone oil on 100–120 mesh Chromosorb W.

Column temperature: 200°C.

Injector temperature: 200°C.

Carrier gas: Helium at 60 ml/minute measured at the detector outlet.

Detector: Nitrogen-selective rubidium bromide thermionic detector operated at 350–370°C with a hydrogen flow rate of 25 ml/minute and air flow rate of 180 ml/minute.

Using the experimental conditions described the retention time of the *O*-methyl (—OCH_3) derivative of ethirimol was 4.2 minutes. An injection of 25 ng of the derivative gave approximately 40% full-scale deflection (FSD) at an attenuation of 8×10^2 (with a noise level of $<0.5\%$ FSD).

ii. *Extraction and Partition Cleanup*

Thoroughly mix the finely chopped or ground sample and weigh a representative portion (100 gm grain and other crops, 20 gm straw) into a macerating jar. Macerate the sample with 200 ml of methanol for 10–15 minutes. Filter through a filter paper (Whatmann no. 1) in a Büchner funnel and wash the crop residuum with further methanol to a final volume of 400 ml (500 ml for straw).

Transfer a portion of the filtrate (80 ml equivalent to 20 gm of grain, etc; 250 ml equivalent to 10 gm of straw) into a round-bottom flask and add 10 ml of water. Concentrate on a rotary evaporator at 50–60°C until no more methanol evaporates and aqueous solution (approximately 10 ml) remains.

Note 1. For soil samples, place a representative portion (20 gm) of the undried soil in a screw-capped bottle and add 50 ml of an acetone–1 *M* aqueous ammonium chloride 1:1 (v/v) solution. Extract by high-speed shaking or rolling for 30 minutes. Filter through a Celite pad (on a

Whatmann no. 1 filter paper) in a Büchner funnel and wash the soil with two 25-ml volumes of acetone–ammonium chloride. Add 5 ml water to the filtrate and concentrate by rotary evaporation at 50–60°C until no more methanol evaporates. Proceed as for crop samples below.

Quantitatively transfer the aqueous solution into a 250 ml separatory funnel using three 5 ml water portions to wash the round-bottom flask. Make the solution acid (pH 1–2) by adding 3–4 drops concentrated hydrochloric acid. Wash the round-bottom flask with 25 ml diethyl ether, adding the washings to the acidified solution, and shake for 1 minute. Discard the diethyl ether. Shake the acid solution with 25 ml hexane for 1 minute and discard the hexane layer.

Neutralize the acid solution (to pH 6–7) by the careful addition of small amounts of solid sodium bicarbonate and extract the ethirimol from the neutralized solution with three 50-ml portions of diethyl ether. Dry the combined ether extracts over 30 gm of anhydrous sodium sulfate and filter through a small plug of sodium sulfate into a 250 ml measuring cylinder. Wash the separatory funnel and sodium sulfate with diethyl ether to a total volume of 150 ml.

Note 2. It is essential, particularly during the first ether partition, to continuously release any pressure increase that occurs in the separatory funnel.

Note 3. To obtain a satisfactory recovery of ethirimol, it is recommended that the above partition procedure be completed in the same day. The final diethyl ether extract may be stored in a refrigerator for several days without loss of ethirimol.

Evaporate one half of the dried diethyl ether extract (equivalent to 10 g of grain or soil) just to dryness at 50–60°C. (For straw samples evaporate the whole of the extract.)

Note 4. No drops of water should remain after evaporation of solvent, particularly if the column chromatographic cleanup is used. If this happens, diethyl ether should be added and the solution dried over anhydrous sodium sulfate before repeating the evaporation to dryness.

Dissolve the residue in the flask by swirling with 1–2 ml of 20% (v/v) methanol in diethyl ether. Transfer the solvent, using a Pasteur pipette, to a 10-ml graduated tube. Repeat with further small volumes of the washing solvent to a total collected volume of 5–6 ml. Evaporate the solution to 1 ml using a stream of clean dry air.

iii. *Methylation and GLC Determination*

Add to each graduated tube 2 ml of a freshly prepared ethereal solution of diazomethane. Plug the tubes with cotton-wool and allow to stand at room temperature for at least 1 hour before evaporating to 1.0 ml using a stream of clean dry air. **CAUTION:** *All operations involving diazomethane must be carried out in a well-ventilated fume hood.* React duplicate 1-ml portions of standard solutions of ethirimol in methanol with diazomethane simultaneously with the sample extracts. Convenient quantities are 5 μg, 10 μg, and 50 μg of ethirimol. Evaporate to 1 ml after standing for at least 1 hour.

Inject a suitable portion (2–4 μl) of the sample extract into the gas chromatograph. Measure the height of the peak at the retention time of the methyl derivative (—OCH_3) of ethirimol. Calculate the weight of ethirimol in the sample by comparison with peak heights obtained from the standard solutions of the methyl derivative injected into the chromatograph before and after the residue samples.

Note 5. In solution ethirimol is best represented as the equilibrium

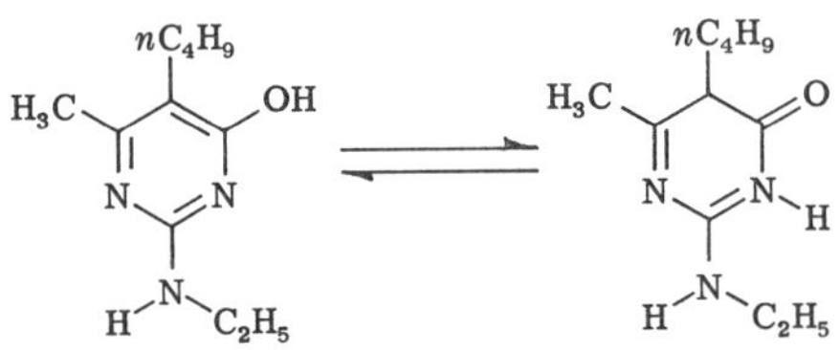

Diazomethane reacts with both tautomers to form the corresponding —OCH_3 and —NCH_3 derivatives. The reaction is repeatable with a consistent ratio of both products formed in the presence of a variety of crop and soil extracts. The —OCH_3 compound is the least polar and more volatile derivative and is used for quantitative determination. Using the experimental GLC conditions detailed above the retention time for the *O*-methyl derivative is 4.2 minutes compared with 11.2 minutes for the *N*-methyl compound.

iv. *Column Chromatographic Cleanup*

Some samples, particularly straw, may contain coextracted materials that interfere with the GLC determination. The following column chromatographic cleanup can be used for these samples.

Add 3 gm of silica gel (see Section b above) in small amounts to a glass chromatography column (1 cm i.d. × 15–20 cm long fitted with a tap and plugged with cotton-wool) half filled with diethyl ether. Allow the excess solvent to drain to the level of the adsorbent. Take a portion (normally 75–150 ml, equivalent to 10 gm crop or soil) of the dried ether extract of the neutralized aqueous solution (Section ii, above) and allow it to percolate through the adsorbent. Elute the column with the mixed solvent methanol–acetone–diethyl ether (1:2:7 v/v) and collect the fraction containing ethirimol. Evaporate to 1.0 ml in a graduated tube and proceed with the methylation and GLC determination as described above.

Note 6. Each batch of silica gel must be calibrated before use. This is done by passing 100 ml of a standard solution (1 μg/ml) of ethirimol in dry diethyl ether through the prescribed column, eluting with the mixed solvent and analyzing 5–10 ml fractions for compound recovery. Using Merck Kieselgel (0.05–0.2 mm), which contained 3.5% (w/w) moisture, the ethirimol emerged in the first 60 ml of eluate.

e. *Recovery and Sensitivity*

Recovery experiments should be carried out by adding known amounts of ethirimol to untreated samples before extraction. The amounts added should be similar to the residue levels expected in the treated samples. Recovery experiments on a large number of grain, straw, apple, cucumber, melon, and soil samples have shown that consistently high recoveries of 70–120% (average of near 100%) can be expected using the described analytical method at the 0.1–5 ppm level.

The sensitivity of the method was found to be 0.05 ppm for grain, apples, cucumbers, melons, and soil, and 0.1–0.2 ppm for straw. The sensitivity depends on the noise level of the chromatograph (arising mainly from the nitrogen detector) and particularly the presence and variation of background material in the sample extract. In the absence of background, the sensitivity was based on four times the noise level of the chromatographic trace at the retention time of the *O*-methyl derivative of ethirimol.

f. *Discussion*

The GLC procedure has been applied successfully to a variety of crops and soils. Most samples can be analyzed to a sensitivity of 0.05 ppm without the need for column chromatographic cleanup. Some cereal straw samples have required the additional cleanup, but generally a sensitivity of 0.1 ppm is readily obtained.

Ethirimol can be chromatographed directly on several polar stationary phases, but its polar nature causes severe tailing. The hydroxy grouping on the pyrimidine ring (of ethirimol and related compounds) is amenable to derivatization with several reagents. Thus, a silyl derivative is utilized for the GLC analysis of technical and formulated materials. The *O*-methyl derivative has been found to be the most convenient for practical residue analysis. It effects appreciable cleanup of the sample extracts and does not cause problems from excess reagent or added impurities. The reaction is repeatable with a consistent ratio of the *O*-methyl and *N*-methyl products formed, and the former more volatile compound is used for quantitation. The rubidium bromide thermionic nitrogen detector used for this analysis has given good results with consistent and repeatable response to nitrogen (to less than 0.5 ng ethirimol) over several years routine use.

REFERENCES

Bagness, J. E., and Sharples, W. E. (1974). *Analyst* **99,** 225–232.

Bebbington, R. M., Brooks, D. H., Geoghegan, K. J., and Snell, B. K. (1969). *Chem. Ind. (London)* p. 1512.

DeBoer, T. J., and Backer, H. J. (1956). *Org. Syn.* **36,** 16–18.

Snell, B. K., Elias, R. S., and Freeman, P. F. H. (1967). British patent 1, 182, 584.

18

Mertect®

J. S. Wood, Jr.

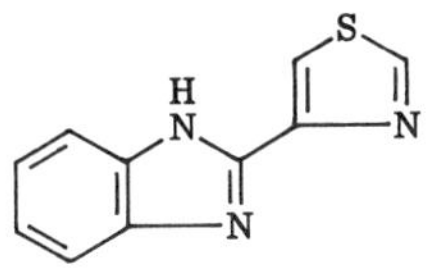

2-(4′-Thiazolyl)benzimidazole

I. GENERAL

A. Empirical Formula

$C_{10}H_7N_3S$ (Mol. wt. 201.26).

B. Alternate Names

Tecto®, Thiabendazole, Thiaben®, Thibenzole®, Bovizole®, Eprofil®, Equizole®, Mintezol®, Omnizole®.

C. Source of Analytical Standard

Merck, Sharp and Dohme Research Laboratories, Rahway, New Jersey.

D. Biological Properties

Mertect® is effective for the control of many phytopathogenic fungi by foliar application at 3–6 oz a.i./acre. It is also effective for postharvest treatment of fruit and vegetables for the control of storage diseases. Activity has been demonstrated against a variety of fungi, including *Aspergillus, Botrylis, Ceratocystis, Cercospora, Colletotrichum, Cytospora, Fusarium, Penicillium, Rhizoctonia, Sclerotinia, Thielaviopsis,* and *Verticillium.*

E. History

2-(4′-Thiazolyl)benzimidazole has been in use as an anthelminthic since 1962 and as an agricultural fungicide since 1968. Its antifungal properties were reported by Robinson *et al.* (1964) and by Staron and Allord (1964). It is covered by U.S. patent 3,017,415 issued to Merck & Co., Inc., 1962.

F. Physical Properties

2-(4′-Thiazolyl)benzimidazole is a white, crystalline solid with melting point 304–305°C (sublimes at 310°C); solubility of 3.84% in water at pH 2.2 (point of maximum solubility, which decreases at higher and lower pH); is soluble in dimethylformamide and slightly soluble in alcohols, esters, and chlorinated hydrocarbons.

G. Chemical Properties

1. METHOD OF SYNTHESIS

2-(4′-Thiazolyl)benzimidazole may be prepared by the reaction of 4-thiazolecarboxamide with *o*-phenylenediamine in polyphosphoric acid.

2. CHEMICAL REACTIONS

2-(4′-Thiazolyl)benzimidazole is stable in aqueous solution regardless of pH.

H. Formulations

Mertect® is available as 40, 60, and 90% wettable powders; as a 40% flowable suspension; and a thermal fumigation tablet containing 7 gm of Mertect®.

I. Toxicity

LD_{50} orally in mice 3.6 gm/kg, orally in rats 3.1 gm/kg; orally in rabbits > 3.8 gm/kg (Robinson *et al.* 1965).

II. ANALYSIS

A. Formulations Analysis

1. REVIEW OF METHODS

Mertect® has generally been determined in formulated products by UV spectrophotometry. Fluorometry, gas chromatography, liquid chromatography, and colorimetric analyses have also been suggested for its analysis.

2. RECOMMENDED METHOD

a. *Principle*

The preferred method for analyzing formulations of Mertect® is UV spectrophotometry. The method is quick, relatively simple, and fairly specific.

b. *Reagents*

Analytical standard, obtainable from Merck, Sharp and Dohme Research Laboratories, Rahway, New Jersey.

Hydrochloric acid, 37%

Hydrochloric acid, 0.1 *N*. Dilute 8.3 ml 37% hydrochloric acid to 1 liter and mix thoroughly.

Reference standards.

Solution A. Weigh exactly 250 mg of standard 2-(4′-thiazolyl)benzimidazole into a 250-ml volumetric flask. Completely dissolve in about 150 ml 0.1 *N* hydrochloric acid. Dilute to the mark with the acid and mix thoroughly. Each milliliter contains 1000 μg of standard.

Solution B. Dilute solution A exactly tenfold and mix well. Each milliliter contains 100 μg of standard.

Solution C. Dilute solution B exactly tenfold and mix well. Each milliliter contains 10 μg of standard.

c. *Apparatus*

Centrifuge capable of centrifuging 50-ml tubes at 2000 rpm.

Flasks, 100, 200, 250, 500, and 1000 ml volumetric.

Centrifuge tubes, 50 ml.

Pipettes, 5 ml and 10 ml volumetric.

Spectrophotometer, Cary model 11, or equivalent recording spectrophotometer, with 1-cm square silica cells.

d. *Experimental Procedure*

i. *Procedure for Suspensions*

To insure proper sampling, thoroughly mix the suspension with a magnetic stirrer before pipetting. Make accurate dilutions based on Table I with 0.1 *N* hydrochloric acid according to the expected concentration. Mix for 10–15 minutes with a magnetic stirrer. Centrifuge if cloudy. With 0.1 *N* hydrochloric acid in both sample and reference cells, zero the instrument at the 303 mm absorbance peak for Mertect®. Replace the acid in the sample cell with reference standard solution C and record the spectrum from 350 mm to 240 mm. Replace the standard with the diluted unknown solution and repeat the measurement. Draw

TABLE I

DILUTION FACTORS FOR MERTECT® SUSPENSIONS

Concentration of suspension (ppm)	Dilution to be made (D)
100–250	20×
250–600	50×
500–1250	100×
1000–2500	200×
2500–6000	500×
5000–10,000	1000×

an extrapolated baseline for both curves as shown in Fig. 1. Read the absorbance at the 303 mm peak A_p and the absorbance at the same wavelength on the extrapolated baseline A_b. Calculate parts per million of Mertect® from the following equation:

$$\text{ppm} = \frac{(A_{px} - A_{bx})}{(A_{ps} - A_{bs})} \times C_s \times D$$

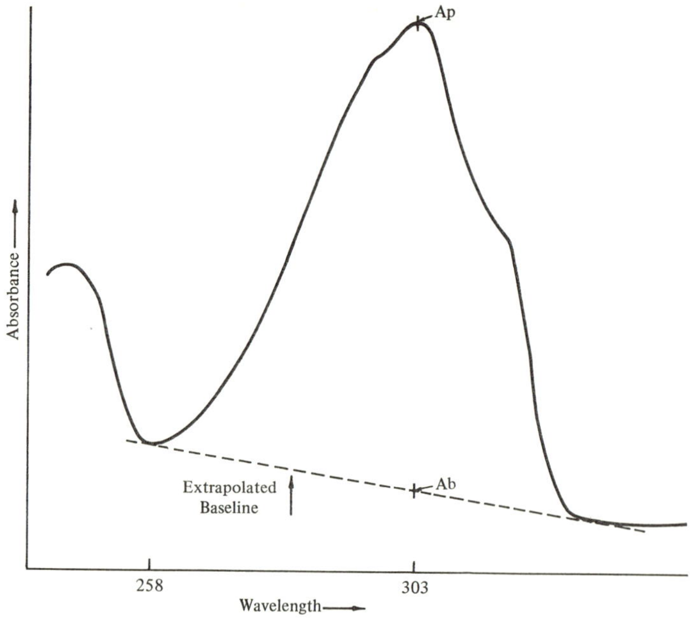

FIG. 1. Ultraviolet absorption spectrum of Mertect® in 0.1 N HCl. A_p is peak absorbance; A_b is extrapolated baseline absorbance at wavelength of the peak absorbance. These values are used to calculate the concentration of formulations (see Section II, A,2,d,i).

TABLE II
DILUTION FACTORS FOR MERTECT® SOLID FORMULATIONS

Percent in formulation	Dilution to be made (D)
2–5	5×
5–10	10×
10–20	20×
20–40	40×
40–60	50×
60–100	100×

where A_{px} = peak absorbance for unknown
A_{ps} = peak absorbance for standard
A_{bx} = absorbance from extrapolated baseline for unknown
A_{bs} = absorbance from extrapolated baseline for standard
C_s = concentration of standard in ppm (10 μg/ml)
D = dilution factor for unknown.

ii. *Solid Formulations*

Weigh exactly 1 gm of the formulation into a 1000-ml volumetric flask. Add 0.1 N hydrochloric acid to the mark and stir 10–15 minutes with a magnetic stirrer. Centrifuge a portion of the solution. Make an exact dilution depending on the formulation according to Table II of the clear centrifugate with 0.1 N hydrochloric acid and mix thoroughly. Proceed with the spectrophotometric measurement and extrapolated baseline drawing as for the suspensions. Calculate the concentration of Mertect® as follows:

$$\%\text{Mertect}^{®} = \left(\frac{A_{px} - A_{bx}}{A_{ps} - A_{bs}} \times C_s \times \text{D}\right) : 10W_x$$

where A_{px}, A_{ps}, A_{bx}, A_{bs}, and C_s are as above. D = dilution factor from Table II; W_x = grams of formulation taken.

B. Residue Analysis

1. REVIEW OF METHODS

Mertect® on fruits, vegetables, and other crops has generally been determined after extraction with solvents such as ethyl acetate and chloroform. Usually, the method of detection is either fluorometry or spectrophotometry. Gas and liquid chromatography have been suggested for its analysis, but the methods have not been completely developed.

2. RECOMMENDED METHOD

a. *Principle*

Mertect® is extracted from an aqueous homogenate of the vegetative matter with ethyl acetate. The extract is cleaned up by a series of solvent–solvent extractions, and the amount of Mertect® in the extract is determined by fluorescence measurement.

b. *Reagents*

Analytical standard, obtainable from Merck, Sharp and Dohme Research Laboratories, Rahway, New Jersey.

Stock solution. Accurately weigh 0.025 gm of analytical standard into a 250-ml volumetric flask. Add about 150 ml of 0.1 *M* hydrochloric acid and agitate until the solid completely dissolves. Dilute to the mark with the 0.1 *M* acid and thoroughly mix. (100 μg per milliliter).

Working standard solutions: (A) Dilute 5 ml of the stock solution accurately to 250 ml with the hydrochloric acid (2 μg per milliliter). (B) Dilute 25 ml of solution A accurately to 100 ml with the 0.1 *M* hydrochloric acid (0.5 μg per milliliter). (C) Dilute 2 ml of solution A accurately to 100 ml with the 0.1 *M* hydrochloric acid (0.04 μg per milliliter). (D) Dilute 5 ml of the stock solution accurately to 250 ml with distilled water (2 μg per milliliter).

Sodium acetate, Reagent Grade, anhydrous.

Sodium chloride, Reagent Grade, anhydrous.

Sodium pyrophosphate, Reagent Grade, decahydrate.

Hydrochloric acid, 37%

Buffer, pH 4.5. Weigh 24.6 gm sodium acetate into a 4-liter beaker. Add by graduate 3 liters distilled water. Stir with a magnetic stirrer to completely dissolve. Adjust the pH to 4.5 with 37% hydrochloric acid and a pH meter.

Buffer, pH 6.0. Weigh 133.8 gm sodium pyrophosphate into a 2-liter beaker. Add enough distilled water to make a volume of 1 liter. Stir with a magnetic stirrer to completely dissolve. Adjust the pH to 6.0 with 37% hydrochloric acid and a pH meter.

Ethyl acetate, glass distilled as from Jackson and Burdick.

Sodium hydroxide, 50% w/v

Sodium hydroxide, 1 M. Add 50 ml 50% sodium hydroxide to 950 ml distilled water and mix thoroughly.

Sodium hydroxide, 0.05 M. Dilute 50 ml 1 *M* sodium hydroxide to 1 liter with distilled water and mix thoroughly.

Hydrochloric acid, 0.1 M. Add 25 ml 37% hydrochloric acid to 2975 ml distilled water and mix thoroughly.

c. *Apparatus*

Magnetic stirrer.

pH meter.

Blender, Waring Blendor or equivalent.

Flasks, 100 ml and 250 ml volumetric.

Centrifuge tubes, 50 ml, stoppered.

Shaker, mechanical

Pipettes, 2 ml, 3 ml, 5 ml, 10 ml, 20 ml, and 25 ml volumetric and 1 ml graduated.

Beakers, 2 and 4 liter.

Balances capable of weighing 0.025 gm, 20 gm, and 100 gm accurately.

Centrifuge capable of centrifuging 50-ml centrifuge tubes at 2000 rpm.

Syringes, 20 ml.

Pipettes, disposable.

Fluorometer, such as Aminco-Bownan or Baird-Atomic model SF-100.

d. *Experimental Procedure*

i. *Sample Preparation*

The selection of vegetative matter must necessarily vary with species. In general, the material should be selected as randomly as possible as to the top, bottom, and side of the container or bunch. A large enough sampling should be taken to be as representative of the whole sample as possible.

Macerate thoroughly a representative 100-gm sample in a blender for 1 minute with an equal weight of pH 4.5 buffer. Quickly weigh 20 gm of the homogenate into a 50-ml centrifuge tube, resuspending the homogenate during the weighing if any settling tendency is noted.

ii. *Solvent–Solvent Extractions*

Weigh about 3 gm sodium chloride into the tube, stopper, and shake for 10 minutes by mechanical shaker. Pipette 10 ml ethyl acetate into the tube, stopper, and shake 10 minutes. Centrifuge for 5 minutes or until completely separated. Move the upper layer by syringe to a second 50-ml tube. Avoid the lower layer completely. Repeat the ethyl acetate extraction twice more, combining the three ethyl acetate layers. Wash the combined ethyl acetate layer twice with 5 ml each of 0.05 M sodium hydroxide, discarding the lower or aqueous phase in each case by syringe or disposable pipette. Add 5 ml of 0.1 *M* hydrochloric acid to the washed ethyl acetate layer, and shake mechanically for 10 minutes and centrifuge for 2 minutes. Move the lower, acid layer as completely as possible to a fresh 50-ml tube, using a disposable pipette. Add 0.3 ml 1 *M* sodium hydroxide, 5 ml pH 6.0 buffer, and exactly 25 ml ethyl acetate to the new tube. Stopper and shake by hand for 1 minute and centrifuge for 5 minutes. Pipette exactly 20 ml of the upper, ethyl acetate layer into a fresh 50-ml tube. Pipette exactly 3 ml of the 0.1 *M* hydrochloric acid into the tube. Stopper and shake by hand for 1 minute. Centrifuge for 5 minutes and move enough of the lower layer to a 1-cm fluorescence cell to read the fluorescence adequately.

iii. *Fluorescence Measurements*

Set the excitation wavelength at 302 nm (uncorrected) and the fluorescence wavelength at 362 nm (uncorrected) on the fluorometer. Zero the instrument with 0.1 *M* hydrochloric acid. Select a standard (A, B, or C) estimated to contain the same general range of material as the unknown, and set the instrument to read 100% (full scale) with said standard. Read the fluorescence as percent *T* for the unknown. If the unknown reads too low, go to a lower standard. If the unknown reads above standard A, dilute the unknown with 0.1 *M* hydrochloric acid so that it reads less than standard A.

iv. *Process Blank and Process Standard*

Carry 10 ml of pH 4.5 buffer (process blank) separately, and exactly 0.5 ml of standard D plus 9.5 ml of pH 4.5 buffer (process standard) through the entire solvent–solvent distribution procedure. Read the fluorescence of the process blank against working standard solution C, set to read 100%, and the process standard against standard solution B set to read 100%. These measurements need be made only often enough to ensure that the analyst can get reproducible results, and in the case of the blank, when new reagents are introduced.

v. *Calculations*

Calculate the process blank as follows:

$$B = \text{Blank concentration equivalent} = 0.0375 F_{\mathrm{B}} C_{\mathrm{SC}}$$

and the process standard as follows:

$$P = \%\ \text{recovery} = 100\left(\frac{0.0750 F_{\mathrm{P}} C_{\mathrm{SB}}}{C_{\mathrm{SD}}} - B\right)$$

and the unknown samples as follows:

$$\text{ppm} = 10\left[\frac{(0.0375 F_{U} C_{\mathrm{SU}} D - B)}{P}\right]$$

where F_B = fluorescence of blank
C_{SC} = concentration (μg/ml) of standard C
F_P = fluorescence of process standard
C_{SB} = concentration (μg/ml) of standard B
C_{SD} = concentration (μg/ml) of standard D
F_U = fluorescence of unknown sample
C_{SU} = concentration (μg/ml) of standard against which unknown was read.
D = dilution of unknown before fluorescence reading if any.

vi. *Recovery and Sensitivity*

Process blanks for the method usually run 0.005–0.0015 ppm. Blanks from control vegetative matter vary with the particular material involved. Blanks of 0.000–0.05 ppm above process blank are common. Thus, the sensitivity generally varies from about 0.002 to 0.1 ppm. A sensitivity of 0.01 ppm is normal. Recovery of both process standards and standards spiked into unknown samples average about 80%. Thus, the average recovery from unknowns corrected for process standard recovery averages 95–100%, with a normal range for a number of assays of 90–110% at fortification levels of 0.02–1 ppm.

e. *Discussion*

As with most residue assays, the quality of the reagents and the cleanliness of the glassware is critical. Working standard solutions are stable for months, but both unknown and standard must be prepared from exactly the same batch of 0.1 *M* hydrochloric acid, as the concentration of acid is critical.

Crops treated with a wax containing Mertect® are extracted directly with ethyl acetate to initiate the assay procedure. Some materials, like banana skins, require special treatment before the homogenization. (For banana skins, the special treatment consists of cutting the skin up into pieces about $\frac{1}{8}$ inch thick across the grain.)

REFERENCES

Aharonson, N., and Ben-Aziz, A. (1973). *J. Ass. Offic. Anal. Chem.* **56,** 1330.

Food and Drug Administration. (1969). "Pesticide Analytical Manual," Sect. 2, pp. 120–242. FDA, Washington, D.C.

Gunther, F. A. (1969). *Residue Rev.* **28,** 83.

Hey, H. (1972). *Z. Lebensm. Unters. -Forsch.* **149,** 79.

Holmes, M. G. (1972). *Pestic. Sci.* **3,** 367.

Kroller, E. (1971). *Deut. Lebensm.-Rundsch.* **67,** 229.

Mestres, R., Campo, M., and Tourte, J. (1970). *Trav. Soc. Pharm. Montpellier* **30,** 193.

Miller, V. L., Gould, C. J., and Csonka, E. (1971). *Plant. Dis. Rep.* **55,** 77.

Miller, V. L., Gould, C. J., Csonka, E., and Jensen, R. L. (1973). *J. Agr. Food Chem.* **21,** 931.
Miller, V. L., Gould, C. J., and Csonka, E. (1974). *J. Agr. Food Chem.* **22,** 90.
Norman, S. M., Fouse, D. C., and Craft, C. C. (1972). *J. Ass. Offic. Anal. Chem.* **55,** 1239.
Rajzman, A. (1974). *Analyst* **99,** 120.
Robinson, H. J., Phares, H. F., and Graessle, O. E. (1964). *J. Invest. Dermatol.* **42,** 479.
Robinson, H. J., Stoerk, H. C., and Graessle, O. E. (1965). *Toxicol. Appl. Pharmacol.* **7,** 53.
Staron, T., and Allord, C. (1964). *Phytiat.-Phytopharm.* **13,** 163.
Szalkowski, C. R., and Kanara, J. (1965). *J. Ass. Offic. Anal. Chem.* **48,** 288.

19

Validacin®

K. Nishi and K. Konishi

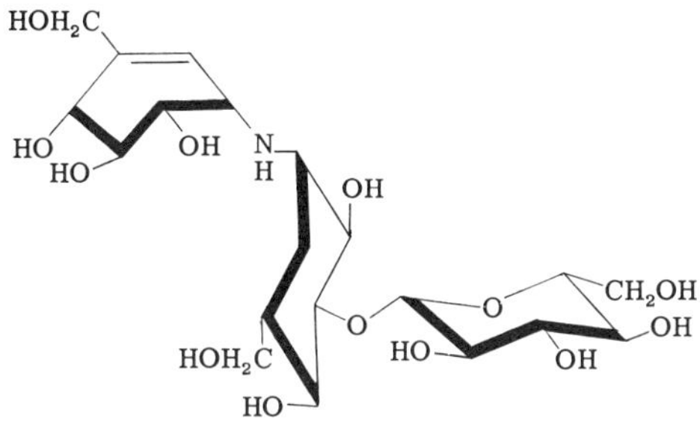

N-[(1*S*)-(1,4,6/5)-3-Hydroxymethyl-4,5,6-trihydroxycyclohex-2-enyl][*O*-β-D-glucopyranosyl-(1 → 3)-(1*S*)-(1,2,4/3,5)-2,3,4-trihydroxy-5-hydroxymethylcyclohexyl]amine

I. GENERAL

A. Empirical Formula

$C_{20}H_{35}NO_{13} \cdot H_2O$ (Mol. wt. 515.52).

B. Alternative Names

Validamycin, validamycin A. Validacin® is the registered trademark of Takeda Chemical Industries, Ltd. for its formulated commercial products containing validamycin A as the active ingredient. The material was tested under the code number T-7545 in Japan.

C. Source of Analytical Standard

Takeda Chemical Industries, Ltd., Osaka, Japan.

D. Biological Properties

Validacin® is a nonsystemic, curative fungicide and is particularly useful as a foliar, seed, and soil treatment at low dosage rates for the control of plant diseases caused by the fungi belonging to Corticiaceae in Basidiomycetes. It is specifically and highly effective against the sheath blight of the rice plant and some soil-borne diseases such as damping-off, root rot, and southern blight of vegetables; soreshin of cotton, black scurf of potatoes; and collar rot of carnation (Wakae and Matsuura, 1973, 1974).

Validacin® has low toxicity to mammals (male and female):

Acute oral LD_{50} to rats and mice	> 20,000 mg/kg
Acute subcutaneous LD_{50} to rats and mice	> 15,000 mg/kg
Acute intravenous LD_{50} to mice	> 13,000 mg/kg
Acute intraperitoneal LD_{50} to mice	> 10,000 mg/kg

No irritation was observed on application of 50% aqueous solution to the eyes and shaved skin of rabbits. In 90-day feeding tests on rats and mice, there were no significant abnormalities at dietary levels up to 10,000 ppm (Onishi and Miyaji, 1973). No adverse effects were found in beagle dogs orally administered at the rates up to 800 mg/kg/day for 4 months (Hosokawa *et al.,* 1974).

Validacin® is nonhazardous to livestock, poultry, and wildlife such as arthropods and fish. TL_m (median tolerance limit) for 48 hours is greater than 40 ppm for carp and greater than 4,000 ppm for killifish. It showed no phytotoxicity to a wide range of plants in various stages nor fungicidal activity against fungi including Corticiaceae, bacteria, and yeasts in a general agar dilution method. Validacin® inhibits the growth of infection hyphae on plants to prevent the development of the disease. It has no tendency to afford drug resistance against antibiotics including itself to human and plant pathogens.

Validacin® is easily biodegradable and decomposes with a half-life of several hours in soil under both paddy and upland conditions. It has no effects on microflora in soil nor does it persist in crops. Orally administered Validacin® is degraded by enteric bacteria into glucose and the amine, which is quantitatively excreted in the feces through the intestinal tract without absorption. Intravenously administered Validacin® is rapidly excreted in the urine (Furuta, 1973).

E. History

Validamycin A is a main component of the validamycin complex (A through F), which was isolated as a new antifungal antibiotic from a new strain of Actinomycetes, *Streptomycetes hygroscopicus* var. *lemoneus* by Takeda Chemical Industries, Ltd., in 1966 (Iwasa *et al.,* 1970, 1971a,b). All of validamycins have the unsaturated aminocyclitol structure in their molecules (Iwasa *et al.,* 1971c; Horii and Kameda, 1972). Among them, validamycin A is most effective against the phytopathogenic fungi and has been developed into a new agricultural fungicide. Its use as a fungicide is covered by Japanese patent nos. 640,249 (1972) and 644,761 (1972), U.S. patent no. 3,754,083 (1973), British patent no. 1,278,289 (1973), French patent no. 6,937,207 (1972), and Swiss patent no. 532,658 (1973).

F. Physical Properties

Pure validamycin A is an odorless, hygroscopic, white powder, which does not have a sharp melting point, softens at about 100°C, and decomposes at about 135°C; $[\alpha]_D^{24}$ 112.5° (c 1, H_2O); pK_a 6.0.

It is soluble in water, methanol, dimethylformamide, and dimethylsulfoxide; slightly soluble in ethanol and acetone; and insoluble in ethyl acetate and diethyl ether.

G. Chemical Properties

Validacin® is marketed in Japan as a 3% water solution and as a acidic aqueous solutions at room temperature. It is also stable under ambient temperature on storage.

H. Formulation

Validacin® is marketed in Japan as a 3% water solution and as a 0.3% dust. It is also commercially available in a variety of combinations with other fungicides and/or insecticides.

II. ANALYSIS

A. Formulation Analysis

1. REVIEW OF METHODS

Three methods have been developed for the analysis of Validacin® formulations. The first is a bioassay method (Iwasa *et al.*, 1971b). The second utilizes the colorimetry of glucose using naphthoresorcinol after the separation by ion exchange resin column chromatography. The third method, which is the recommended one, is a gas chromatographic procedure.

2. RECOMMENDED METHOD

a. *Principle*

The formulated sample is extracted with water, evaporated under reduced pressure, and analyzed by gas chromatography using a flame ionization detector after the preparation of validamycin A trimethylsilyl (TMSi) derivative.

b. *Reagents*

Pyridine, Reagent Grade.

Internal standard solution. Dissolve 30 mg melezitose (Difco Laboratories, Inc.) in 100 ml distilled water.

Trimethylchlorosilane (TMCS), Reagent Grade.

N,O-Bis(trimethylsilyl)acetamide (BSA), Reagent Grade.

Silylating reagent. Mix 2 volumes of pyridine, 1 volume of BSA, and 1 volume of TMCS.

Extracting solution, 0.1% cationic surfactant (Colomine A, Laurylpicolinium chloride, Kao Atlas, or equivalent) aqueous solution.

Validamycin A, analytical standard obtained from Takeda Chemical Industries, Ltd., Osaka, Japan. Validamycin A shows one peak when gas chromatography is used for this analytical method and one spot at R_f value 0.3 on a thin-layer chromatogram on silica gel G with the mixed solvent of *n*-propanol–acetic acid–water (4 : 1 : 1 v/v).

c. *Apparatus*

Shaker.

Rotary vaccum evaporator.

Gas chromatograph, Yanagimoto G-80 equipped with a flame ionization detector or equivalent.

Microliter syringe, Gintan Termo or equivalent.

d. *Gas Chromatographic Operating Parameters*

Column: 1.5 m × 3 mm glass.

Packing: 1% Silicone OV-17 on 80–100 mesh Chromosorb G, AW, DMCS.

Detector oven temperature: 320°C.

Column temperature: 280°C.

Injection temperature: 320°C.

Gas: Helium, 8 ml/minute; hydrogen, 20 ml/minute; air, 0.8 liter/minute.

Recorder chart speed: 5 mm/minute.

e. *Experimental Procedure*

i. *Standards*

(a) *Standard Solution.* Weigh 110–120 ± 0.1 mg of validamycin A into a 200-ml volumetric flask (W_1 mg). Dissolve and dilute to volume with water.

(b) *Preparation of the Validamycin A TMSi Derivative.* Measure 1 ml portion from the standard solution into a 25-ml pear-shaped flask, and add 1 ml of the internal standard solution and 3 ml of pyridine, respectively. After evaporation at 40°–60°C, add 1 ml of pyridine and again evaporate in a similar manner as above. Put the pear-shaped flask in a vacuum dryer to dry for 1 hour under reduced pressure at 60°C. Add 400 μl of the silylating reagent to the residue, seal the flask tightly and heat it for 30 minutes at 70°C to prepare the standard validamycin A TMSi derivative solution.

ii. *Samples*

Weigh enough sample to contain 10–20 mg of the active ingredient accurately (±0.1 mg), transfer into a 200-ml Erlenmeyer flask (W_2 mg), and add 50 ml of the extracting solution. Shake for 30 minutes and filter with a Toyo Roshi no. 5 filter paper. Measure accurately 2 ml of the filtrate into a 20-ml pear-shaped flask with a ground glass stopper after discarding the 10 ml of the first filtrate. Add 1 ml of the internal standard solution and then prepare the sample TMSi derivative solution in the same manner as above.

iii. *Measurement*

Using a microliter syringe, inject a 1-μl portion of the standard or the sample TMSi derivative solution into a gas chromatograph and measure the peak height. To carry out the analysis, inject a 1-μl portion in the following order: (1) standard, (2) sample, (3) standard, (4) sample, and (5) standard. The retention time for the validamycin A TMSi derivative is about two times as long as that for the internal standard (melezitose TMSi derivative). Typical gas chromatograms for the standard and the sample are shown in Fig. 1.

iv. *Calculation.*

Divide the peak height of the validamycin A TMSi derivative by that of the internal standard to determine the ratio R. Calculate the average R for the sample injections and the average R for the standard injections.

$$\text{Validamycin A content (\%)} = \frac{\text{average } R \text{ (sample)}}{\text{average } R \text{ (standard)}} \times \frac{W_1}{W_2} \times 12.5$$

3. DISCUSSION OF METHOD

As validamycin A TMSi derivative decomposes upon contact with metal at high temperature, its solutions should be injected directly onto a glass column of the gas chromatograph. The columns should be main-

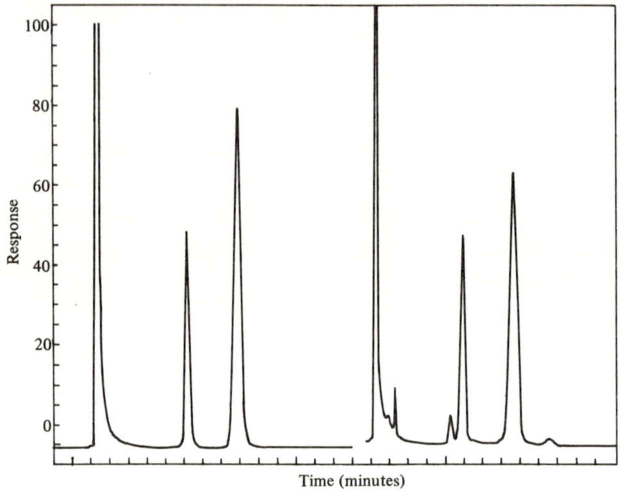

FIG. 1. Typical gas chromatograms for Validacin® standard and sample.

tained at 320°C for at least 10 hours for conditioning in order to obtain reproducible values.

B. Residue Analysis

1. REVIEW OF METHOD

One method has been developed to determine Validacin® in soil and crops. The method utilizes gas chromatography using a flame ionization detector after cleanup process.

2. RECOMMENDED METHOD

a. *Principle*

Residues of Validacin® are extracted with an acidic aqueous methanol, and interfering substances are removed by a cleanup procedure using cation-exchange resin column chromatography. However, it is necessary to digest with α-amylase in the case of rice before extraction. This chapter describes the residue analysis in soil.

b. *Reagents*

Internal standard solution. Dissolve 10 mg melezitose in 100 ml distilled water.

Aqueous methanol. Mix 3 volumes of methanol (Reagent Grade) and 7 volumes of distilled water.

1 N sulfuric acid. Dilute 30 ml sulfuric acid (Reagent Grade) to 1 liter with distilled water.

0.1 M barium hydroxide. Dissolve 32 gm barium hydroxide octahydrate (Reagent Grade) to 1 liter with distilled water. Allow to stand over night and filter with a filter paper.

1 M Pyridine. Dilute 80 gm pyridine to 1 liter with distilled water.

Other reagents are the same as those in the formulation analysis (Section II,A).

Ion exchange resin. Prepare a slurry of strong cation Dowex 50 W × 2 (Dow Chemical Company) as follows. Place 100 gm resin in a glass-stoppered Erlenmeyer flask, add water to slurry, and wash several times to neutralize. Plug loosely the lower end of a chromatographic column (250 mm × 15 mm) with a small amount of glass wool. Add 10–20 ml distilled water to the column and pour 30 ml of slurry of the resin gently into the column to obtain a settled bed. Keep the column bed covered with water to prevent drying and subsequent channeling. Wash the resin with 150 ml 4 *N* hydrochloric acid, 150 ml 2 *N* sodium chloride, 150 ml 4 *N* hydrochloric acid, and finally with water until a neutral effluent is obtained.

c. *Apparatus*

Chromatographic column, 15 × 250-mm glass fitted with a stopcock.

Other apparatus are the same as those in the formulation analysis (Section II,A).

d. *Gas Chromatographic Operating Parameters*

The Yanagimoto G-80 gas chromatograph is operated under the same conditions as in Section II,A.

e. *Experimental Procedure*

i. *Standards*

(a) *Standard Solution.* Weigh accurately 40 ± 0.1 mg validamycin A into a 100-ml volumetric flask. Dissolve and dilute to volume with distilled water. Again dilute the solution with distilled water to appropriate concentration for the residue analysis of Validacin®.

(b) *Preparation of the Validamycin A TMSi Derivative.* Measure a 1-ml portion from the appropriate diluted solution of validamycin A standard (1–200 μg/ml) into a 10-ml pear-shaped flask and add 1 ml of the internal standard solution. After evaporation at 40–60°C, put the pear-shaped flask in a vacuum dryer to dry for 1 hour under reduced pressure at 60°C. Add 300 μl of the silylating reagent to the residue, seal the flask tightly, and heat it for 30 minutes at 70°C.

ii. *Samples*

Weigh 20 gm soil into a 100-ml centrifuge tube, add 80 ml of the aqueous methanol and 2 ml 1 *N* sulfuric acid, and shake for 1 hour. Centrifuge for 5 minutes at 3000 rpm and filter the supernate through glass wool. Repeat this procedure with the same volume of the aqueous methanol and 1 *N* sulfuric acid. Transfer the extract into a 300-ml round-bottom flask and evaporate methanol with a rotary vacuum evaporator. Add 0.1 *M* barium hydroxide to the residual solution to neutralize and centrifuge for 5 minutes at 3000 rpm. Pour the supernate onto the top of the column and rinse out with wash water obtained from addition of 10 ml of distilled water to the residue after centrifugation. Wash the ion-exchange resin with 100 ml distilled water, discard this effluent, replace the receiver and elute with 200 ml of 1 *M* pyridine. Collect this eluate into a 500-ml pear-shaped flask. Add 1 ml of the internal standard solution and concentrate the eluate to 3–5 ml at 50°C with rotary vacuum evaporator. Transfer into a 10-ml pear-shaped flask and evaporate carefully under reduced pressure. Dry sample and prepare TMSi derivative in a similar manner as above.

iii. *Determination*

Using a microliter syringe, inject a 3-μl portion of the TMSi derivative solution into a gas chromatograph. The analytical procedure is the same as in Section II,A. A typical chromatogram is shown in Fig. 2.

iv. *Calculation*

$$\text{Validamycin A (ppm)} = \frac{\text{average } R \text{ (sample)}}{\text{average } R \text{ (standard)}} \times \frac{\text{weight (standard taken) } (\mu\text{g})}{\text{weight (sample) (gm)}}$$

f. *Recoveries*

Recovery data obtained by using the gas chromatographic method are shown in Table I. These data were obtained by adding known amounts of validamycin A to the various soils in Japan immediately before extraction.

g. *Sensitivity*

The determination and detection limits of a sample are affected by the noise level of the gas chromatograph and the sample size of soil. The

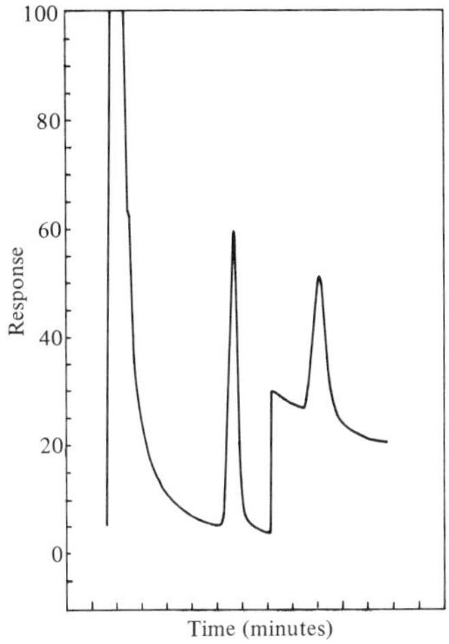

FIG. 2. Gas chromatogram for the extract from the alluvial soil fortified with 1 ppm of validamycin A before extraction.

lower limit of detection for this entire method is approximately 0.05 ppm. The peaks for untreated soils tested were below the 0.01-ppm level.

h. *Applicability of the Gas Chromatographic Method*

The gas chromatographic method has been used for a wide variety of crops in combination with a cleanup procedure using the ion exchange resin column chromatography.

3. DISCUSSION OF METHOD

Validacin® can be easily extracted from a soil sample and separated from the residues of other pesticides by aqueous acidic methanol extraction, ion exchange resin column chromatography and gas chromatography cleanup procedures. This method is highly sensitive and can be performed even by a flame ionization detector because an amino group

TABLE I
RECOVERIES OF VALIDAMYCIN A-FORTIFIED SOIL

Soil	Validamycin A added (ppm)	No. of determination	% Recovered	
			Range	Average
Alluvial	0.05–20.0	7	67–81	75
Volcanic ash	0.05–20.0	8	73–95	86

and many hydroxyl groups of validamycin A are trimethylsilylated. Barium hydroxide is added in the extracting solvent to remove the sulfate ion and other interfering substances and to reduce the amount of the ion exchange resin needed.

References

Furuta, H. (1973). *Jap. Pestic. Inform.* **15,** 28.

Horii, S., and Kameda, Y. (1972). *J. Chem. Soc., Chem. Commun.* 747.

Hosokawa, S., Ogiwara, S., and Murata, Y. (1974). *J. Takeda Res. Lab.* **33,** 119.

Iwasa, T., Yamamoto, H., and Shibata, M. (1970). *J. Antibiot.* **23,** 595.

Iwasa, T., Higashide, E., Yamamoto, H., and Shibata, M. (1971a). *J. Antibiot.* **24,** 107.

Iwasa, T., Higashide, E., and Shibata, M. (1971b). *J. Antibiot.* **24,** 114.

Iwasa, T., Kameda, Y., Asai, M., Horii, S., and Mizuno, K. (1971c). *J. Antibiot.* **24,** 119.

Onishi, S., and Miyaji, T. (1973). *Med. J. Osaka Univ.* **23,** 215.

Wakae, O., and Matsuura, K. (1973). *Abstr. Papers Int. Congr. Plant Pathol., 2nd, 1973,* p. 129, Minneapolis, Minnesota.

Wakae, O., and Matsuura, K. (1974). *Proc. Intersec. Congr. Int. Ass. Microbiol. Soc., 1st, 1974,* Vol. 3, p. 620, Tokyo, Japan.

20

Vitavax®*

G. M. Stone

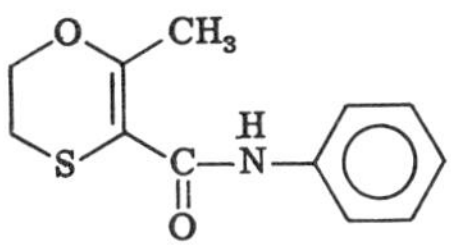

5,6-Dihydro-2-methyl-1,4-oxathiin-3-carboxanilide

I. GENERAL

A. Empirical Formula

$C_{12}H_{13}NO_2S$ (Mol. wt. 235).

B. Alternate Names

D735, carboxin, carbothiin, DCMO.

C. Source of Analytical Standard

Uniroyal Chemical, Division of Uniroyal, Inc., Elm Street, Naugatuck, Connecticut (Attention G. M. Stone).

D. Biological Properties

Vitavax is a systemic fungicide and is effective as a seed treatment for control of plant pathogenic fungi on a variety of crops.

The acute oral LD_{50} of technical grade Vitavax for albino rats is 3820 mg/kg of body weight. Albino rats and beagle dogs fed technical Vitavax for 2 years showed no pathological effects at the highest rate tested, 600 ppm. The growth of rats at the 600 ppm daily intake level was retarded. The acute dermal LD_{50} of technical grade Vitavax for albino rabbits is in excess of 8000 mg/kg of body weight. Formulated Vitavax has a similar low toxicity to rabbits. Eye irritation tests have shown that formulated Vitavax is a severe eye irritant and the wearing of goggles is recommended.

The LD_{50} for the 75% wettable powder Vitavax formulation is 4.5

* Vitavax is a registered trade name; no. 0862225.

ppm for rainbow trout and 2.9 ppm for bluegill sunfish. The LC_{50} is greater than 5620 ppm for bobwhite quail and greater than 4640 ppm for mallard ducks. Both rates were the highest tested.

E. History

Vitavax was developed by Uniroyal, Ltd., Guelph, Ontario, and Uniroyal Chemical, Division of Uniroyal, Inc., Bethany, Connecticut, and its use first reported in 1966 (von Schmeling and Kulka, 1966). U.S. patents 3,249,499, 3,393,202, and 3,454,391 have been issued.

F. Physical Properties

Vitavax is an off-white crystalline solid with two crystal structures. The melting points are 91.5–92.5°C and 98.0–100°C. The infrared spectra of the two forms in solution are identical. Vitavax has a solubility of 0.017 gm per 100 gm of water, 15 gm per 100 gms of benzene, 60 gm per 100 gms of acetone, and 21 gm per 100 gm of methanol.

G. Chemical Properties

1. METHOD OF SYNTHESIS

Vitavax is synthesized by reacting acetoacetanilide with sulfuryl chloride to give 2-chloroacetoacetanilide, which is reacted with 2-mercaptoethanol. The resulting intermediate dehydrates to form Vitavax in the presence of an acid catalyst.

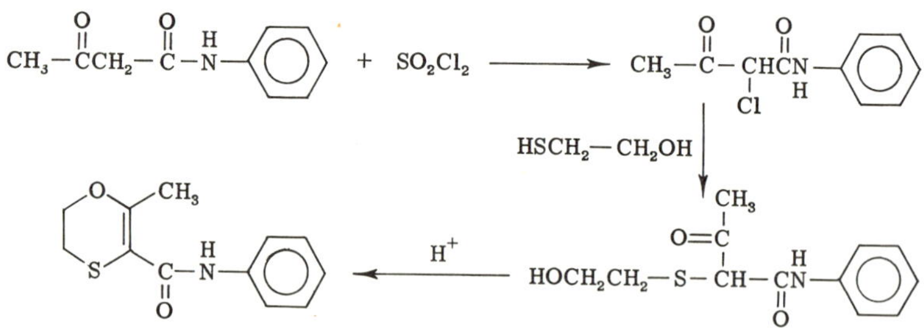

H. Formulations

Vitavax is marketed in the form of a 75% wettable powder, a 34% active flowable (liquid suspension), and as a 25% wettable powder formulation. It is also combined with thiram in dry and flowable formulations.

II. ANALYSIS

A. Formulation Analysis

1. RECOMMENDED METHOD—USED TO ANALYZE TECHNICAL AND FORMULATED VITAVAX

a. *Principle*

Vitavax is assayed by infrared spectroscopy based on measurements at 5.97, 6.30, and 7.75 μm in comparison with a Vitavax standard.

b. *Reagents*

Benzene, Reagent Grade.

c. *Apparatus*

Grating spectrophotometer, Perkin Elmer 421, matched 0.2 mm sodium chloride cells.

d. *Experimental Procedure*

i. *Standards*

A sample of 0.2 gm of pure Vitavax is dissolved in benzene in a 25-ml volumetric flask diluted to the mark. The reference cell is filled with benzene and the spectrum is run from 5.75 to 8 μm at slow speed.

ii. *Samples*

The optimum concentration for a cell path of 0.2 mm is 0.20 gm Vitavax per 25 ml benzene. Therefore, for a 75% formulation, 0.27 gm is weighed; for 34% formulation, 0.6 gm is weighed. The fillers in the formulation are allowed to settle, and then the spectrum is obtained using 0.2-mm cells with benzene in the reference cells.

e. *Calculations*

$$\Delta A_{5.97\ \mu m} = A_{5.97\ \mu m} - A_{6.13\ \mu m} \tag{1}$$

$$\Delta A_{6.30\ \mu m} = A_{6.30\ \mu m} - A_{6.13\ \mu m} \tag{2}$$

$$\Delta A_{7.75\ \mu m} = A_{7.75\ \mu m} - A_{7.90\ \mu m} \tag{3}$$

The above baseline absorptions for both samples and standard are obtained. The ratio of the sample to standard absorptions for all three absorption bands is calculated, and the weight of standard used is multiplied by each ratio. Finally, this value is multiplied by 100 and divided by the sample weight to obtain the percentage of Vitavax at each wavelength.

For 5.97 μm band:

$$\frac{\Delta A_{\text{unknown}}}{\Delta A_{\text{standard}}} \times \frac{\text{weight}_{\text{standard}}}{\text{weight}_{\text{unknown}}} \times 100 = \%\ \text{Vitavax at } 5.97\ \mu\text{m}$$

The calculation is repeated at 6.30 and 7.75 μm, then an average is taken of the three Vitavax percentage values.

2. ALTERNATIVE METHOD—USED TO ANALYZE TECHNICAL VITAVAX*

a. *Principle*

Vitavax is oxidized by H_2O_2 to its sulfoxide, and the excess H_2O_2 is back titrated with sodium thiosulfate.

b. *Reagents*

Hydrogen peroxide, 15%.
Potassium iodide, 50%.
Sulfuric acid, 3 *N*.
Glacial acetic acid.
Starch indicator.
Sodium thiosulfate, 0.5 *N*.

c. *Apparatus*

Erlenmeyer flask, 250 ml.
Pipette, 2 ml.

d. *Experimental Procedure*

A sample of 1.0 gm of Vitavax is weighed into an Erlenmeyer flask and dissolved in 20 ml glacial acetic acid with slight warming on a steam bath. After cooling, 2 ml 15% hydrogen peroxide is pipetted into the flask and swirled to mix. The flask is covered and allowed to stand 1 hour at room temperature. To the mixture, 10 ml 3 *N* H_2SO_4 is added followed by 10 ml 50% KI solution. After mixing, the mixture is titrated with 0.5 *N* sodium thiosulfate solution. A blank is also run omitting the sample. For a sample of about 1.03 gm, titers run between 20 and 30 ml and the blank between 40 and 45 ml.

e. *Calculation*

$$\% \text{ Vitavax} = \frac{(B - T)N \times 11.75}{W}$$

where B = milliliters of 0.5 *N* sodium thiosulfate used for blank
T = milliliters of 0.5 *N* sodium thiosulfate used for sample
N = normality of 0.5 *N* sodium thiosulfate
W = weight of the sample.

3. ALTERNATIVE METHOD—USED TO ANALYZE TECHNICAL VITAVAX

The procedure (Puchalsky, 1969) is based on the room-temperature oxidation of sulfide to sulfoxide by hydrogen peroxide in acetic acid.

* M. Kulka and H. Thomson, private communication (1965).

After 30 minutes, the acetic acid and unused hydrogen peroxide are removed in a rotary evaporator. The sulfoxide is then titrated potentiometrically (as a base) with perchloric acid in acetic anhydride.

B. Residue Analysis

1. REVIEW OF METHODS

Studies with ^{14}C-labeled Vitavax show that Vitavax is systemic but is not translocated freely. The residues are found mainly in the stem just above ground level and in the first leaves. The sulfoxide of Vitavax is the major component of residues in plants. No ^{14}C-residue occurs in mature seed, therefore, the residue tolerance is based on the sensitivity of the method.

The method developed for Vitavax in seed and plants at harvest consists of extraction of Vitavax and the sulfoxide from the crop, cleanup by solvent partitioning, caustic hydrolysis of the Vitavax and distillation of the aniline, and cleanup preceding gas chromatographic detection using a specific nitrogen detector (Sisken and Newell, 1971). The method has been used for small grains, cottonseed, corn, and peanuts and for plants at harvest time. The method is sensitive to 0.2 ppm.

Young plants have residues ranging from 0.5 ppm to several parts per million and may be analyzed by a colorimetric procedure (Lane, 1970). In this test, fresh macerated plants are digested in caustic solution to liberate aniline from Vitavax and its sulfoxide. The distilled aniline is reacted with *p*-dimethylaminobenzaldehyde, and the dye is measured spectrometrically. As plants mature the increasing interference levels make determination of less than 1 ppm impractical by this test.

A third method has been developed by D. S. Gunew of ICI Australia Limited (private communication, 1974) in which the Vitavax is determined by gas chromatography. This method of detection was used in studying the fate of Vitavax in plants (Chin, *et al.*, 1970), and is of particular value when the parent compound as such is to be measured. It can not be used for total residues, since it does not measure the sulfoxide (Chin *et al.*, 1970).

2. RECOMMENDED METHOD

a. *Gas–Liquid Chromatographic Procedure*

i. *Principle*

Vitavax and its metabolite, the sulfoxide, are extracted from the ground plant tissue and the extract is cleaned up. Vitavax and the sulfoxide are hydrolyzed to liberate aniline in a caustic reducing medium.

The aniline is distilled, concentrated, and determined by gas chromatography using a nitrogen detector.

ii. *Reagents*

Water purified by distillation and redistilled from alkaline permanganate.

Hexane extracted with 10 *N* H_2SO_4 before use.

Diethyl ether extracted with 10 *N* H_2SO_4 before use.

Titanium trichloride, 20% solution (Fisher).

Sodium hydroxide, 50% (J. T. Baker).

Methanol, toluene, benzene, acetone (Baker & Adamson).

Vitavax reference sample (Uniroyal Chemical, Division of Uniroyal, Inc., Naugatuck, Connecticut).

iii. *Apparatus*

Microcoulometer, Dorhmann C-200-A, or Coulson, N detector.

Blender, Waring or equivalent.

Soxhlet extractor and thimbles, 43 × 123 mm.

Steam generator.

iv. *Experimental Procedure*

(a) *Sample Preparation and Analysis.* A 30-gm sample of seed or dry plant part is ground with 100 ml methanol in a blender for 2 minutes, the methanol is decanted through a funnel into the Soxhlet thimble in an extractor attached to a 300-ml round-bottomed flask. A glass stopper is placed under the thimble in the Soxhlet extractor to promote efficient drainage. The seed in the blender is treated with an additional 75 ml of methanol and blended for 2 more minutes. Both seed and methanol are transferred to the Soxhlet thimble. Soxhlet extraction at reflux temperature is continued for 3 hours. Excessive heating should be avoided because it promotes scorching, giving erratic blanks due to the thermal degradation of natural components of the extract.

The methanol extract is cooled and transferred to a 1-liter separatory funnel. The flask is rinsed with two 25-ml portions of water. The rinsings are added to the separatory funnel, followed by 2 ml saturated NaCl solution. The methanol–water mixture is extracted twice with 50 ml portions of *n*-hexane, which are discarded. Then 70 ml water is added, and the contents of the funnel extracted with 100 ml chloroform, followed by four successive extractions with 30 ml portions of chloroform. The chloroform extracts are combined in a 250-ml separatory funnel with 15 gm anhydrous Na_2SO_4 powder, and the separatory funnel shaken. After 1 minute, the chloroform is decanted from the separatory funnel through Whatman no. 4 filter paper into a 500-ml round-bottomed flask. The Na_2SO_4 in the separatory funnel is rinsed with two 15-ml portions of chloroform, which are filtered through the filter paper into the flask. The flask, connected to a vacuum rotary evaporator, is half submerged in warm water and the chloroform solution concentrated to about 2 ml. Then 30 ml of toluene is added to the flask and the solution concentrated again until about 3 ml of liquid remain.

After concentration, 2 ml of 20 mesh zinc granules, 5 ml of titanium trichloride, and 100 ml of 50% NaOH are added to the flask, which is placed in a heating mantle for insulation, connected to a steam generator, and steam distilled until 100 ml of distillate is collected.

The distillate is transferred to a separatory funnel, brought to a pH of 1 with 10 *N* H_2SO_4 (using 1 to 11 range pH paper), and extracted with 20 ml benzene, which is discarded. The aqueous phase is then rendered alkaline to pH 11 with 50% NaOH solution and is extracted with 40 ml diethyl ether, followed by three extractions with 20-ml portions of diethyl ether. The 40 ml extract is drained into a 50-ml beaker containing 6 gm anhydrous Na_2SO_4 powder and stirred to remove most of the water present. The ether is decanted into a 30-ml beaker with 30 μl of 3 *N* H_2SO_4 is added. The beaker is placed in a dish of warm water (not greater than 30°C) and the ether evaporated under a fine jet of nitrogen gas. The 20 ml ether extracts are successively dried with the same Na_2SO_4 in the first beaker, and are added to the ether under the jet. When the ether is about 5 ml in volume, it is transferred, portionwise, to a tared conical tube and the evaporation is continued. The beaker is rinsed with 1 ml acetone, the rinsing added to the conical tube, and evaporated to 0.4–0.7 ml (ether free). Forty microliters of 3 *N* NaOH are added and the tube is reweighed. Since the sample solution is mostly water, the weight in milligrams is assumed to be equal to the volume in microliters. Ten microliters from the tube are injected into the gas chromatograph to determine any aniline present in the sample.

The analytical results are corrected for reagent blanks, which vary from 0.04 to 0.1 ppm. The reagent blank is run starting with the methanol used to extract the sample going to the final concentration, with no sample of plant tissue present. A standard curve is run using aniline.

(b) *Gas Chromatographic Columns and Conditions.* Two detectors have been used successfully: the Dorhmann C-200-A microcoulometer with a T-400 titration cell and Coulson N-cell. The columns used are:

Column I. 18 feet × $\frac{1}{10}$ inch i.d., type 304 stainless steel packed with 70–80 mesh Anakrom SD, coated with 4% Carbowax 20M, Igepal CO-880 and 1% KOH.

Column temperature programmed from 100°C to 200°C at 6°/minute.

Injection port: 150°C.

Carrier gas: Ultrapure hydrogen moistened by passing over damp silica gel, 55 ml/minute at GC column temperature of 100°C, and 40 ml/minute at 200°C.

Column II. 8 feet × $\frac{1}{4}$ inch o.d. SS packed with 80–100 Gas Chrom Q coated with 40% Castorwax.

Column temperature: 170°C.

Injection port: 230°C.

Carrier gas: H_2 90 ml/minute.

c. *Calculation*

The residue is calculated as Vitavax from

$$\text{ppm} = \frac{\text{area of unknown peak}}{\text{area of aniline standard peak}} \times 0.060 \times 2.53 \times \frac{\text{total volume}}{\text{volume injected}} \times \frac{1}{\text{sample weight}}$$

The 0.060 is from the 0.060 μg aniline standard injection.

d. *Recoveries*

Recovery data on various crops are given in Table I. These data are obtained by adding 0.2 ppm of Vitavax to the ground crop before extraction.

e. *Sensitivity*

Sensitivity of the method is 0.2 ppm.

3. COLORIMETRIC METHOD

a. *Procedure for Green Plant Tissue*

i. *Principle*

Vitavax and its metabolite, the sulfoxide, are hydrolyzed, and the aniline is distilled. The distillate is cleaned up by extraction with hexane. The aniline is then extracted into 20% acetic acid and determined

TABLE I
RECOVERY DATA ON VARIOUS CROPS

Crop	No. of determinations	Recovery average (%)
Wheat	4	90
Barley	4	103
Peanuts	4	91
Cotton	10	84

colorimetrically after reaction with *p*-dimethylaminobenzaldehyde. The crop interference forms a red color, the aniline a yellow color.

ii. *Reagents*

Antifoam A, paste form, Dow Corning Corporation.

DAB reagent (Ehrlich's), 2 gm *p*-dimethylaminobenzaldehyde dissolved in 100 ml 4% HCl solution.

Titanium(ous) trichloride, 20% aqueous (Fisher SO-T-43).

Zinc granules, 10 mesh.

Vitavax standard solution, 100 μg/ml acetone.

iii. *Apparatus*

Alar steam distillation head (Edgerton *et al.,* 1967). A slightly adapted head, using a Graham type coil condensing tube of about five loops to replace the original 45° angle condensing tube, is employed (Macalaster Bicknell Co., 181 Henry Street, New Haven, Connecticut).

Spectrophotometer equipped with 50-mm silica cells.

iv. *Experimental Procedure*

(a) *Sample Preparation and Analysis.* The weight of macerated crop sample, the volume of NaOH solution required, and the suitable flask size are given in Table II. To the flask, these additional items are also added: 5 ml titanium trichloride solution, 2 ml granular zinc, and 1 ml Antifoam A.

To 150 ml of distillate, 5 gm NaCl is added, and the solution then extracted with 50 ml hexane in a 250-ml separator. The aqueous phase is transferred to a 250-ml beaker and, by dropwise addition of 85% H_3PO_4, the pH is adjusted to 6.5 using a pH meter. The distillate is extracted two times with 30-ml portions of benzene and the recovered benzene extracts are combined. The benzene is extracted two times with 50 ml water. Retaining the benzene layer, it is extracted with 12 ml 20% acetic acid solution. The pH is adjusted to 3.0 or lower. Two minutes is allowed for any benzene droplets or fog to clear from the lower aqueous phase before transferring the aqueous phase to a 25-ml mixing cylinder (be sure no benzene is drawn off with the aqueous phase to cause a foggy solution). Two milliliters DAB reagent are added to the mixing cylinder and mixed well.

TABLE II

DISTILLATION CONDITIONS FOR GREEN PLANTS

Anticipated residue (ppm)	Sample (gm)	Flask size (liters)	NaOH solution
3–20	10	1	500 ml of 35%
0.5–2.0	50	1	500 ml of 35%
0.2–0.4	100	2	500 ml of 35%

After 30 minutes, 1 ml glacial acetic acid is added and the pH adjusted to 3.5 by dropwise addition of 50% NaOH solution.

The volume is adjusted to 16 ml with distilled water, if necessary (this is sufficient volume to fill the 50-mm silica cell). All volumes of color solutions are recorded for use in the calculation. The color solutions must be crystal clear for measurement. If a color solution is foggy at this point, filter it through Whatman no. 1 paper.

(b) *Color Measurement.* A cell blank is prepared by adding 12 ml 20% acetic acid solution, 2 ml DAB reagent, and 1 ml glacial acetic acid to a 25-ml cylinder and shaking well before adjusting the pH to 3.5 and the volume to 16 ml as given in the procedure. The absorbance of crop analyses or standard analyses is measured against the cell blank in 50-mm cells at 440 and 490 nm on a spectrophotometer. If the absorbance measured at 440 nm is greater than 1.5 units, remeasure the solutions in 1-cm silica cells and multiply by 5 to fit the calculation.

b. *Standard Curve*

Five hundred milliliters 35% NaOH solution, 2 ml zinc granules, 5 ml 20% titanium trichloride solution, 5 ml concentrated NH_4OH (to stimulate the amines and NH_4OH in a crop analysis), and finally 0, 20, 60, or 80 μg of Vitavax standard solution are added to a 1-liter flask. One hundred fifty milliliters of distillate is collected and the analysis completed as described above. The net absorbance versus micrograms of Vitavax is plotted, and from the average straight line through the points, K is determined (average value about 60).

c. *Calculation*

$$\text{net absorbance (50-mm cells)} = (A_{440\text{nm}} - A_{490\text{nm}}) \times \frac{\text{total ml color solution}}{\text{16 (the standard color volume and capacity of the 50-mm cell)}}$$

$$\text{ppm} = \text{net absorbance} \times K/\text{sample weight}$$

where $K = \mu\text{g/net absorbance} = \text{about } 60$

d. *Recoveries*

Recovery data on various green tissue samples are given in Table III. These data are obtained by adding Vitavax to the ground tissue before the hydrolysis step.

e. *Sensitivity*

Sensitivity of this method is 0.2 ppm on green tissue, 1 ppm on seed or dry tissue.

TABLE III
RECOVERY DATA ON VARIOUS GREEN TISSUE SAMPLES

Crop	No. of determinations	Average recoveries (%)
Barley	7	94
Cotton	5	99
Peanuts	7	85
Wheat	6	88

4. SPECIFIC METHOD FOR VITAVAX

a. *GLC Method for the Parent Compound*

i. *Principle*

Specific determination of Vitavax by gas–liquid chromatography is useful (Chin *et al.*, 1970), since Vitavax can be separated from its degradation products and determined quantitatively in plant extracts with a short, inert GLC column and a specific sulfur detector. D. S. Gunew of ICI Australia Limited has developed a similar method for the unchanged Vitavax molecule, which is determined by gas chromatography using a flame photometric detector in the sulfur mode. Vitavax is extracted from the crop, and the extract is subjected to a column cleanup prior to the final determination.

ii. *Reagents*

See Section II,B,2,a,ii above.

iii. *Apparatus*

(a) *Gunew Method.*
Gas chromatograph, Hewlitt-Packard model 7610, Tracor flame photometric detector, S mode.
Glass column, 40 cm × 2.5 cm, fitted with Teflon tap.
Other apparatus as in Section II,B,2,a,iii, above.

(b) *Chin Method.*
Gas chromatograph, Dorhmann M-100 with sulfur titration cell, S-200.

iv. *Experimental Procedure*

(a) *Sample Preparation and Analysis.* The Gunew method sample preparation is similar to the recommended method (Section II,B,2,a,iv, above). The 50-gm sample is ground and extracted. The methanol extract,* with salt added, is extracted with hexane and then chloroform. The chloroform is dried with sodium sulfate and evaporated to dryness using a rotary evaporator.

At this point, the method is different from that in Section 2, above, since the residue is taken up in 5 ml benzene transferred to a silica gel column. The column is washed with 100 ml benzene and then eluted with 80 ml acetone, discarding the first 10-ml fraction. The volume of acetone eluate is reduced to 1 ml and rediluted to 5 ml using methanol. A 5-μl portion is injected into the gas chromatograph.

A calibration curve is run using 0 to 10 ppm Vitavax in methanol. Five microliters of each standard is injected into the gas chromatograph. The peak height versus the square root of the concentration in ppm is plotted on log–log graph paper, resulting in a straight line.

(b) *Gas Chromatographic Columns and Conditions.*

I. The conditions recommended by D. Gunew method are:

Column: 4 feet × $\frac{1}{4}$ inch o.d. glass U-shaped, packed with 100–120 mesh Gas Chrom Q coated with 5% OV-1.

Injection temperature: 240°C.
Column temperature: 220°C.
Transfer line temperature: 240°C.
Detector temperature: 200°C.
Photomultiplier tube voltage: 750 V.
Carrier gas: Nitrogen.
Gas flow
Hydrogen: 130 cc/minute.
Oxygen: 15 cc/minute.
Air: 60 cc/minute.
Detector: Flame photometric, S mode.

II. The column conditions recommended by Chin *et al.* (1970) are:

Column: 8 inches × $\frac{1}{4}$ inch o.d. glass tube packed with Anakron ABS coated with 15% silicone grease

Injection temperature: 220°C.
Column temperature: 220°C.
Furnace temperature: 660°C.
Carrier gas
Helium: 150 cc/minute.
Oxygen: 35 cc/minute.

* For dry grain, acetone may be used as the extracting solvent. After extracting 3 hours, the extract is evaporated to dryness, then the residue is redissolved in 5 ml benzene to add to the silica column. Acetone extracts less oil from dry crops than methanol and makes the cross extraction with *n*-hexane or chloroform unnecessary.

Detector: Dorhmann M-100 with sulfur titration cell S-200.

(c) *Calculation.* The height of the Vitavax peak is measured from the calibration graph the equivalent concentration of Vitavax is determined. Dividing this figure by 10 gives the concentration of Vitavax in parts per million when a 50-gm sample is taken.

REFERENCES

Chin, W. T., Stone, G. M., and Smith, A. E. (1970). *J. Agr. Food Chem.* **18,** 709–712.

Edgerton, L. J., Rocky, M. L., Arnold, H., and Lisk, D. J. (1967). *J. Agr. Food Chem.* **15,** 812.

Lane, J. R. (1970). *J. Agr. Food Chem.* **18,** 409–412.

Puchalsky, C. B. (1969). *Anal. Chem.* **41,** 843–844.

Sisken, H. R., and Newell, J. E. (1971). *J. Agr. Food Chem.* **19,** 738–741.

von Schmeling, B., and Kulka, M. (1966). *Science* **142,** 659.

Part IV

Herbicides

21

Balan®

W. S. JOHNSON AND R. FRANK

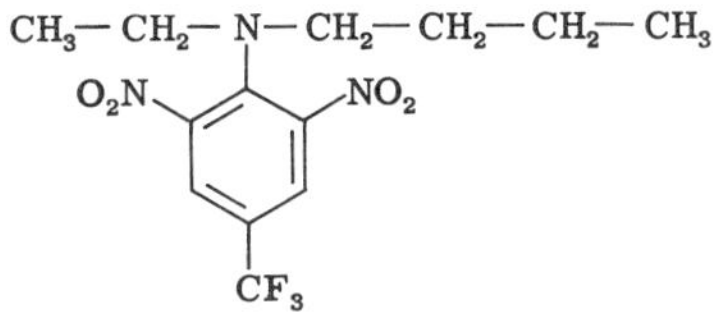

N-Butyl-*N*-ethyl-α,α,α-trifluoro-2,6-dinitro-*p*-toluidine

I. GENERAL

A. Empirical Formula

$C_{13}H_{16}F_3N_3O_4$ (Mol. wt. 335.3).

B. Alternative Names

Benefin, benfluralin, Quilan.® Balan® and Quilan® are registered trademarks of Elanco Products Company, Division of Eli Lilly and Company.

C. Source of Analytical Standard

Elanco Products Company, 640 S. Alabama Street, Indianapolis, Indiana.

D. Biological Properties

Balan® is a selective pre-emergence herbicide that controls a wide variety of undesirable grasses and broadleaf weeds. A partial list of tolerant plants includes lettuce (direct seeded and transplant), established turf grass, peanuts, transplanted tobacco, red clover, alfalfa, and other small-seeded legumes.

The acute oral LD_0 of technical Balan® for adult male and female rats is greater than 10 gm/kg and for newborn rats the LD_{50} is 0.8 gm/kg. For the dog, chicken, and rabbit the acute oral LD_0 is greater than 2 gm/kg. No hazards have been found to result from skin contact or inhalation of the technical material. Adult male and female rats and dogs were fed or dosed with Balan® for 3 months at levels of 1250 and 500 ppm, respectively. Gross and histologic examination of these animals indicated there were no changes that could be attributed to the Balan® treatment. Balan® dissolved in acetone and added to aquaria containing bluegill fingerlings has an LD_{50} in the range of 375 ppb.

E. History

Balan® is one of a class of substituted dinitroanilines whose herbicidal properties were first reported by Alder *et al.* (1960). Its use as a selective preemergence annual grass and broadleaf herbicide for use on established turf grass and agronomic crops was first reported by Ford (1965).

F. Physical Properties

Pure Balan® is a yellow-orange crystalline solid that melts at 67–68°C. It is readily soluble in organic solvents such as acetone and xylene. Its solubility in water is less than 1 ppm at 25°C. The vapor pressure of Balan® was determined to be 7.8×10^{-5} mm Hg at 25°C, and the boiling point is 121–122°C at 0.5 mm Hg. The ultraviolet absorption spectrum of a dilute methanolic solution exhibits maxima at 275 and 376 nm, the latter giving a molar absorptivity of 2388.

G. Chemical Properties

1. METHOD OF SYNTHESIS

Balan® is prepared commercially by nitration of 4-chloro-α,α,α-trifluorotoluene, followed by reaction with ethyl-*n*-butylamine.

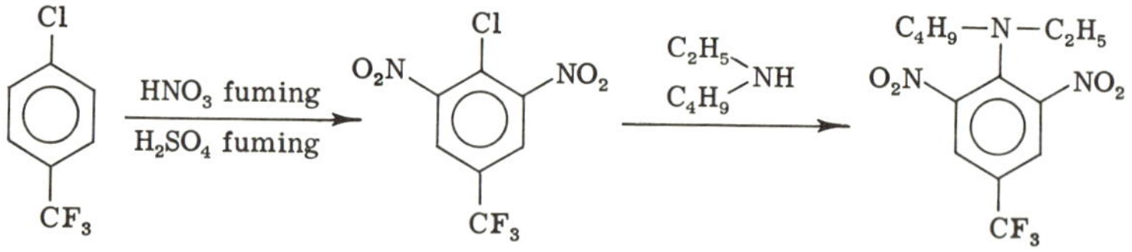

2. CHEMICAL STABILITY

Balan® is subject to decomposition by ultraviolet radiation.

H. Formulation

Balan® is formulated as a 1.5 pound per gallon liquid concentrate and as 2.5% granules. It is also available as a 93% technical material and 50% milled concentrate.

II. ANALYSIS

A. Formulations Analysis

1. REVIEW OF METHODS

Two methods of analysis are recommended, one being a spectrophotometric measurement and the other a gas chromatographic measurement. Fuzeri (1971) describes a formulation assay method by gas chromatography using an electron capture detector. Hambleton (1971) has assayed Balan® formulations by gas chromatography with temperature programming using a flame ionization detector.

2. SPECTROPHOTOMETRIC METHOD

a. *Principle*

Balan® is measured by determining the absorbance of solutions in hexane at 376 nm. Technical material and liquid formulations are dissolved directly in hexane. Solid formulations are extracted with chloroform, the solvent removed from the extract and the residue dissolved in hexane. Interfering colored impurities are removed by Florisil column chromatography using hexane as eluant.

b. *Reagents*

Florisil, 100–200 mesh, Floridin Company, Berkeley Springs, West Virginia.

Sodium sulfate, anhydrous granular, analytical reagent.

n-Hexane, analytical reagent.

Balan® standard, 1.25 mg/ml in *n*-hexane.

Chloroform, analytical reagent.

c. *Apparatus*

Chromatographic column, 10 mm × 400 mm fitted with stopcock.

Chromatographic column, 25 mm × 400 mm fitted with stopcock.

Soxhlet extraction apparatus.

Rotary vacuum evaporator, Rinco or equivalent.

Spectrophotometer, Model DU, Beckman Instrument Co., or equivalent.

d. *Experimental Procedure*

i. *Preparation and Standardization of Florisil*

Determine the loss on drying of the Florisil as received by heating for 6 hours at 110°C. Add sufficient water to a weighed amount of Florisil to give a total moisture content of 10%. Mix well and store the mixture in a sealed container.

Insert a pledget of glass wool in the bottom of a 25 mm × 400 mm glass chromatographic column. Add, with constant tapping, 2.5 gm sodium sulfate, 5.0 gm prepared Florisil, and 2.5 gm sodium sulfate. Open the stopcock, add 10 ml *n*-hexane and allow to drain until the liquid level reaches the top of the upper layer of sodium sulfate. Transfer 5.0 ml of the Balan® standard (1.25 mg/ml) to the column and allow the liquid to drain to the top of the upper sodium sulfate layer. Rinse the wall of the column three times with 4-ml portions of *n*-hexane. Allow each rinse to run into the column before adding the next. Elute the column with *n*-hexane until the intense yellow band is almost to the bottom of the Florisil layer. Discard the eluate to this point. Place a 100-ml volumetric flask under the column and continue the elution until the yellow band is completely removed from the column. Dilute to volume with *n*-hexane. Normally, 60–90 ml of eluate should be collected. If less than 60 ml is required, reduce the moisture content of the Florisil slightly. If the volume collected is greater than 90 ml, increase the moisture content of the Florisil. The amount of Balan® recovered from the column may be determined by comparing the absorbance value at 376 nm for the column eluate with the value obtained for 5.0 ml of Balan® standard diluted to 100 ml.

ii. *Sample Preparation*

(a) *Technical Balan® and Emulsifiable Concentrate.* Accurately weigh a sample equivalent to 125 mg of Balan®. Transfer the sample to a 100-ml volumetric flask and dilute to volume with *n*-hexane.

(b) *50% Milled Concentrate.* Accurately weigh a sample equivalent to 125 mg of Balan® and transfer to a 100-ml volumetric flask. Add chloroform to the mark and mix for 1 minute. Allow to stand undisturbed for 30 minutes and then transfer a 10-ml portion to a boiling flask. Evaporate the sample to dryness on a rotary vacuum evaporator and then dissolve the residue in 10 ml hexane.

(c) *2.5% Granules.* Insert a pledget of glass wool in the bottom of a 10 mm × 400 mm glass column and add about $\frac{1}{2}$ inch of sodium sulfate. Accurately weigh a sample of granular formulation equivalent to 125 mg of Balan® and transfer to the column. Gently tap the column to pack the sample and then layer $\frac{1}{2}$ inch of sodium sulfate over the top. Open the stopcock and set a 100-ml volumetric flask in place to collect the eluate. Add chloroform to the column and adjust the flow rate to approximately 10 ml/minute. Pass sufficient chloroform over the column until 100 ml has been collected. Stopper the flask and mix well. Transfer a 10-ml portion of the extract to a boiling flask and evaporate to dryness on a rotary vacuum evaporator. Dissolve the residue in 10 ml *n*-hexane.

iii. *Florisil Chromatography*

Prepare Florisil columns as described in Section i. Transfer a 5.0-ml portion of the *n*-hexane solutions of Balan® standard and sample to separate columns and allow the liquid to drain to the top of the upper sodium sulfate layer. Rinse the walls of the column three times with 4-ml portions of *n*-hexane and allow each rinse to run into the column before adding the next. Proceed with the elution as previously described and collect the volume of eluate determined in the standardization of the Florisil. Add sufficient *n*-hexane to the eluate to make 100 ml and mix well.

iv. *Measurement*

Measure the absorbance of the standard and sample solutions in a 1-cm cuvette at 376 nm using *n*-hexane as the reference solvent. The analyst should determine the linear response of the spectrophotometer and employ a standard response curve in making calculations if necessary.

v. *Calculation*

$$\% \text{ Balan}^{\circledR} = \frac{A_{\text{sample}}}{A_{\text{standard}}} \times \frac{\text{concentration of standard (mg/ml)} \times 200{,}000}{\text{sample weight (mg)}}$$

vi. *Discussion*

The spectrophotometric method has a coefficient of variation of 1.1%.

3. GAS CHROMATOGRAPHIC METHOD

a. *Principle*

Balan® is measured by flame ionization gas chromatography with the column oven operated at isothermal conditions. Technical material is dissolved directly in chloroform, and formulated products are extracted with or diluted directly with chloroform. The chloroform solution is injected into the gas chromatograph. Diethyl phthalate is used as an internal standard.

b. *Reagents*

Chloroform, analytical reagent.
Balan® analytical standard, 1.0 mg/ml in chloroform.
Diethyl phthalate internal standard, 1.1 mg/ml in chloroform.
Sodium sulfate, anhydrous, analytical reagent.

c. *Apparatus*

Glass column, 10 mm × 400 mm fitted with a stopcock.
Rotary vacuum evaporator, Rinco or equivalent.

Gas chromatograph, Hewlett-Packard model 402, equipped with a flame ionization detector, model 7621A automatic liquid sampler, and model 7123A 1-mV recorder, or equivalent GC system. The chromatographic column is 120 cm × 3 mm i.d. glass tubing packed with 5% Carbowax 20M on 100–200 mesh Chromosorb W-HP (Applied Science Laboratories, State College, PA). The column is conditioned at 240°C under carrier gas purge for 16 hours prior to use.

d. *Experimental Procedure*

i. *Sample Preparation*

(a) *Technical Material, 50% Milled Concentrate, and Emulsifiable Concentrate.* Accurately weigh a sample equivalent to 100 mg Balan® and transfer to a 100-ml volumetric flask. Dilute to volume with chloroform and mix well.

(b) *2.5% Granules.* Place a glass wool pledget in the bottom of a 10 mm × 400 mm glass column and add about ½ inch of anhydrous sodium sulfate. Accurately weigh a sample of granular formulation equivalent to 100 mg of Balan® and transfer to the column. Tap the column gently to pack the granules and then layer ½ inch of sodium sulfate on the top of the sample. Open the stopcock and add 100 ml chloroform to the column. Adjust the flow to give a flow rate of about 10 ml/minute, and collect the effluent in a 100-ml volumetric flask. Add sufficient chloroform through the column to bring the liquid level to the mark on the flask, stopper, and mix well.

ii. *Gas Chromatography*

Flash heater temperature: 240°C.
Column temperature: 180°C.
Detector temperature: 230°C.
Nitrogen carrier flow: 35 ml/minute.
Hydrogen flow: 35 ml/minute.
Oxygen flow: 200 ml/minute.
Electrometer: 1.28×10^{-9} AFS (range 10, attenuation 32).

Pipette 1.0 ml of the chloroform solution of the sample and 1.0 ml of the diethyl phthalate into a small vial. Stopper the vial and mix well. Prepare a vial of standard Balan® solution in the same manner. Inject approximately 2 μl of the sample and standard into the gas chromatograph. Measure the peak response (area or peak height) for the Balan® and diethyl phthalate peaks.

iii. *Calculations*

Calculate the response ratio R for the Balan® and diethyl phthalate peaks:

$$R = \frac{\text{peak response Balan}^{\circledR}}{\text{peak response diethyl phthalate}}$$

Calculate the percent Balan® in the sample by the formula:

$$\%\ \text{Balan}^{\circledR} = \frac{R_{\text{sample}}}{R_{\text{standard}}} \times \text{concentration of standard (mg/ml)} \times \frac{100\ \text{ml}}{\text{sample weight (mg)}} \times 100$$

Prior to the measurement of experimental samples, the analyst should determine the linear response of the gas chromatograph and employ a standard response curve in making calculations if necessary. The Balan® standard should be injected frequently during the course of the measurement of a series of samples to insure the instrument response has not changed.

e. *Discussion*

The gas chromatographic method has a coefficient of variation of about 1.5%. Balan® has also been successfully measured by gas chromatography using the following column packing–internal standard combinations: 3% JXR–*n*-octacosane; 5% XE-60–lindane; and 2% OV-17–eicosane.

B. Residue Analysis

1. RECOMMENDED METHOD

a. *Principle*

Balan® is quantitatively extracted from plant tissue and soil with methanol. The extract is purified by liquid–liquid extraction with methylene chloride and by column chromatography over Florisil. Detection and measurement are made by gas–liquid chromatography using an electron capture detector.

b. *Reagents*

Analytical standard stock solution. Accurately weigh 10 mg of analytical standard, transfer quantitatively to a 200-ml volumetric flask with benzene, and dilute to volume (50 μg/ml). Store in a refrigerator and prepare fresh every 3 months. Dilutions should be protected from light and be prepared fresh weekly.

Methanol, analytical reagent.

Sodium chloride, analytical reagent, 5% in deionized water.

Methylene chloride, analytical reagent, redistilled before use.

n-Hexane, analytical reagent, redistilled before use.

Florisil, 100–200 mesh, Floridin Co., Berkeley Springs, West Virginia, deactivated and standardized as follows: Determine the water content of the Florisil as received by heating to 600°C for 4 hours. Add sufficient water to an appropriate amount of Florisil to yield a total water content of 8.5–9.0%. After addition of the water, stir well with a glass rod to break up lumps and tumble in a closed container for 1 hour. Allow to stand for an additional 2 hours prior to use.

Prepare a Florisil column by placing 10–15 ml of hexane in a 14-mm i.d. glass chromatographic column. Insert a glass wool pledget and tamp in place with a glass rod to eliminate trapped air. Add 4.8 gm deactivated Florisil and allow to settle. Stir with a glass rod to eliminate air bubbles and start to drain the column. Wash down the sides of the column with hexane to remove adhering Florisil. Carefully layer 2–3 cm anhydrous sodium sulfate on the surface of the Florisil. Continue to drain the hexane until the liquid layer reaches the top of the sodium sulfate layer.

Add 100 μg of Balan® dissolved in 5 ml hexane to the column and start to drain. Rinse the flask with two successive 5-ml portions of hexane, allowing each to go into the column before the next addition. When the last hexane rinse reaches the top of the sodium sulfate, add 65 ml hexane to the column, allow to drain at a flow rate of 3–5 ml/minute. Note the volume of solvent required to move the leading edge of the yellow band to within 1 inch of the bottom of the Florisil layer. Discard this amount of eluate as a forerun. Continue the elution of the yellow band and record the volume of hexane required to completely remove the yellow color from the column. This volume plus an additional 15 ml will be collected in the column chromatography cleanup step to insure complete removal of the compound.

Benzene, analytical reagent, redistilled before use.

c. *Apparatus*

Blender, Omni-mixer (Ivan Sorvall Co.) or equivalent.

Gas chromatograph, Hewlett-Packard model 5713A, equipped with a model 18731A Ni-63 linear EC detector, model 7621A automatic liquid sampler, and a model 7123A 1-mV recorder. The chromatographic column is 120 cm × 3 mm i.d. glass tubing packed with 5% XE-60 on 80–100 mesh Chromosorb W-HP (Applied Science Laboratories, State College, PA). The column is conditioned at 240°C. for 16 hours prior to use.

d. *Experimental Procedure*

i. *Sample Preparation*

Weigh a representative finely ground 25-gm sample of plant tissue or soil into a quart Mason jar. Add sufficient methanol to the plant sample to make 200 ml total volume after allowance is made for the water content of the sample. Add 200 ml methanol to soil samples. Blend the sample at moderate speed for 5 minutes on an Omni-mixer. Filter a sufficient volume of the extract by gravity through Whatman No. 1 filter paper to provide a 20-ml portion. Transfer the 20-ml portion to a 125-ml separatory funnel containing 40 ml 5% sodium chloride solution. Extract the aqueous methanol solution twice with 20-ml portions of methylene chloride, passing each extract through sodium sulfate into a 125-ml boiling flask. Evaporate the methylene chloride on a rotary vacuum evaporator using a water bath at 50°C or less. Remove the flask from the evaporator immediately upon complete evaporation of the methylene chloride. Prepare a Florisil column as described under Florisil standardization. Dissolve the residue in the flask in 5 ml hexane, transfer to the column, and start to drain. Rinse the flask with two successive 5-ml portions of hexane, allowing each to go into the column before the next addition. Add an additional 65 ml of hexane to the column and allow to drain at a flow rate of 3–5 ml/minute. Collect the volume of effluent as determined by the Florisil standardization. Remove the solvent by rotary vacuum evaporation using a 50°C water bath.

Recovery samples are prepared by adding 1.25 μg of Balan® in methanol to 25 gm of control plant tissue, or 2.5 μg of Balan® to 25 gm of control soil.

ii *Gas Chromatography*

Column temperature: 140°C.
Detector temperature: 300°C.
Injection temperature: 250°C.
Electrometer attenuation: setting to provide 30–40% of full scale deflection on injection of 0.3–0.4 ng of Balan®. For the gas chromatograph described, the cell current is constant 10^{-9} A, and the signal

measured is the pulse frequency required to maintain the standing current.

Dilute 10.0 ml of the standard Balan® solution (50 μg/ml) to 50 ml with benzene. Pipette 0.05, 0.10, 0.20, 0.40, and 0.50 ml of this solution (10 μg/ml) into 20-ml volumetric flasks and dilute to volume with benzene. Inject 2 or 3 μl of each solution into the gas chromatograph. Plot peak response versus nanogram injected to obtain the standard response curve. Check the response with each set of samples assayed, and frequently inject a standard dilution (e.g., 0.1 μg/ml) during the course of the assay to determine whether instrument response has changed.

Dissolve the crop extract in 1.0 ml benzene and the soil extract in 2.0 ml benzene. Inject the same volume of extract as used for the response curve. One microliter is equivalent to 2.5 mg of the crop tissue and 1.25 mg of soil.

iii. *Recovery and Sensitivity*

The procedure as described gives a recovery of about 85% after fortification of crops at 0.05 ppm and 0.10 ppm in soils. Recovery of Balan® at lower levels indicates the assay is sensitive to 0.01 ppm.

iv. *Interferences*

A thin-layer chromatographic (TLC) procedure is described to provide a means for further purification of the sample extract and for confirmation of Balan® residues found in crops or soils when desired. Evaporate the benzene extract used for gas chromatographic measurement to dryness and reconstitute in 0.5 ml hexane. Apply 100 μl of the hexane solution to a precoated silica gel TLC plate (0.25 mm, Brinkmann Instrument Co.). Marker zones with 25 μg of Balan® are spotted on the plate to locate the R_f for the sample zone. Develop the plate once in an unlined tank containing carbon tetrachloride. The development time is about 3 hours for 15 cm solvent travel.

Scrape the zone in the sample lane corresponding to the R_f of the marker zones. Transfer the scrapings to a 1-dram vial and add 0.4 ml benzene. Agitate gently for a few minutes and allow the silica gel to settle. Inject a portion of the supernatant benzene into the chromatograph. The TLC sample is now at twice the dilution of the original sample, and the dilution factor must be taken into account when calculating the amount of residue found.

v. *Sample Calculation*

The number of nanograms of Balan® present in the number of milligrams of plant tissue or soil injected into the gas chromatograph may be obtained from the standard curve. The value obtained may be corrected to reflect assay efficiency as determined from recovery samples.

e. *Discussion*

All evaporations should be performed by thin-film rotary vacuum evaporation to prevent loss of the compound. Use of a current of air for evaporation must be avoided. Balan® is unstable to light and should be protected from undue exposure during the assay. Routine analysis should be completed within one working day; however, if the analysis must be interrupted, the samples must be stored overnight in hexane or benzene protected from light.

Crop samples having a high oil content may leave an oily residue in the flask upon evaporation of the methylene chloride. The oil may change the elution pattern on the Florisil. To remove the oil from the sample, dissolve the extract in 20 ml hexane and extract twice with acetonitrile. Combine the extracts, evaporate on the rotary evaporator, and proceed with the column chromatography.

References

Alder, E. F., Wright, W. L., and Soper, Q. F. (1960). *Proc. N. Cent. Weed Contr. Conf.* **17,** 23.

Ford, D. H. (1965). *Proc. Annu. Calif. Weed Contr. Conf., 7th, 1965* p. 129.

Fuzeri, M. (1971). *J. Ass. Offic. Anal. Chem.* **54,** 711.

Hambleton, L. G. (1971). *J. Ass. Offic. Anal. Chem.* **54,** 125.

22

Devrinol®

G. G. PATCHETT, D. J. BROOKMAN, AND J. E. RODEBUSH

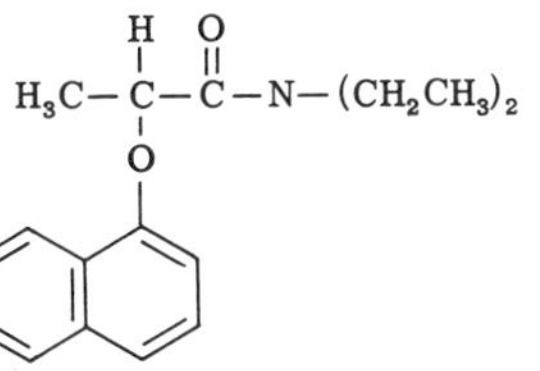

2-(α-Naphthoxy)-*N*,*N*-diethyl propionamide

I. GENERAL

A. Empirical Formula

$C_{17}H_{21}NO_2$ (Mol. wt. 271.36).

B. Alternative Names

Stauffer Code R-7465, Orion® (French). Devrinol® and Orion® are the registered trademarks of the Stauffer Chemical Company.

C. Source of Analytical Standard

Stauffer Chemical Company, Richmond Research Center, 1200 South 47th Street, Richmond, California.

D. Biological Properties

Devrinol® is a selective herbicide that controls most annual grass weeds and many annual broadleaf weeds when applied to the soil before emergence of the weeds. Effective reduction in nutsedge populations also occurs at recommended rates. Devrinol® is registered for use in established or newly planted almonds, apricots, cherries, grapes, nectarines, oranges, peaches, plums, prunes, and tomatoes. Other tolerant crops include asparagus (established), cole crops, eggplant, English pea, Irish potatoes, oilseed rape, peanuts, peppers, radish, safflower, sunflower, tobacco, turnip, watermelon, many established ornamentals, and ground covers. Devrinol® may be applied to the soil surface prior to weed emergence where ample rain occurs soon after application; however, best results are obtained when it is incorporated in the soil to a depth of 1–2 inches. Devrinol® may also be applied to established weeds if used with a contact herbicide and rain follows soon after application.

Devrinol® exhibits very little soil movement. Microbial decomposition occurs in moist loam soils, resulting in a half-life of 8–12 weeks. Moderate photodecomposition occurs under high sunlight intensity if the herbicide remains on the soil surface.

Devrinol® is relatively nontoxic to animals, with an acute oral LD_{50} to male and female rats greater than 5000 mg/kg and an acute dermal LD_{50} to rabbits greater than 4640 mg/kg. Devrinol® is nonirritating to the eyes of rabbits.

E. History

Devrinol® was originally synthesized and developed at the laboratories of the Western Research Centers of the Stauffer Chemical Company (Tilles *et al.*, 1969; Baker *et al.*, 1973).

F. Physical Properties

Pure Devrinol® is a white crystalline solid with a melting point of 74.8–75.5°C. Its solubility in water is 73 ppm at 20°C. Devrinol® is soluble to the extent of 10% (w/v) or more in most organic solvents, including toluene, ethanol, benzene, chloroform, methyl isobutyl ketone, and acetone. The vapor pressure of pure Devrinol® is 0.004 μm at 25°C and 0.1 μm at 50°C.

G. Chemical Properties

1. Method of Synthesis

Devrinol® can be prepared by reacting 2-(α-naphthoxy)propionic acid with *N,N*-diethylamine in the presence of phosphorus oxychloride.

2. Chemical Stability

Devrinol® is stable to hydrolysis in aqueous, acidic, or basic solutions. At pH values of 4, 7, and 10, no decomposition was observed after 9 weeks. Devrinol® is fairly stable with respect to thermal degradation. At 90°C, its half-life was 71 days; at 110°C its half-life was 14 days.

H. Formulations

Devrinol® is formulated as a wettable powder (designated 50-W) containing 50% by weight of active ingredient. Devrinol® is also formulated as an emulsive liquid (designated 2-E) containing 2 pounds of the active ingredient per gallon.

II. FORMULATION ANALYSIS

1. RECOMMENDED METHOD

a. *Principle*

Devrinol® is determined by gas chromatography using diisooctyl adipate as internal standard. A liquid emulsifiable formulation is diluted to the appropriate concentration with mixed solvents. A wettable powder formulation is extracted with mixed solvents known to quantitatively extract the active ingredient; sufficient solvent is used to produce the appropriate concentration of Devrinol® in the extract.

b. *Apparatus*

Gas chromatograph, Hewlett-Packard model 7620 equipped with a flame ionization detector.

Column, 1.5 feet × 0.25 inch o.d. (0.16 inch i.d.) Pyrex glass packed with 12% SE-30 on 60–80 mesh Gas Chrom Q.

c. *Experimental Procedure*

Weigh, to the nearest 0.1 mg, approximately 200 mg of the Devrinol® 50-W formulation or 450 mg of the 2-E formulation into a 1-ounce narrow-mouth screw-cap bottle. Pipette into the bottle exactly 15.00 ml of an extracting solvent consisting of 80:15:5 (v/v) carbon disulfide–chloroform–acetone, and containing 8 mg diisoctyl adipate per milliliter. Cap thc bottlc with a Polyseal-lined lid, and shake it briefly to dissolve thc 2-E formulation. Shake the bottle for 15–30 minutes for the 50-W formulation using a wrist-action shaker.

Prepare a calibration solution in a similar manner, substituting analytical standard Devrinol® for the formulation; use 100 mg of the standard material weighed to the nearest 0.1 mg. Perform successive gas chromatographic injections of the calibration solution; operate the gas chromatograph using the following conditions:

Column oven temperature: 225°C.
Injection port temperature: 225°C.
Detector temperature: 240°C.
Carrier gas (nitrogen): 80 ml/minute.

Hydrogen: 30 ml/minute.
Air: 300 ml/minute.
Volume injected: 1.5 μl.

Under these conditions the approximate retention times for Devrinol® and diisooctyl adipate are 2.7 and 5.7 minutes, respectively.

When successive injections of the calibration solution yield area or peak height ratios (Devrinol® versus diisooctyl adipate) reproducible to within 1% relative, begin injections of the sample solutions. Make replicate injections of each sample solution; follow each series with injections of the calibration solution.

Calculate percentage of Devrinol® as follows:

$$\%(\text{weight})\ \text{Devrinol}^{\circledR} = \frac{R \times A_D}{W_S \times A_I}$$

where $R = (W_D' \times A_I' \times P)/A_D'$ = response factor

and P = weight percent purity of Devrinol® analytical standard
W_D' = weight of standard Devrinol® in calibration solution
A_D' = area or peak height of Devrinol® in calibration solution
A_I' = area or peak height of diisooctyl adipate in calibration solution
A_D = area or peak height of Devrinol® in sample solution
A_I = area or peak height of diisooctyl adipate in sample solution
W_S = weight of sample.

2. DISCUSSION OF METHOD

The precision obtainable with this method is 1% relative, based on the level of Devrinol®. The method is linear when the level of Devrinol® in a sample is within 10% of that in the calibration solution.

III. RESIDUE ANALYSIS

1. RECOMMENDED METHOD

a. *Principle*

Most crop samples are extracted with benzene, cleaned up using a combination of one or two liquid–solid chromatographic procedures, and quantitatively determined by gas chromatography using the Coulson electrolytic conductivity detector in the nitrogen mode.

b. *Reagents*

Solvents, benzene, acetonitrile, ethyl acetate, hexane and acetone, all distilled in glass; methanol, Reagent Grade.

Alumina, aluminum oxide, Woelm, acidic, activity grade I.

Silicic acid, Mallinckrodt SilicAR CC-4, 200–325 mesh, pH4.

Activated carbon, Kensco no. K3211, acid-washed Nuchar, Kensington Scientific Corp., Emeryville, California.

Sodium sulfate, anhydrous, Reagent Grade.

c. *Apparatus*

Gas chromatograph, Tracor model MT-220 equipped with a Coulson electrolytic conductivity detector, model C-321.

GLC column, 110 cm × 2 mm i.d. Pyrex, packed with 10% OV-17 on 80–100 mesh Gas Chrom Q, Applied Science Laboratories, Inc., State College, Pennsylvania.

Waring Blendor, 1-quart capacity.

Chromatography tubes, Pyrex, 15 cm long × 11 mm i.d., constricted lower end to hold Pyrex wool plugs. An 80-ml pear-shaped reservoir is fitted to the top of the tube with a neoprene rubber stopper.

Compressed air, regulated for use at 0–10 psi.

Centrifuge tubes, 40 ml, graduated, ₮ 14/15 ground glass neck.

Rotary evaporator, water aspirator vacuum system, with ₮ 14/15 ground glass adaptor, and 35–40°C water bath.

2. EXPERIMENTAL PROCEDURE

a. *Sample Extraction*

i. *Oily Crops (Nuts, Avocados)*

Weigh 100 gm of shelled nuts or avocados into a Waring Blendor and blend them with 400 ml of acetonitrile for 5 minutes. Filter the mixture through a suitable filter paper.

ii. *Soil and Green Tobacco*

Weigh 400 gm of a representative sample of soil or 100 gm of tobacco into a 1-quart Waring Blendor jar. Add 400 ml of methanol–water 60:40 (v/v) to the sample to obtain a 1.0 gm/ml sample/solvent ratio. Blend this mixture for 2 minutes in the blender, pour the mixtures into a centrifuge bottle, and centrifuge it for 15 minutes at 1800 rpm. Tobacco samples may be filtered directly without centrifuging. Decant the liquid phase into a funnel fitted with a no. 7 Whatman filter paper. Collect at least 200 ml of the filtrate.

iii. *All Other Crops and Tissues*

A suitable portion of the prepared sample is extracted with benzene for 5 minutes in a Waring Blendor. The recommended benzene ratios are listed in Table I. After blending, allow the phases to separate, and filter the mixture through a suitable filter paper. For rapeseed, add 15 ml 10%

TABLE I
SAMPLE EXTRACTION RATIOS

Crop	Weight (gm)	Water (ml)	Benzene (ml)
Tobacco, dried	25	25	200
Cottonseed	50	–	500
Rapeseed	50	–	500
Cotton plants	100	–	500
All other fresh samples	100	–	330

aluminum sulfate, and allow the mixture to stand overnight prior to the filtration step.

b. *Cleanup*

Table II lists the cleanup procedures which are recommended for the different sample types. Descriptions of these procedures follow.

i. *Acetonitrile Evaporation*

For samples extracted with acetonitrile, transfer a portion containing 5 gm of sample to a 40-ml centrifuge tube and evaporate the solution to dryness on a rotary evaporator. Dissolve the residue in 5 ml benzene by shaking the sample vigorously 1 minute. Centrifuge the sample for 5 minutes at 1800 rpm. Remove the benzene with a pipette and add the sample to the alumina column described in Section v, below. Repeat the benzene partition step twice more, adding each extract to the column. Follow these with an additional 10 ml benzene. Add 25 ml of 15:85 ethyl acetate–benzene to the column and discard all eluates. Elute the Devrinol® with 20 ml of 1:1 acetone–benzene. Evaporate this eluate to dryness on a rotary evaporator, then reconstitute the residue with 0.25 ml benzene and retain this solution for gas chromatographic analysis.

ii. *Hexane–Acetonitrile Partition*

Transfer a portion of the benzene extract containing 5 gm of sample into a 200-ml round-bottom flask, and remove the solvent under vacuum using the rotary evaporator. Transfer the residual oil to a 40-ml centrifuge tube with 15 ml hexane followed by 10 ml acetonitrile. Shake the mixture vigorously and centrifuge it for 10 minutes at 2000 rpm. Transfer the clear acetonitrile phase to a clean 40 ml tube, add 30 ml benzene, and proceed to the alumina column no. 2 cleanup in Section vi, below.

iii. *Methanol–Water–Benzene Partition*

Measure 200 ml of the filtrate from Section ii, above, into a 500-ml separatory funnel. For tobacco use a 6-gm portion (usually 24 ml). For

TABLE II
Cleanup Procedures[a]

	ACN evaporation	Hexane–ACN	Methanol partition	Nuchar–SilicAR CC-4	Alumina no. 1	Alumina no. 2	Florisil		ACN evaporation	Hexane–ACN	Methanol partition	Nuchar–SilicAR CC-4	Alumina no. 1	Alumina no. 2	Florisil
Almonds	+				+			Kidney		+				+	
Apples					+			Lemons				+	+		
Apricots				+	+			Liver		+				+	
Avocado	+				+			Muscle						+	
Bananas					+			Nectarines				+	+		
Bell peppers				+	+			Oranges				+	+		
Blackberries				+	+			Peaches				+	+		
Blueberries				+	+			Pears					+		
Broccoli				+	+			Pecans	+				+		
Brussels sprouts				+	+			Pistachio nuts	+				+		
Cabbage				+	+			Plums				+	+		
Cauliflower				+	+			Potatoes					+		
Cherries				+	+			Quince					+		
Corn ears, plants				+	+			Rapeseed		+				+	+
Cottonseed						+		Rutabagas				+	+		
Cottonplants						+		Soil			+				
Cranberries					+			Sugarbeets					+		
Eggplant				+	+			Sugarcane					+		
Fat		+				+		Tomatoes				+	+		
Figs				+				Tobacco, green			+	+	+		
Filberts	+				+			Tobacco, dried		+		+	+		
Grapefruit				+	+			Walnuts	+				+		
Grapes				+	+										

[a] "+" indicates inclusion of specific cleanup step described in text.

soil, add 5 drops concentrated hydrochloric acid and 10.0 ml benzene. For tobacco, add 12 ml benzene. Stopper the funnel and shake it vigorously for 1 minute. Add 300 ml water, shake the funnel for another 30 seconds, and let it stand until the benzene separates out and the aqueous phase is moderately clear (about 1 hour). Draw off the aqueous phase and discard it. For soils, draw off the benzene phase into a 1-ounce amber bottle, cap the bottle, and retain the sample for gas chromatography. For tobacco, proceed to the Nuchar–SilicAR column using 10 ml extract from this step.

iv. *Nuchar–SilicAR CC-4 Column*

Add 3.0 gm 1:19 Nuchar–SilicAR CC-4 to a chromatography tube containing a plug of Pyrex wool, and top it with about an inch of granular anhydrous sodium sulfate. To the top of the column add a portion containing 5 gm sample in 20–40 ml benzene. Force the solution into the column with about 3 psi pressure and follow it with three 4-ml washes of benzene. Discard the eluate. Elute the Devrinol® with 25 ml of ethyl acetate–benzene 15:85 (v/v), and collect the eluate in a 40-ml centrifuge tube.

This is the only cleanup required for figs. The sample is then evaporated to dryness, reconstituted with 0.25 ml benzene and reserved for gas chromatographic analysis.

v. *Alumina Column No. 1*

Prepare the column by adding 4.0 gm alumina to a chromatography tube containing a plug of Pyrex wool, and top it with about an inch of anhydrous sodium sulfate. To the top of the column, add the ethyl acetate–benzene 15:85 eluate from the Nuchar–SilicAR column (Section iv, above). Force the sample onto the column with a slight amount of air pressure, and discard the eluate. For Brussels sprouts and cauliflower only, rinse with 30 ml acetonitrile–benzene 25:75 (v/v). Elute the Devrinol® with 20 ml of 1:1 acetone–benzene (v/v) and collect the eluate in a 40-ml centrifuge tube. Evaporate the eluate to dryness on a rotary evaporator, reconstitute the residue with 0.25 ml benzene, and retain this solution for gas chromatographic analysis.

All other crops that are extracted with benzene and that do not require the Nuchar–SilicAR CC-4 column cleanup described above are cleaned up using only the alumina column. These include bananas, pome fruit, cranberries, sugar cane, and potatoes. Add a portion of the benzene extract containing 5 gm of sample to the column. Add an additional 10 ml benzene, and continue with 25 ml 15:85 ethyl acetate–benzene (v/v), and 20 ml 1:1 acetone–benzene (v/v) as described for acetonitrile extracts in Section i, above.

vi. *Alumina Column No. 2*

Add 8 gm alumina to a chromatography tube containing a plug of Pyrex wool, and top it with about an inch of granular anhydrous sodium sulfate. To the top of the column add a portion containing 5 gm of sample in 20–40 ml of benzene or 1:3 acetonitrile–benzene (v/v). Force the solution into the column with about 1 psi pressure and follow the solution with three 4-ml washes of 1:3 acetonitrile–benzene (v/v). Add an additional 20 ml of 1:3 acetonitrile–benzene (v/v) to the column and discard the eluate. Elute the Devrinol® into a 40-ml centrifuge tube with 30 ml of 1:1 acetone–benzene (v/v). Evaporate all the solvent from the sample under vacuum using the rotary evaporator. Reconstitute the residual with 0.25 ml benzene and proceed to the gas chromatographic analysis. Rapeseeds require additional cleanup using the Florisil column.

vii. *Florisil Column*

Add 3.5 gm Florisil to a chromatography tube containing a plug of Pyrex wool, and top it with 1 inch of granular anhydrous sodium sulfate. To the top of the dry column, add the residue from the preceding step dissolved in 20 ml benzene. Apply about 3 psi pressure to force the liquid through the column, and add consecutively three 4-ml and one 8-ml washes of benzene. Add 40 ml of 1:19 ethyl acetate–benzene (v/v) to the column, and apply pressure. Discard the first 15 ml of eluate, and collect the Devrinol® in the next 25 ml. Evaporate all the solvent under vacuum using the rotary evaporator, dissolve the residue in 0.25 ml benzene, and proceed to the gas chromatographic analysis.

Florisil adsorptivity may vary with different lots, and the Devrinol® elution zone must be established for each lot. The 5% ethyl acetate eluent should be adjusted in concentration to effect the desired Devrinol® elution rate. The Devrinol® is eluted in a 5- to 10-ml zone, but a 25-ml fraction should be collected to allow for minor zone shifts.

c. *Gas Chromatography*

Column oven temperature: 220°C.
Injection port temperature: 225°C.
Transfer line temperature: 250°C.
Furnace temperature: 820°C.
Helium carrier gas flow: 70 ml/minute.
Hydrogen flow: 50 ml/minute.
Conductivity bridge voltage: 30 V.
Conductivity bridge attenuation: 1×.
Sample size: 10 μl.

Under these conditions Devrinol® should have a retention time of about 4 minutes and a peak height of 1–3 centimeters for 10 ng of Devrinol®. The benzene solutions of standards and samples are injected alternately, and the peak heights are measured for calculations of Devrinol® in the sample. Devrinol® in the standards should be about the same quantity as found in the samples for greatest accuracy. Calculations are made using the following equation:

$$\text{ppm Devrinol}^{\circledR} = \frac{H_S \times C}{H_R \times 20}$$

where H_S = peak height of sample
H_R = peak height of reference standard
C = concentration of Devrinol® in reference standard solution (usually 1.0 ppm)
20 = concentration of sample in final solution (5 gm/0.25 ml).

d. *Detection Limit*

The detection limit is 0.05 ppm of Devrinol® for all the sample types listed in Table III.

e. *Recovery*

Analytical recovery is 80–100% for most sample types as listed in Table II.

TABLE III
RECOVERY OF 0.05 PPM DEVRINOL®

Sample type	Number of samples	Percent recovered (average)	Sample type	Number of samples	Percent recovered (average)
Almonds	11	89	Grapes	8	79
Apples	14	90	Kidney, quail	1	122
Apricots	2	96	Lemons	3	90
Avocado	1	60	Liver, quail	1	90
Bananas	1	68	Muscle, quail	1	86
Bell peppers	3	90	Muscle, chicken	1	86
Blackberries	1	96	Nectarines	1	88
Blueberries	1	100	Oranges	3	95
Broccoli	3	91	Peaches	2	93
Brussels sprouts	1	90	Pears	3	88
Cabbage	2	93	Pecans	2	93
Cauliflower	2	93	Pistachio nuts	1	50
Cherries	1	88	Plums	1	82
Corn ears	1	86	Potatoes	6	93
Corn plants	1	86	Quince	1	93
Cotton plants	1	98	Rapeseed	3	82
Cottonseed	1	88	Rutabagas	2	84
Cranberries	2	82	Soil	25	101
Eggplant	1	76	Sugarbeets	2	92
Fat, quail	1	81	Sugarcane	2	92
Figs	3	101	Tomatoes	26	92
Filberts	2	88	Tobacco, green	3	81
Fish	4	80	Tobacco, dried	1	77
Grapefruit	5	93	Walnuts	2	94

f. *Interferences*

All registered pesticides containing nitrogen have significantly different retention times than Devrinol® and therefore they do not interfere.

3. DISCUSSION

A thoroughly efficient cleanup is required for Devrinol® residue analyses due to the abundance of naturally occurring nitrogen compounds that would otherwise interfere with the gas chromatographic determination. Most crops therefore require the two-column cleanup that is recommended for all fresh fruit and vegetable samples. These procedures are described in Sections I,J,2,b,iv and I,J,2,b,v, and are considered to be the basic cleanup methods for Devrinol®. They may also be applicable to rapeseeds, but have not yet been tested.

REFERENCES

Baker, D. R., Tilles, H., and Dewald, H. C. (1973). U.S. Patent 3,718,455.
Tillis, H., Baker, D. R., and Dewald, H. C. (1969). U.S. Patent 3,480,671.

23

Dinitramine

H. C. NEWSOM

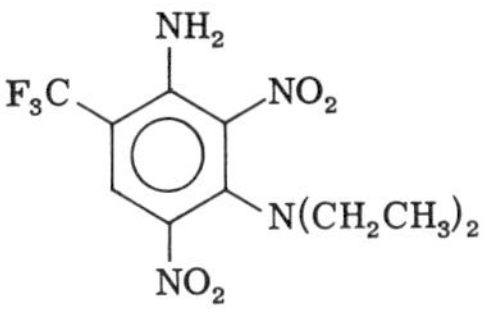

N^3,N^3-Diethyl-2,4-dinitro-6-trifluoromethyl-*m*-phenylenediamine

I. GENERAL

A. Empirical Formula

$C_{11}H_{13}F_3N_4O_4$ (Mol. wt. 322.25).

B. Alternative Names

USB 3584. The emulsifiable concentrate is Cobex®.

C. Source of Analytical Standard and Internal Standard

U.S. Borax Research Corporation, 412 Crescent Way, Anaheim, California.

D. Biological Properties

1. HERBICIDAL ACTIVITY

Dinitramine is a preplant herbicide incorporated $1\frac{1}{2}$–2 inches into the seed bed at a rate of $\frac{1}{3}$–$\frac{2}{3}$ pound per acre, depending on soil type. It is a selective herbicide, finding application in cotton, dry beans, peanuts, and soybeans. Many annual grasses and broadleaf weeds are controlled, including barnyard grass, brachiaria, carpetweed, crabgrass, fall panicum, Florida pusley, foxtail species, goosegrass, johnsongrass, junglerice, lambsquarters, pigweed, purslane, Texas panicum, and smartweed. In addition, one-half to two-thirds population reduction is achieved on annual morning glory, annual ragweed, prickly sida, and velvetleaf.

2. TOXICITY

Test	Species	Result
Acute oral	Rats	LD_{50} 3000 mg/kg
Acute oral	Mallard ducks	LD_{50} 10,000 mg/kg
Acute oral	Bobwhite quail	LD_{50} 1200 mg/kg
Acute dermal	Rabbits	LD_{50} >6800 mg/kg
Eye irritation	Rabbits	Not an irritant
Primary skin	Rabbits	Not an irritant

E. History

Coinventors are D. L. Hunter, W. G. Woods, J. D. Stone, and C. W. LeFevre in U.S. Patents 3,617,252 (1971) and 3,764,623 (1973), assigned to United States Borax & Chemical Corporation.

F. Physical Properties

Dinitramine is a yellow crystalline solid; m.p. 98–99°C; vapor pressure 3.6×10^{-6}mm at 25°C (extrapolated); specific gravity 1.52 g/cc at 25°C (crystal density as determined by ASTM D153-54; 1965). Solubility in water is 1 ppm at 24.5°C, and solubility in organic solvents (wt. % at room temperature):

Hexane	2	Methyl isoamyl ketone	41
Carbon tetrachloride	7	Ethyl acetate	46
Ethanol	10	Dioxane	55
Diethyl ether	24	Acetone	57
Chloroform	29	Tetrahydrofuran	60
Benzene	36	Dimethylformamide	60
Acetonitrile	41	Dimethylsulfoxide	61

G. Chemical Properties

1. METHOD OF SYNTHESIS

Dinitramine is prepared from α,α,α-trifluoro-2,4-dichlorotoluene by dinitration followed by displacements with diethylamine and ammonia, as shown in Eqs. (1)–(3).

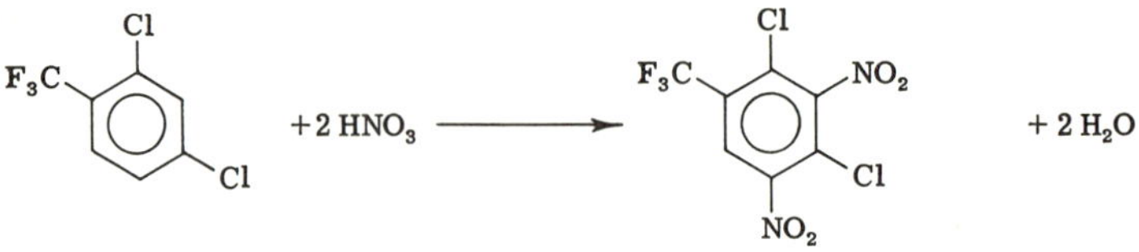

(1)

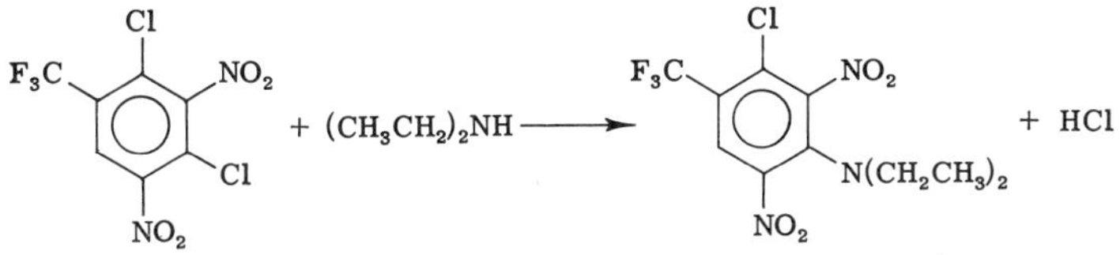

(2)

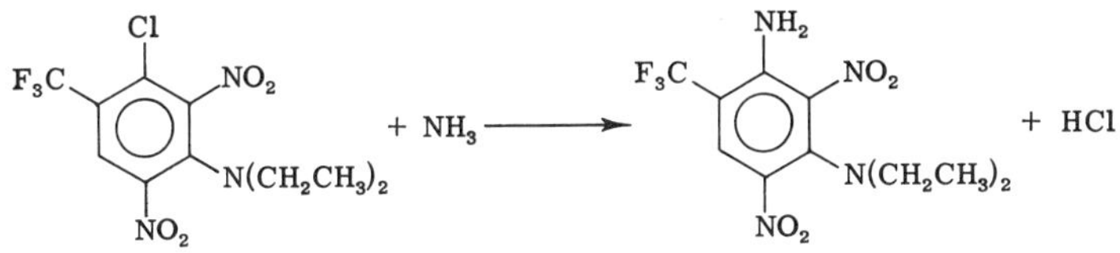

(3)

2. CHEMICAL REACTIONS

Dinitramine melts without decomposition, but when heated in a thermal analyzer, an exotherm is observed near 200°C that is accompanied by conversion of the red liquid to a black liquid with gas evolution. Exposure of a dilute aqueous solution to mild ultraviolet radiation (e.g., sunlight) caused rapid photolytic degradation to two benzimidazoles and their dihydroxy precursor (Newsom and Woods, 1973). Similar compounds were observed in soil degradation (Smith *et al.*, 1973) and in metabolism by specific microorganisms (Laanio *et al.*, 1973). Room temperature hydrolytic stability has been demonstrated over several months at pH 5, 7, and 9. The electron-withdrawing effect of the two nitro groups reduces the basicity of the amino functions and contributes to the general inert chemical reactivity of the molecule.

H. Formulation

Dinitramine is available as a 25% emulsifiable concentrate containing 2 pounds of active ingredient per gallon. The name of the formulated product is Cobex®.

II. ANALYSIS

A. Formulation Analysis

1. REVIEW OF METHODS

Total nitrogen analysis is not appropriate for determination of dinitramine in the formulated Cobex®, since several process impurities are of similar structure with nitro and amino groups.

2. RECOMMENDED METHOD

a. *Principle*

The preferred method for analysis of dinitramine in the formulated material is gas chromatography using an internal standard. A QF-1 column effectively separates dinitramine and the standard from impurities and other components of the formulation, and detection may be by either thermal conductivity or flame ionization.

b. *Procedure*

Weigh out 1.0 gm of the 2 pound/gallon formulated product (Cobex®) and 0.25 gm pure N^3,N^3-dipropyl-2,4-dinitro-6-trifluoromethyl-*m*-phenylenediamine (internal standard), each to the nearest 0.1 mg. Add 5 ml of acetone and dissolve completely. Inject approximately 5 μl into the gas chromatograph* using the following conditions:

Column: $\frac{1}{8}$ inch × 6 feet QF-1, 5%.
Flow rate: 30 cc/minute.
Chart speed: 1 inch/minute.
Program: Isothermal.
Column temperature: 195°C.
Detector temperature: 280°C.
Injector temperature: 220°C.

The active dinitramine and internal standard peaks have respective retention times of 4 minutes and 6.5 minutes. Determine the area of the peaks by digital integrator, disk integrator, or by cutting and weighing. Perform the following calculations:

$$\frac{W_{\mathrm{std}}}{W_{\mathrm{D}}} \times \frac{P_{\mathrm{D}}}{P_{\mathrm{std}}} \times 100 = \text{uncorrected \% purity}$$

Where W_{std} = weight (gm) of internal standard powder (N^3,N^3-dipropyl-2,4-dinitro-6-trifluoromethyl-*m*-phenylenediamine)
W_{D} = weight (gm) of dinitramine formulation
P_{std} = area of GC internal standard peak
P_{D} = area of GC dinitramine peak.

To improve the accuracy, weigh out three mixtures for each sample of Cobex® to be analyzed and inject each solution in duplicate. Determine the average and the standard deviation for the six results.

A response factor is then determined by analyzing a sample of analytically pure dinitramine by the same procedure as above, using approximately 0.3 gm each of dinitramine and internal standard. The "uncor-

*This volume is appropriate for thermal conductivity detection; if a flame ionization detector is used, the volume injected should be 1–2 μl.

rected purity" obtained for the pure material is used to calculate an instrument response factor, as follows:

$$F = 1/\text{uncorrected purity}$$

This factor is usually 1.00–1.03 for thermal conductivity detection and 1.1–1.2 for flame ionization detection.

A second set of three samples of Cobex® is then weighed out and analyzed as before. The average of the two uncorrected purity determinations is calculated. The final assay is obtained by multiplying the average uncorrected purity by the response factor:

$$\text{Assay (\%)} = \text{F} \times \text{average uncorrected purity}$$

The average of twelve results should show a standard deviation σ of less than 0.4%.

B. Residue Analysis*

1. REVIEW OF METHODS

The structure of the dinitramine molecule is such that free electrons are readily picked up and are stabilized in the extended π-cloud system. Consequently, electron capture gas chromatography offers very high sensitivity and good selectivity for dinitramine analysis. A confirmatory method is available that employs alkali flame ionization detection. The methods have been described by Newsom and Mitchell (1972).

2. RECOMMENDED METHOD

a. *Principle*

The method involves methanol extraction, methylene chloride partitioning, Florisil column chromatography, and measurement by electron-capture gas chromatography. It is applicable to soil, forage, and crop samples, and it gives good recoveries down to 0.01 ppm without subtraction of blanks. An alternative method utilizes the same extraction and cleanup procedure, but with alkali flame ionization detection. The latter method also is quantitative at 0.01 ppm.

b. *Apparatus*

A gas chromatograph, such as the Varian Aerograph model 1740 equipped with a tritium foil electron capture detector is required. The column is 5 feet × $\frac{1}{8}$ inch S.S. packed with 3% QF-1 on 100–120 mesh Varaport-30. Prepurified nitrogen flow is 25–35 ml/minute, and column and inlet temperatures are 200° and 225°C, respectively. Detector temperature is 200–210°C.

*Reprinted from *J. Agr. Food Chem.* **20,** 1222–1224 (1972), with permission of copyright owner.

c. *Reagents*

Dinitramine standard solution, 0.1 ppm in Nanograde benzene, is prepared fresh every week and protected from light. Methylene chloride and *n*-hexane are redistilled. Methanol, ethyl ether, and acetonitrile are Analytical Reagent Grade.

d. *Experimental Procedure*

i. *Florisil Standardization*

Batches of 500 gm of 60–100 mesh Florisil are heated to 130°C for 48 hours, and then moistened by thoroughly mixing with 15 gm water and kept in a sealed jar.

The Florisil cleanup column is prepared as follows: To a 22-mm i.d. tube fitted with a Teflon stopcock and a glass wool pledget, add Florisil with light tapping until a 7.5-cm column is formed. Add 2.5 cm of anhydrous sodium sulfate on top of the Florisil. Wash the column with 100 ml *n*-hexane, keeping a layer of hexane on the column. Add 200 μg dinitramine (200 μl of a 1000 ppm benzene solution) to a flask containing 5 ml of 95:5 hexane–ethyl ether. Transfer to the column and start to drain, collecting the eluate. Rinse the flask with four successive 5-ml portions of hexane–ether, allowing each to go into the column before the next addition. Add additional hexane–ether to the column and follow the visible movement of the dinitramine. Note the volume of solvent mixture required to move the dinitramine as a yellow band to the bottom of the Florisil column. This volume of forerun is to be discarded in the sample cleanup procedure following. Continue to elute and record the volume of hexane–ether required to remove the dinitramine from the column. This volume plus an additional 50 ml will be collected in the cleanup procedure to ensure complete recovery of the dinitramine. Each fresh batch of Florisil should be standardized in this manner.

ii. *Sample Extraction*

For plant tissue, grind a representative sample of frozen tissue in a Hobart food cutter in the presence of dry ice. When thawed and free of dry ice, weigh 100 gm into a quart Mason jar. Extract with 400 ml of methanol for 10 minutes using an Omni-mixer. In the case of soil samples, extract 100 gm of thawed, air-dried soil with 400 ml methanol in a Mason jar for 30 minutes on a platform shaker. In either case, after extraction, filter through Whatman no. 1 paper into a glass bottle.

iii. *Cleanup Procedure A*

This procedure is suitable for soil samples as well as soybean and some other plant tissues. Transfer 100 ml of the extract (representing a

25-gm sample) to a 1-liter separatory funnel. Add 500 ml 5% sodium chloride solution to the funnel and mix. Add 50 ml methylene chloride to the funnel and shake vigorously for 1 minute. Allow the layers to separate. Drain the methylene chloride layer through anhydrous sodium sulfate into a 300-ml round-bottomed flask. Repeat the extraction with two more 50-ml portions of methylene chloride, collecting the extracts as before. Wash the sodium sulfate with 25 ml methylene chloride. Evaporate the combined methylene chloride extracts using a flash evaporator and a 40–50°C water bath. Remove the flask as soon as the solvent is completely evaporated to prevent loss of dinitramine. Add 5 ml 95:5 hexane–ether mixture to the flask, swirl to dissolve contents, and transfer to the Florisil column. Start collecting the eluate. Rinse the flask with four 5-ml portions of hexane–ether, allowing each portion to go into the column before the next addition. Elute with hexane–ether. Retain the fraction of the eluate as determined in the standardization of the Florisil. Normally the first 50–55 ml is discarded, and the next 100 ml saved in a 200-ml round-bottomed flask. Evaporate to dryness as before. Dissolve the residue in 2 ml of Nanograde benzene and transfer to a small vial. Stopper and protect the solution from light with aluminum foil.

Inject a 2-μl portion into a gas chromatograph operated under the previously described conditions. Measure the area under the peak at the retention time of dinitramine (height × width at $\frac{1}{2}$ height is adequate).

iv. *Cleanup Procedure B*

A variation of the Florisil column elution is required for cotton plant tissue to remove an interference at the GC retention time of dinitramine, as follows: After the methylene chloride extraction and evaporation, dissolve the residue in 5 ml of a 1:1 mixture of hexane–benzene (instead of hexane–ether) and transfer to the Florisil column. Rinse the flask with four more 5-ml portions and elute the column with the same solvent mixture. *Note:* The Florisil must be previously standardized with 1:1 hexane–benzene to determine which fraction of eluate to retain. Normally, the first 50–55 ml are discarded and the next 100 ml saved in a 200-ml round-bottomed flask. Evaporate to dryness as before and take up the residue in 2 ml of Nanograde benzene for injection into the GC.

v. *Modification*

The addition of an oil removal step to procedures A and B is required for plant tissues that have a high oil content, e.g., soybeans and cottonseed. The residue from the methylene chloride evaporation is dissolved in 25 ml of *n*-hexane and partitioned with two 25-ml volumes acetonitrile. (If problems with phase separation occur, addition of one drop of concentrated hydrochloric acid helps alleviate the difficulty.) The acetonitrile extract is evaporated to dryness and the residue dissolved in 5 ml solvent and transferred to the Florisil column according to procedure A for soybean seeds and procedure B for cottonseed.

vi. *Standard Curve*

Determine the GC linear response range for dinitramine using standard solutions in benzene by plotting peak area (square millimeters) (height × width at $\frac{1}{2}$ height) versus dinitramine injected on log–log coordinates. A linear range of 200–2000 pg is typical. If a sample extract contains dinitramine, the observed peak area must be within this linear range to obtain reliable data. Depending on dinitramine concentration in the sample, some adjustment in volume of the final concentrate or of the portion injected may be necessary. Some variations in detector sensitivity are observed with time. For best results, two points on the line (e.g., 200 and 1000 pg) should be established every few hours. A reasonable approximation is obtained by determining one point and applying the same slope as previously established (for the same detector and conditions).

vii. *Calculations*

Measure the area of the dinitramine peak obtained from the sample and compare with the standard curve. The residue in the original sample is calculated as follows:

$$\text{Dinitramine (ppm)} = \frac{D \times V \times 10^{-6}}{W \times I}$$

where D = dinitramine corresponding to observed peak area in picograms
V = volume final concentrate in microliters
W = weight of analyzed sample (25 gm)
I = volume injected into chromatograph in microliters.

TABLE I
RECOVERY OF DINITRAMINE ADDED TO UNTREATED SOILS

Soil type	Added ppm	Recovery (average) (%)
Sandy loam	0.012	87
Sandy loam	0.04	97
Clay loam	0.2	85
Silty clay loam	0.04	120

TABLE II
RECOVERY OF DINITRAMINE ADDED TO UNTREATED SOYBEAN AND COTTON PLANTS AND SEEDS

Crop	Added ppm	Recovery (average) (%)
Soybean plants	0.01	88
Soybean seed	0.01	98
Cotton plants	0.005	80
Cotton plants	0.01	102
Cotton seed	0.01	97

e. *Confirmatory Method*

Values obtained by the above procedure can be verified by use of an alkali flame ionization detector. All extraction and cleanup procedures are carried out as previously described, giving a final benzene concentrate for gas chromatographic analysis. The flow rates of nitrogen, air, and hydrogen in the chromatograph are adjusted to maximize the signal-to-noise ratio, typically about 30, 250, and 35 cc/minute, respectively. The sample peak area produced at the dinitramine retention time is compared with a log–log standard curve determined separately for this detector. Due to variation in detector sensitivity, a standard should be injected every few hours for calibration.

f. *Discussion*

The method is quantitative to 0.01 ppm. Some sample types (e.g., low organic soils) present so little GC background that levels of dinitramine much lower than 0.01 ppm are readily determined. Tables I and II list the recoveries, ranging from 85 to 120%, which were obtained from spiked samples of untreated soil and plant tissue.

REFERENCES

Laanio, T. L., Kearney, P. C., and Kaufman, D. D. (1973). *Pestic. Biochem. Physiol.* **3,** 271.

Newsom, H. C., and Mitchell, E. M. (1972). *J. Agr. Food Chem.* **20,** 1222.

Newsom, H. C., and Woods, W. G. (1973). *J. Agr. Food Chem.* **21,** 598.

Smith, R. A., Belles, W. S., Shen, K. W., and Woods, W. G. (1973). *Pestic. Biochem. Physiol.* **3,** 278.

United States Borax & Chemical Corporation (1971). U.S. Patent 3,617,252.

United States Borax & Chemical Corporation (1973). U.S. Patent 3,764,623.

24

Isopropalin

W. S. JOHNSON AND O. D. DECKER

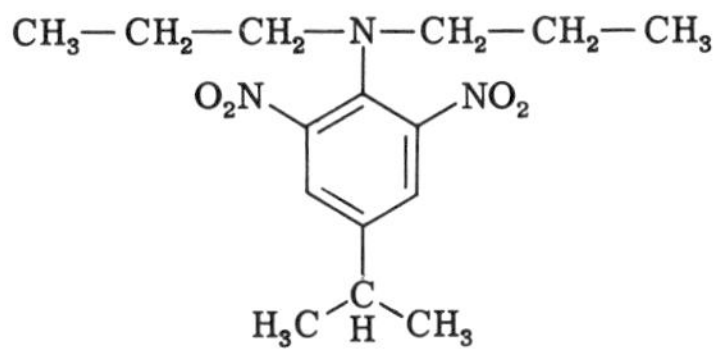

2,6-Dinitro-*N*,*N*-dipropylcumidine

I. GENERAL

A. Empirical Formula

$C_{15}H_{23}N_3O_4$ (Mol. wt. 309.4).

B. Alternative Name

Paarlan®, a registered trademark of Elanco Products Company, Division of Eli Lilly and Company

C. Source of Analytical Standard

Elanco Products Company, 640 South Alabama Street, Indianapolis, Indiana.

D. Biological Properties

Isopropalin is a selective, preemergence, soil incorporated herbicide that controls a wide variety of grasses and broadleaf weeds. Tolerant crops include direct-seeded and transplant tomatoes, direct-seeded and transplant peppers, and all types of transplant tobacco.

The acute oral LD_{50} of technical isopropalin for mice and rats is greater than 5 gm/kg. The LD_0 for the rabbit, chicken, and dog is not less than 2 gm/kg. Technical isopropalin fed in the diet of rats and young dogs at 250 ppm for 3 months, and in the diet of 1-day-old chicks for 3 weeks produced no adverse effects. Rabbits exhibited only very slight skin irritation when isopropalin was applied at a dosage of 2 gm/kg. The

LC_{50} of technical isopropalin for fathead minnows is 50–150 parts per billion by the standard 96-hour bioassay.

E. History

Isopropalin is one of a class of substituted dinitroanilines whose herbicidal properties were first reported by Alder *et al.* (1960). Its use as a selective, preemergence herbicide for the control of annual grasses and broadleaf weeds in agronomic crops has been described by Shoop (1969), Guse (1969), and Keaton and Webster (1970).

F. Physical Properties

Pure isopropalin is an orange liquid that is readily soluble in organic solvents such as acetone, hexane, and chloroform. The solubility in water is less than 0.5 ppm at 25°C. The vapor pressure of isopropalin was found to be 3×10^{-5} mm Hg at 24.6°C.

The infrared spectrum of isopropalin between salt plates exhibits absorption maxima at 2950, 2860, 1615, 1530, 1465, 1200, and 1105 cm^{-1}. The ultraviolet absorption spectrum of a dilute hexane solution exhibits maxima at 230 and 386 nm, the latter giving a molar absorptivity of 1710.

G. Chemical Properties

1. METHOD OF SYNTHESIS

Isopropalin is prepared by nitration of *p*-isopropylphenol, followed by chlorination using $POCl_3$, and by amination with di-*n*-propylamine.

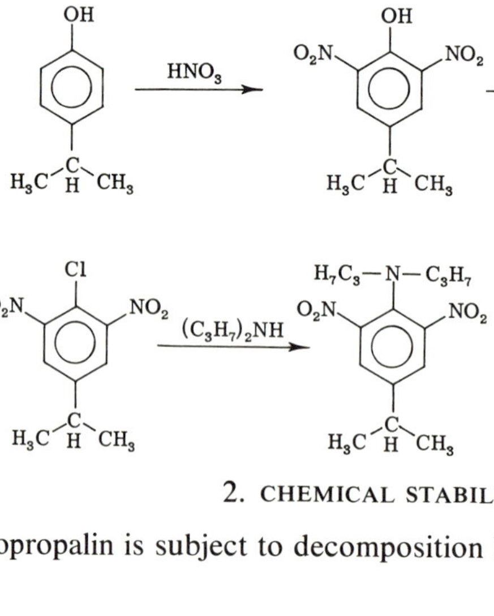

2. CHEMICAL STABILITY

Isopropalin is subject to decomposition by ultraviolet radiation.

H. Formulations

Isopropalin is formulated as a 6-pound/gallon emulsifiable concentrate. Technical material containing approximately 95% isopropalin is also available.

II. ANALYSIS

A. Formulation Analysis

1. REVIEW OF METHODS

Two methods are available for the determination of isopropalin. Decker and Frank (1974) assayed isopropalin colorimetrically by measuring its absorption in *n*-hexane at 386 nm following removal of colored impurities. Isopropalin may also be assayed by flame ionization gas chromatography without prior purification of the sample.

2. RECOMMENDED METHOD

a. *Principle*

Isopropalin is measured by determining its absorbance in *n*-hexane at 386 nm. Colored impurities in the sample are removed by Florisil column chromatography.

b. *Reagents*

Sodium sulfate, anhydrous, analytical reagent.

n-Hexane, analytical reagent.

Florisil, 100–200 mesh, deactivated, Floridin Company, Berkeley Springs, West Virginia.

Isopropalin standard, 0.80 mg/ml in *n*-hexane.

c. *Apparatus*

Chromatographic columns, 14 mm × 240 mm, fitted with stopcock and 250-ml solvent reservoir.

Spectrophotometer.

Rotary vacuum evaporator, Rinco or equivalent.

d. *Experimental Procedure*

i. *Deactivation and Standardization of Florisil*

Determine the apparent moisture content of the Florisil by heating at 600°C for 4 hours. Add sufficient deionized water to the Florisil as received to provide a total moisture content of 10–10.5%. Mix for 1 hour, then allow the Florisil to stand overnight before use. Store in a well-closed container.

Insert a pledget of glass wool in the bottom of the chromatographic column. Add, in the following order with constant tapping, 3.5 gm sodium sulfate, 5.0 gm deactivated Florisil, and 3.5 gm sodium sulfate. With the stopcock open, add 20 ml *n*-hexane and allow to drain until the liquid reaches the top of the upper layer of sodium sulfate. Transfer a 10-ml portion of isopropalin standard solution to the column and allow the liquid to drain until the level reaches the top of the upper sodium sulfate layer. Rinse the walls of the column with small portions of *n*-hexane and allow the rinses to run into the column. Elute the column with *n*-hexane until the main yellow band is approximately 1 inch from the bottom of the Florisil packing, and discard the eluate to this point. Place a 100-ml volumetric flask under the column, continue the elution, and note the volume of *n*-hexane required to completely remove the yellow band from the column. Properly deactivated Florisil should allow for complete elution of the isopropalin with 70–90 ml *n*-hexane after collection begins. If less than 70 ml of eluate is required, reduce the moisture content of the Florisil. If more than 90 ml is required, increase the moisture content.

The amount of isopropalin recovered in the Florisil standardization may be determined by diluting the eluate to 100 ml and comparing the absorbance value at 386 nm of this solution with the value obtained for 10.0 ml of isopropalin standard diluted to the same volume.

ii. *Sample and Standard Preparation*

(a) *Technical Material or Emulsifiable Concentrate.* Accurately weigh a sample equivalent to 80 mg isopropalin into a 100-ml volumetric flask, dilute to volume with *n*-hexane, and mix well. Transfer a 10-ml portion of the *n*-hexane solution to a prepared Florisil column and wash the sample onto the column as previously described. Proceed with the elution and collect the volume of eluate determined in the standardization of the Florisil. Add sufficient *n*-hexane to the flask to make 100 ml and mix well.

(b) *Standard Preparation.* Transfer a 10-ml portion of isopropalin standard solution (0.80 mg/ml) to a prepared Florisil column and proceed with the elution as described for the sample preparation, above. Dilute the eluate to 100 ml with *n*-hexane and mix.

iii. *Measurement*

Measure the absorbance of the standard and sample solutions in a 1-cm cuvette at 386 nm using *n*-hexane as the reference solvent. The analyst should determine the linear response of the spectrophotometer and employ a standard response curve in making calculations, if necessary.

iv. *Calculation*

$$\% \text{ isopropalin} = \frac{A_{\text{sample}}}{A_{\text{standard}}} \times \frac{\text{concentration of standard (mg/ml)} \times 100{,}000}{\text{sample weight (mg)}}$$

If necessary, correction for the purity of the isopropalin standard should be made in the calculation of results.

v. *Discussion*

The Florisil column cleanup adequately removes interfering colored impurities in the technical grade material and emulsifiable concentrate formulation. The precision of the colorimetric method is approximately ±1.5% at the 95% confidence level.

3. GAS CHROMATOGRAPHIC METHOD

a. *Principle*

The technical material or emulsifiable concentrate sample is dissolved in chloroform and analyzed by gas chromatography using a flame ionization detector. Dipentyl phthalate is used as the internal standard.

b. *Reagents*

Chloroform, analytical reagent.
Isopropalin standard, 1.0 mg/ml in chloroform.
Dipentyl phthalate, 1.0 mg/ml in chloroform.

c. *Apparatus*

Gas chromatograph, Hewlett-Packard model 402, equipped with a flame ionization detector, model 7621A automatic liquid sampler, and model 7123A 1-mV recorder or equivalent GC system. The chromatographic column is 120 cm × 3 mm i.d. glass tubing packed with 2% OV-17 (w/w) on 80–100 mesh Chromosorb W-HP (Applied Science Laboratories, State College, Pennsylvania). The column is conditioned at 290°C for 16 hours under carrier gas purge prior to use.

d. *Experimental Procedure*

i. *Sample Preparation*

Accurately weigh a sample of technical material or emulsifiable concentrate equivalent to 100 mg isopropalin into a 100-ml volumetric flask. Dilute to volume with chloroform and mix.

ii. *Gas Chromatography*

Flash heater temperature: 260°C.
Column temperature: 200°C.
Detector temperature: 260°C.
Nitrogen carrier flow: 35 ml/minute.
Hydrogen flow: 30 ml/minute.
Air flow: 300 ml/minute.
Electrometer: 1.28×10^{-9} AFS (range 10, attenuation 32)

Isopropalin retention time: approximately 3.5 minutes.
Dipentyl phthalate retention time: approximately 5 minutes.

Pipette a 1.0-ml portion of the sample solution into a small vial containing 1.0 ml of the internal standard solution. Prepare a standard solution by pipetting 1.0 ml of isopropalin standard solution, 1.0 mg/ml, into a small vial containing 1.0 ml of the internal standard solution. Stopper the vials securely and mix. Inject approximately 2 μl of the sample and standard solutions into the gas chromatograph. Measure the peak response for the isopropalin and dipentyl phthalate peaks.

iii. *Calculations*

Calculate the peak response ratio *R* for the isopropalin and dipentyl phthalate peaks:

$$R = \frac{\text{peak response isopropalin}}{\text{peak response dipentyl phthalate}}$$

Calculate the percent isopropalin in the sample by the formula:

$$\%\ \text{isopropalin} = \frac{R_{\text{sample}}}{R_{\text{standard}}} \times \frac{\text{concentration of standard (mg/ml)} \times 2000}{\text{sample weight (mg)}}$$

If necessary, correction for the purity of the isopropalin standard should be made in the calculation of results.

Prior to the measurement of experimental samples, the analyst should inject a series of known concentrations of isopropalin with internal standard and use a standard response curve in the calculations, if necessary. The isopropalin standard should be injected frequently during the course of the measurement of a series of samples to insure the instrument response has not changed.

iv. *Discussion*

The precision of the method is approximately ±2% at the 95% confidence level. No interferences have been encountered in the assay of commercial formulations. Although the gas chromatographic method can be used for the assay of technical isopropalin, the colorimetric method is recommended since better precision of measurement can be obtained.

B. Residue Analysis

1. RECOMMENDED METHOD

a. *Principle*

Isopropalin is quantitatively extracted from plant tissue and soil with methanol. The extract is purified by liquid–liquid extraction with methylene chloride and by column chromatography over Florisil. Detection and measurement are made by gas–liquid chromatography using an electron capture detector.

b. *Reagents*

Analytical standard stock solution. Accurately weigh 10 mg of analytical standard, transfer quantitatively to a 200-ml volumetric flask with benzene, and dilute to volume (50 μg/ml). Store in a refrigerator and prepare fresh every 3 months. Dilutions should be prepared fresh weekly and must be protected from light.

Methanol, analytical reagent.

Sodium chloride, analytical reagent, 5% in deionized water.

Methylene chloride, analytical reagent, redistilled before use.

n-Hexane, analytical reagent, redistilled before use.

Florisil, 100–200 mesh, Floridin Co., Berkeley Springs, West Virginia, deactivated and standardized as follows:

> Determine the water content of the Florisil as received by heating to 600°C for 4 hours. Add sufficient water to an appropriate amount of Florisil to yield a total water content of 8.5–9.0%. After addition of the water, stir well with a glass rod to break up lumps and tumble in a closed container for 1 hour. Allow to stand for an additional 2 hours prior to use.
>
> Prepare a Florisil column by placing 10–15 ml hexane in a 14-mm i.d. glass chromatographic column. Insert a glass wool pledget and tamp in place with a glass rod to eliminate trapped air. Add 4.8 gm deactivated Florisil and allow to settle. Stir with a glass rod to eliminate air bubbles and start to drain the column. Wash down the sides of the column with hexane to remove adhering Florisil. Carefully layer 2–3 cm anhydrous sodium sulfate on the surface of the Florisil. Continue to drain the hexane until the liquid layer reaches the top of the sodium sulfate layer.

Add 100 μg isopropalin dissolved in 5 ml hexane to the column and start to drain. Rinse the flask with two successive 5-ml portions hexane, allowing each to go into the column before the next addition. When the last hexane rinse reaches the top of the sodium sulfate, add 65 ml of hexane to the column, allow to drain at a flow rate of 3–5 ml/minute. Note the volume of solvent required to move the leading edge of the yellow band to within 1 inch of the bottom of the Florisil layer. Discard this amount of eluate as a forerun. Continue the elution of the yellow band and record the volume of hexane required to completely remove the yellow color from the column. This volume plus an additional 15 ml will be collected in the column chromatography cleanup step to insure complete removal of the compound.

Benzene, analytical reagent, redistilled before use.

c. *Apparatus*

Blender, Omni-mixer (Ivan Sorvall Co.) or equivalent.

Gas Chromatograph, Hewlett-Packard model 5713A, equipped with a model 18731A Ni-63 linear EC detector, and a model 7123A 1-mV recorder. The chromatographic column is 120 cm × 3 mm i.d. glass tubing packed with 5% Carbowax 20M on 80–100 mesh Chromosorb W-HP (Applied Science Laboratories, State College, Pennsylvania). The column is conditioned at 240°C for 16 hours prior to use.

d. *Experimental Procedure*

i. *Sample Preparation*

Weigh a representative finely ground 25-gm sample of plant tissue or soil into a quart Mason jar. Add sufficient methanol to the plant sample to make 200 ml total volume after allowance is made for the water content of the sample. Add 200 ml methanol to soil samples. Blend the sample at moderate speed for 5 minutes on an Omni-mixer. Remove the jar and allow the insolubles to settle. Filter a sufficient volume of the extract by gravity through Whatman No. 1 filter paper to provide a 20-ml portion. Transfer the 20-ml portion to a 125-ml separatory funnel containing 40 ml 5% sodium chloride solution. Extract the aqueous methanol solution twice with 20-ml portions of methylene chloride, passing each extract through sodium sulfate into a 125-ml boiling flask. Evaporate the methylene chloride on a rotary vacuum evaporator using a water bath at 50°C or less. Remove the flask from the evaporator immediately upon complete evaporation of the methylene chloride. Prepare a Florisil column as described under Florisil standardization. Dissolve the residue in the flask in 5 ml hexane, transfer to the column, and start to drain. Rinse the flask with two successive 5-ml portions of hexane, allowing each to go into the column before the next addition. Add an additional 65 ml hexane to the column and allow to drain at a flow rate of 3–5 ml/minute. Collect the volume of effluent as determined by the Florisil standardization. Remove the solvent by rotary vacuum evaporation using a 50°C water bath.

Recovery samples are prepared by adding 1.25 μg isopropalin in methanol to 25 gm of control plant tissue, or 2.5 μg isopropalin to 25 gm of control soil.

ii. *Gas Chromatography*

Column temperature: 190°C.
Detector temperature: 300°C.
Injection temperature: 250°C.
Electrometer attenuation: Setting to provide 30–40% of full scale deflection on injection of 0.3–0.4 ng isopropalin. For the gas chromatograph described, the cell current is constant 10^{-9} A and the signal measured is the pulse frequency required to maintain the standing current.

Dilute 10.0 ml of the standard isopropalin solution (50 μg/ml) to 50 ml with benzene. Pipette 0.05, 0.10, 0.20, 0.40, and 0.50 ml of this solution (10 μg/ml) into 20-ml volumetric flasks and dilute to volume with benzene. Inject 2 or 3 μl of each solution into the gas chromatograph. Plot peak response versus nanograms injected to obtain the standard response curve. Check the response with each set of samples assayed, and frequently inject a standard dilution (e.g., 0.1 μg/ml) during the course of the assay to determine whether instrument response has changed.

Dissolve the crop extract in 1.0 ml benzene and the soil extract in 2.0 ml benzene. Inject the same volume of extract as used for the response curve. One microliter is equivalent to 2.5 mg of the crop tissue or 1.25 mg of soil.

iii. *Recovery and Sensitivity*

The procedure as described gives a recovery of 85 to 100% after fortification of crops at 0.05 ppm and 0.10 ppm in soils. Recovery of isopropalin at lower levels indicates the assay is sensitive to 0.01 ppm.

iv. *Interference Due to Other Pesticides*

Toxaphene and BHC have been found to interfere with the gas chromatographic measurement of isopropalin when using a 5% Carbowax column. These two compounds will appear in the final benzene extract if they were present in the original sample. Interference due to toxaphene, but not BHC, can be eliminated by using a 5% XE-60 GLC column at about 190°C.

A thin-layer chromatographic (TLC) procedure using alumina plates will eliminate the interference due to toxaphene and BHC. If a peak appears in the gas chromatogram at the retention time of isopropalin from samples whose treatment history is unknown, the TLC step and perhaps the alternate GLC column should be employed before the peak is identified as isopropalin.

To carry out the TLC procedure, evaporate the sample to dryness and reconstitute in 0.5 ml hexane. Apply 100 μl (one-fifth portion) to a precoated alumina TLC plate (Type T, Brinkmann Instrument Co.). Marker zones with 25 μg of isopropalin spotted on the plate are used to locate the R_f for the sample zone. Develop the plate twice in an unlined tank containing *n*-hexane. Scrape the zone in the sample lane corresponding to the R_f of the marker zone. Transfer the scrapings to a 1-dram vial and add 0.4 ml benzene. Agitate the vial and allow the alumina to settle. Inject a portion of the supernatant benzene into the gas chromatograph. The TLC sample is now twice the dilution of the original crop sample, and the dilution factor must be taken into account when calculating residue found in the original sample.

v. *Sample Calculation*

The number of nanograms of isopropalin present in the number of milligrams of plant tissue or soil injected into the gas chromatograph may be obtained from the standard curve. The value obtained may be corrected to reflect assay efficiency as determined from recovery samples.

e. *Discussion*

Isopropalin is unstable to light and should be protected from undue exposure during the assay. If the assay is not completed within one working day, the extract may be stored overnight under refrigeration in any of the solvents used. Isopropalin is relatively volatile, and undue losses can occur if extracts are left on the rotary vacuum evaporator too long or if an air stream is used for evaporation of solvents.

Crop samples having a high oil content may leave an oily residue in the flask upon evaporation of the methylene chloride. The oil may change the elution pattern on Florisil. To remove the oil from the sample, dissolve the extract in 20 ml hexane and extract twice with acetonitrile. Evaporate the acetonitrile on the rotary evaporator and continue with the column chromatography.

REFERENCES

Alder, E. F., Wright, W. L., and Soper, Q. F. (1960). *Proc. N. Cent. Weed Contr. Conf.* **17,** 23–24.

Decker, O. D., and Frank, R. (1974). *J. Ass. Offic. Anal. Chem.* **57,** 645.

Guse, L. R. (1969). *Proc. N. Cent. Weed Contr. Conf.* **24,** 44–45.

Keaton, J. A., and Webster, H. J. (1970). *Proc., S. Weed Sci. Soc.* **23,** 106–114.

Shoop, G. J. (1969). *Proc. N. Cent. Weed Contr. Conf.* **24,** 19–20.

25

Karbutilate

R. F. Cook

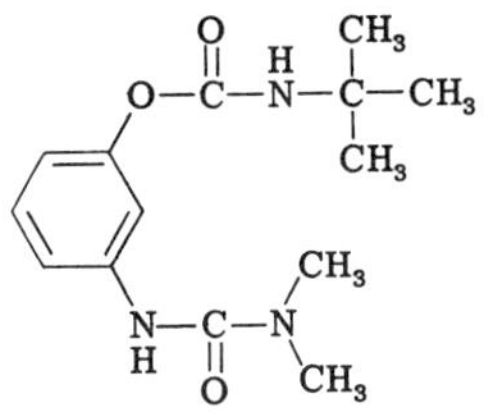

m-(3,3-Dimethylureido)phenyl-*tert*-butylcarbamate

I. GENERAL

A. Empirical Formula

$C_{14}H_{21}N_3O_3$ (Mol. wt. 279.3).

B. Alternate Names

Tandex®*, NIA 11092.

C. Source of Analytical Standard

Formulations Section, Agricultural Chemical Division, FMC Corporation, 100 Niagara Street, Middleport, New York.

D. Biological Properties

Karbutilate is a nonselective broad-spectrum herbicide. Formulations of karbutilate are especially well suited for the control of annual weeds, hard-to-kill perennial weeds, grasses, vines, and brush on noncrop land.

The acute oral LD_{50} of technical karbutilate to rats is 3000 mg/kg. The acute oral LD_{50} of a karbutilate 80% wettable powder formulation to male rats is 28,200 mg/kg and to female rats is 18,800 mg/kg. At dermal doses up to 15,400 mg/kg of body weight, no deaths or adverse

*® Registered Trademark of Agricultural Chemical Division, FMC Corporation.

behavioral reactions were observed in rabbits treated with a 50% aqueous paste of karbutilate. The inhalation LC_{50} for rats was found to be >3.7 mg/liter. Neither the technical karbutilate or the 80% wettable powder formulation was found to be irritating to the eyes or skin.

Fish were exposed, under static conditions, to various concentrations of technical karbutilate for 96 hours. The LC_{50} value obtained for bluegills was 75 ppm and for trout was 135 ppm. Wildlife toxicity data were obtained for technical karbutilate on the basis of 8-day oral tests. The LC_{50} value obtained for bobwhite quail was >10,000 ppm and for mallard ducks was >4640 ppm.

E. History

Karbutilate was first synthesized by chemists in the laboratories of FMC Corporation. The compound per se is covered by U.S. patent no. 3,532,738. Herbicidal compositions based on karbutilate and the method of their use are covered by U.S. patent no. 3,434,822. Processes for the manufacture of karbutilate are described in U.S. patent nos. 3,801,625 and 3,488,376. The first registered use for karbutilate in the United States was as a noncrop herbicide in 1969.

F. Physical Properties

Appearance: Off-white powdered solid.
Odor: Nearly odorless.
Density: 1.175 20/20°C.
Melting point: Pure 176–176.5°C; technical 160°C minimum.
Solubility at 25°C

Water	325 ppm
Organic solvents	% w/w
Acetone	<3
Dimethylsulfoxide	20–25
Dimethylformamide	20–25
Isophorone	<3
Isopropanol	<3
Xylene	<3

Flammability: Not flammable
Explosive hazard: Not explosive at room temperature storage
Corrosive action: Noncorrosive

G. Chemical Properties

1. METHOD OF SYNTHESIS

Karbutilate is prepared by interaction of *tert*-butyl isocyanate (I) and 1-(3-hydroxyphenyl)-3,3-dimethylurea (II). Compound II is dis-

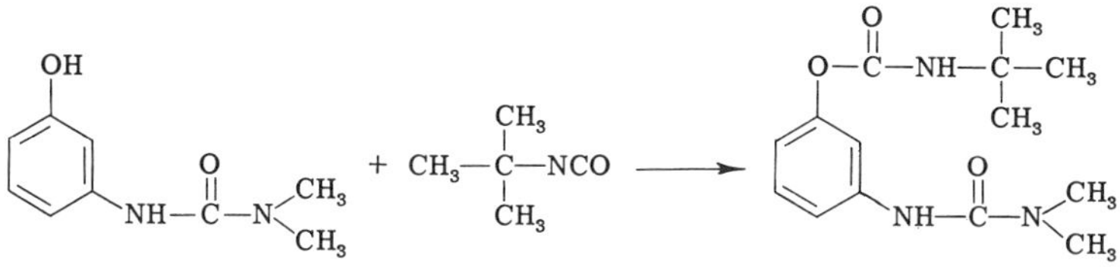

(II) (I) (III)

solved in an appropriate solvent and agitated while I is added slowly. A carbamylation catalyst, such as a tertiary amine, aids in initiating the reaction. Upon completion of the addition of I, the slurry is heated and stirred until the reaction is completed. The solid karbutilate (III) is isolated by filtration, washed thoroughly, and dried under vacuum. The product is ready for formulation without further purification.

2. STABILITY

At 5°C, karbutilate is stable in pH 5 and pH 7 solutions for at least 45 days. At pH 9, karbutilate is hydrolyzed to the phenol of karbutilate and has a half-life of 6.8 days.

At 22°C, karbutilate is also stable in pH 5 solution after 45 days, but is hydrolyzed to the phenol of karbutilate at pH 7 and 8. At pH 7, 55% of karbutilate remains after 45 days, and at pH 8, karbutilate has a half-life of 4.6 days.

The hydrolysis of karbutilate to the phenol of karbutilate follows a pseudo first-order reaction. The rate constant for the reaction is 0.14 day^{-1} at 22°C and pH 8 and 0.018 day^{-1} at 22°C and pH 7.

H. Formulations

Formulations of karbutilate that are commercially available include a 4% granular, 10% granular, and an 80% wettable powder.

II. ANALYSIS

A. Formulation Analysis

1. REVIEW OF METHODS

Infrared spectrophotometry is the recommended method for the quantitative determination of the karbutilate content in technical and formulated products. Gas chromatography is not a recommended technique because of the thermal instability of karbutilate.

2. RECOMMENDED METHOD—INFRARED SPECTROPHOTOMETRY

a. *Principle*

Intact karbutilate is determined by comparing the absorbance maximum of an unknown sample to the absorbance maximum of a standard at 1745 cm^{-1} (Agricultural Chemical Division, TM10-68). Chloroform is used for sample extraction and the media for absorbance measurement.

b. *Reagents*

Analytical standard, Tandex®. Reference standard material is available from Formulation Group, Agricultural Chemical Division, FMC Corporation, Middleport, New York.

Chloroform, certified A.C.S. spectroanalyzed.

Sodium sulfate, anhydrous powder.

c. *Apparatus*

Infrared spectrophotometer, Perkin & Elmer, model 257 or equivalent.

Cells, two 0.2-mm NaCl-sealed cells, matched and calibrated. (International Crystal Labs, Irvington, New Jersey.)

Flask, 25 ml volumetric.

Bottle, 4 ounce screw-cap bottle.

Syringe, tuberculin, 5 ml (Luer, B-D Yale).

Syringe, tuberculin, $\frac{1}{4}$ ml (Luer, B-D Yale).

d. *Experimental Procedure*

i. *Operating Conditions*

Instrument: Perkin & Elmer, model 257, infrared spectrophotometer.

Mode: Normal.

Speed: Slow (50 cm^{-1}/minute).

Slit: 6.

Two 0.2 mm NaCl sealed cells, matched and calibrated.

ii. *Sample Preparation*

(a) *Karbutilate Analytical Standard and Technical.* Karbutilate standard about 0.2500 gm; karbutilate technical about 0.2500 gm.

Weigh the appropriate amount of sample to the nearest 0.1 mg into a 25-ml volumetric flask. Dilute to volume with spectral grade chloroform. Stopper, and mix thoroughly. Analyze the clear solution.

(b) *Tandex® Products.* Tandex® 80 WP about 0.3000 gm; Tandex® 4 G about 5.00 gm.

Weigh accurately the appropriate amount of Tandex® sample as listed above into a 4-ounce screw-cap bottle. Add accurately by pipette 25 ml of spectral grade chloroform. Cap tightly and mechanically shake for 30 minutes. Analyze the clear supernatant liquid (filter quickly or centrifuge if necessary). If water is suspected by a cloudiness in the solution, shaking of the supernate over about 0.2 gm sodium sulfate for 15 minutes is recommended.

iii. *Operating Procedure*

A 5 ml tuberculin syringe is filled and flushed clean with spectral grade chloroform several times, then used to gently flush clean each NaCl cell with chloroform. A $\frac{1}{4}$ ml tuberculin syringe is filled and flushed with chloroform several times, then used to fill each NaCl cell with air-bubble-free chloroform. The cells are stoppered with the recorder pen now adjusted to about 95% transmittance (a baseline scan of chloroform versus chloroform is run from about 1925 cm^{-1} to 1650 cm^{-1} at slow speed) normal mode and a slit setting of 6. This gives an idea of how well matched in path length the cells are. The sample side is flushed and refilled, as described above, with a solution of karbutilate standard. Again, a 1925 cm^{-1} to 1650 cm^{-1} scan is made.

Additional samples are introduced into the sample cell after pre-flushing the cell with the new sample. The standard solution should be run at least at the beginning and the end of an analysis series to ensure consistency of response. The reference cell continues to hold chloroform. The cells are flushed with chloroform, then dry nitrogen, and finally placed in a desiccator.

e. *Calculations*

Calculations are based upon the Tandex® absorbance maximum at 1745 cm^{-1}. The Tandex® peak height at 1745 cm^{-1} of unknown sample is compared with that of the standard under identical conditions.

$$\%\ \text{Tandex}^{®} = \frac{\text{peak height unknown sample}}{\text{peak height standard} \times F} \times 100$$

where

$$F = \frac{\text{unknown sample weight/25 ml}}{(\text{standard weight/25 ml}) \times \text{standard purity}}$$

Peak height is determined as the difference in absorbance between the chloroform versus chloroform baseline at 1745 cm^{-1} and the peak at 1745 cm^{-1} (see Fig. 1).

Precision

Average deviation	ca. ±0.7%
Standard deviation	ca. ±1.0%
95% confidence limits	ca. ±2.0%

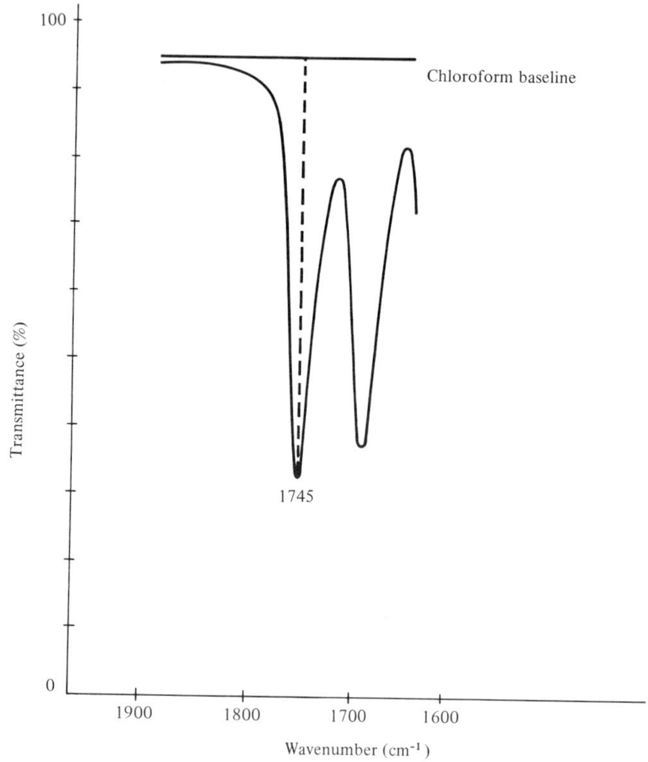

FIG. 1. Infrared spectrum of Tandex®.

B. Residue Analysis

1. REVIEW OF METHODS

High-pressure liquid chromatography (HPLC) was used as the instrumentational basis for developing a residue method capable of determining karbutilate and its major degradation products in soil and water. The HPLC system using an ultraviolet detector provided adequate selectivity and sensitivity. In addition, karbutilate could be determined intact, since thermal decomposition was not a factor in the use of the HPLC system.

Metabolism studies on karbutilate in soil by Knaak and Robinson (1971) have shown the major mechanism for degradation to be demethylation. That is, karbutilate (I) is demethylated by soil microorganisms to monomethyl karbutilate (II) and, to a lesser degree, further demethylated to desmethyl karbutilate (III). Soil metabolism by hydrolysis to phenolic products accounts for less than 20% of the total dose applied as karbutilate. In addition, the phenolic metabolites are immediately and firmly bound to the soil media. Extraction of the phenolic metabolites from soil for the purpose of quantitative analysis is not practical. In addition, since the phenolic metabolites are not phytotoxic, there is little necessity in being able to analyze for them.

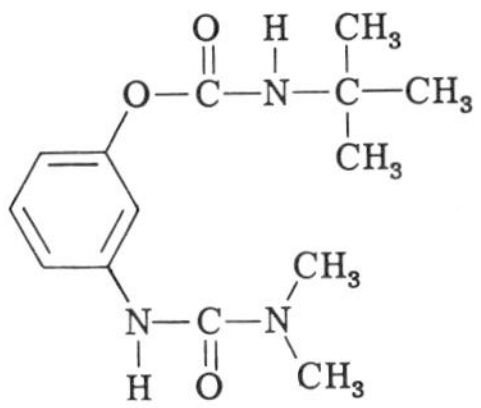

(I)

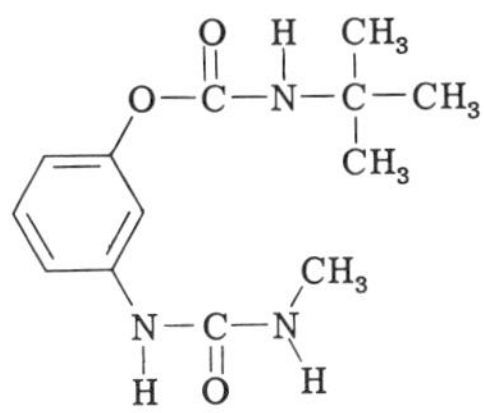

(II)

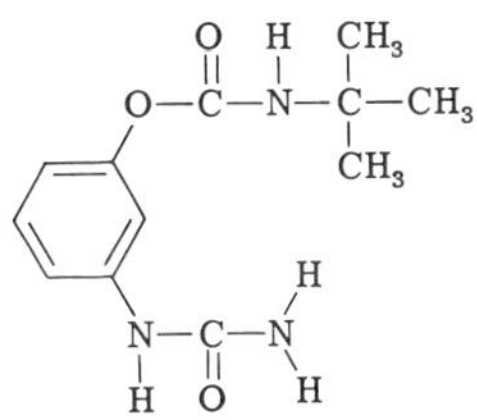

(III)

Hydrolysis to the phenol of karbutilate (IV) has been found to be the major route of degradation in water. The more alkaline the water and the higher the temperature, the more rapid the degradation (see Section I,G,2). By proper selection of the HPLC column and adjustments in the

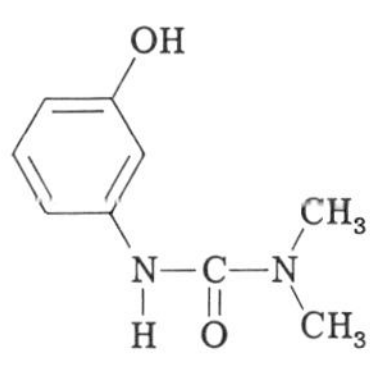

(IV)

concentration of the mobile phase, all the above metabolites and degradation products can be determined intact (i.e., without structural modification).

2. RECOMMENDED METHOD

a. *Principle*

Extraction of soil samples was accomplished by blending the soil with a highly polar, water-miscible solvent system. Separation of soil particles from the extraction system was accomplished centrifugally with subsequent partition of the residues into a water-immiscible organic solvent. Determination of karbutilate down to 0.1 ppm and the major urea–carbamate residues down to 0.2 ppm can be accomplished by concentration of the organic extraction solution and analysis by HPLC using a UV detector (254 nm).

Extraction of water samples involves making the sample acidic and then partitioning the urea–carbamate residues as well as the phenolic residues into a water immiscible organic solvent. Residue determination is by concentration of the extracts and injection into the HPLC.

b. *Reagents*

All chemicals used were Analytical Reagent Grade or better. Solvents were laboratory redistilled in glass or purchased as a "pesticide grade" (Burdick and Jackson). Note that all the chloroform used in the method was purchased with 1% ethanol as a stabilizer.

Analytical standards, Residue Section, Agricultural Chemical Division, FMC Corporation, 100 Niagara Street, Middleport, New York.

Celite 545 (Fisher Scientific Corporation).

Micro-Porasil (10 μm) prepacked column, 4 mm i.d. × 30 cm (Waters Associates, Catalog No. 27477).

Neutral Norit (Fisher Scientific Corporation).

Sodium Sulfate (Van Water and Rogers Scientific).

c. *Apparatus*

Centrifuge, 2000 rpm capability at maximum load (IEC model K or equivalent).

Centrifuge bottles, 250 ml.

Centrifuge tube, graduated 13 ml capacity with No. 13 ₮ ground glass stopper.

Chromatography columns, 22 mm i.d. × 600 mm.

Erlenmeyer flask, 500 and 250 ml with neck.

Filter paper, Whatman no. 42.

Food chopper, Hobart model 4114 or equivalent.

Graduated cylinder, 100 and 250 ml.

Grinding mill.

Liquid chromatograph, Waters Model ALC 202 (or equivalent) equipped with an UV detector (254 nm) and a 10 mV recorder.

Millipore filter, 47 mm Pyrex (model XX10 047 03).

Millipore filter holder, 47 mm.

Nucleopore membrane filter (0.1 μm) 47 mm, G.E.

pH Meter, Beckman Phsar II (or equivalent).

Rinco Rotary Evaporator (model 1007).
Round-bottom flask, 500 ml with 24/40 ₮ joint.
Separatory funnel, 1000 ml.
Sieve, no. 10, 2 mm screen.
Waring Blendor.

d. *Procedure*

i. *Sample Preparation*

Air-dry soil samples for at least 24 hours. Break up large lumps of soil using a mortar and pestle or hammer. Screen samples to remove stones. Pulverize soil samples using a grinding mill and thoroughly mix using a V-Blender or riffle. Store all samples (soil or water) at −20°C until needed for analysis.

ii. *Extraction*

(a) *Soil.* Place a 100 gm subsample of the soil as prepared above into a blender jar. Add 100 ml methanol and 50 ml distilled deionized water. Blend at high speed for 3 minutes. Transfer the mixture to a 250-ml centrifuge bottle. Centrifuge for 30 minutes at 2000 rpm. Decant solution into a graduated cylinder and record the recovered volume. Filter the water–methanol solution through a nucleopore filter (0.1 μm) into a 500-ml Erlenmeyer flask and rinse the filter cake with 25 ml distilled deionized water. Transfer the filtrate to a 1000-ml separatory funnel and extract twice with 150 ml chloroform. Wash the chloroform with 150 ml distilled deionized water and dry with sodium sulfate. Filter the chloroform through filter paper into a 500-ml round-bottom flask and concentrate to about 5–10 ml on a rotary evaporator. Transfer the concentrated chloroform solution to a graduated centrifuge tube.

(b) *Water.* Adjust the pH of 200 ml water to 5 using hydrochloric acid. Extract the acidified solution twice with 200 ml ethyl acetate. Concentrate the ethyl acetate to about 5 ml on a rotary evaporator and transfer to a graduated centrifuge tube.

iii. *Analysis*

Concentrate the samples in the graduated centrifuge tubes to less than 1 ml using a gentle stream of nitrogen. Dilute the sample to exactly 1 ml with chloroform. Inject a suitable portion, e.g., 10 μl, into a liquid chromatograph equipped with a UV detector (254 nm) operating under the following conditions:

(a) *Soil—Karbutilate and Monomethyl Karbutilate Analysis.*
Column: 30 cm × 4 mm i.d. prepacked microporasil.

Mobil phase: 3% v/v absolute ethanol–dichloroethane.
Temperature: Ambient.
Pressure: 1000 psi
Flow rate: 2.4 ml/minute.
Chart speed: 5 minutes/inch.
Detector: UV Photometer fixed at 254 nm.
Absolute setting: 0.04.

With the instrument equilibrated under these conditions, a retention time, i.e., from the point of injection to the peak center, of 3.75 minutes for karbutilate and 7.75 minutes for monomethyl karbutilate should be obtained.

(b) *Soil—Desmethyl Karbutilate Analysis.* Use all the same HPLC conditions as listed in Section iii,a, except for the mobil phase concentration, which should be 7% v/v absolute ethanol–dichloroethane. Injection of the soil extract already analyzed above (Section iii,a) under these conditions will allow the determination of desmethyl karbutilate at a retention time of 4.5 minutes.

(c) *Water—Karbutilate, Karbutilate Phenol and Monomethyl Karbutilate Analysis.* Use all the same HPLC conditions as listed in Section iii,a, except increase the absorbance setting to 0.02. Retention times of 3.75 minutes should be obtained for karbutilate, 5.0 minutes for karbutilate phenol, and 7.75 minutes for monomethyl karbutilate.

iv. *Calculations*

Quantitation was accomplished by comparing the response (peak height) of an unknown sample to the response (peak height) for a known amount of standard material. The following formula can be used to determine the residue content in parts per million for a given sample.

$$\text{ppm} = \frac{H \times R_f}{G}$$

where H = unknown sample response in centimeters of peak height
R_f = response factor for the standard expressed in micrograms per centimeter of peak height
and G = gram sample size injected for analysis.

Note that the gram sample size injected needs to account for the amount of sample recovered from the centrifuging step, as well as the final fraction injected, two easily overlooked points.

The response for karbutilate and its metabolites was found to be linear over several absorbance range settings. All standard solutions of karbutilate and its metabolites were prepared and maintained in chloroform. No stability problems were encountered over 3 months' room temperature storage.

v. *Sensitivity*

The minimum amount considered to be acceptable for accurate quantitation was 0.075 μg for karbutilate and karbutilate phenol and 0.150 μg for monomethyl karbutilate and desmethyl karbutilate. These quantities yield peak height responses of about 1.5 cm (6% scale deflection). Detectability, the point at which a response begins to become distinguishable from the maximum background noise level, was considered to be one-half the amounts needed for accurate quantitation.

Therefore, the method described for soil (100 gm sample) will allow quantitation down to a limit of 0.1 ppm for karbutilate, 0.2 ppm for monomethyl karbutilate (see Fig. 2), and 0.2 ppm for desmethyl karbutilate (see Fig. 3). The method described for water (200 gm sample) will allow the quantitation down to a limit of 0.01 ppm for karbutilate, 0.01 ppm for karbutilate phenol, and 0.02 ppm for monomethyl karbutilate (see Fig. 4). Sample background is a limiting factor on the sensitivity for soil method. No interference has been encountered with the water method. Method sensitivity is primarily limited by the response of the compounds at 254 nm and instrument background noise for water samples.

vi. *Recovery*

A wide variety of soil types (e.g., clay loam, sandy loam, silt loam, loam, loamy sand) have been analyzed using the above method. Fortifications over a wide range of values have yielded the following average method recoveries: karbutilate, 87%; monomethyl karbutilate, 92%; desmethyl karbutilate, 83%. Analysis by the above procedure of fortified water samples has shown method recoveries of about 95% should be obtained for karbutilate, phenolic hydrolysis product of karbutilate, and monomethyl karbutilate.

Variation (1 σ) in the above average recoveries can be expected to be $\pm 10\%$ for any given group of samples. All fortifications were made by adding a water-miscible standard solution (acetone) of the compounds of interest directly to sample prior to the addition of the initial extraction solvent.

e. *Discussion*

i. *Method*

In applying the soil procedure, it has been found that samples with a high silt content do not centrifuge completely clear at 2000 rpm. That is, some fine particles remain suspended. This creates a problem by plug-

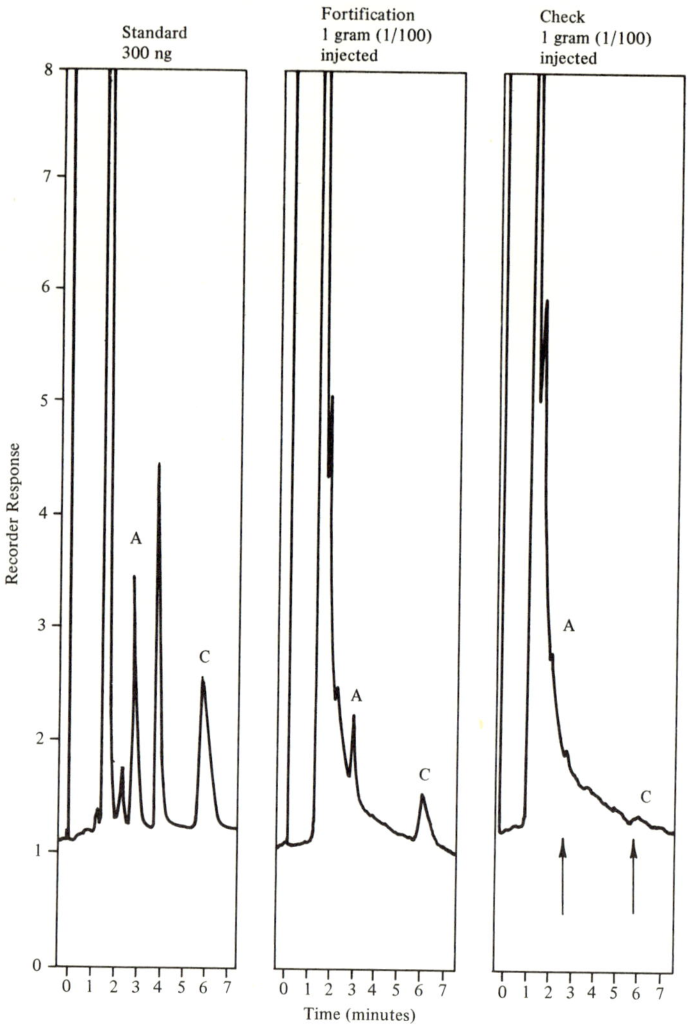

FIG. 2. Typical liquid chromatograms of soil extracts; (A) karbutilate, (C) monoethyl karbutilate.

ging the nucleopore filter or eventually becoming deposited in the liquid chromatograph. To eliminate this problem, soil samples with suspended solids (cloudy appearance) were centrifuged a second time at up to 7000 rpm.

Some variation in response has been noted between individual prepacked microporasil columns. It is also important that final injection

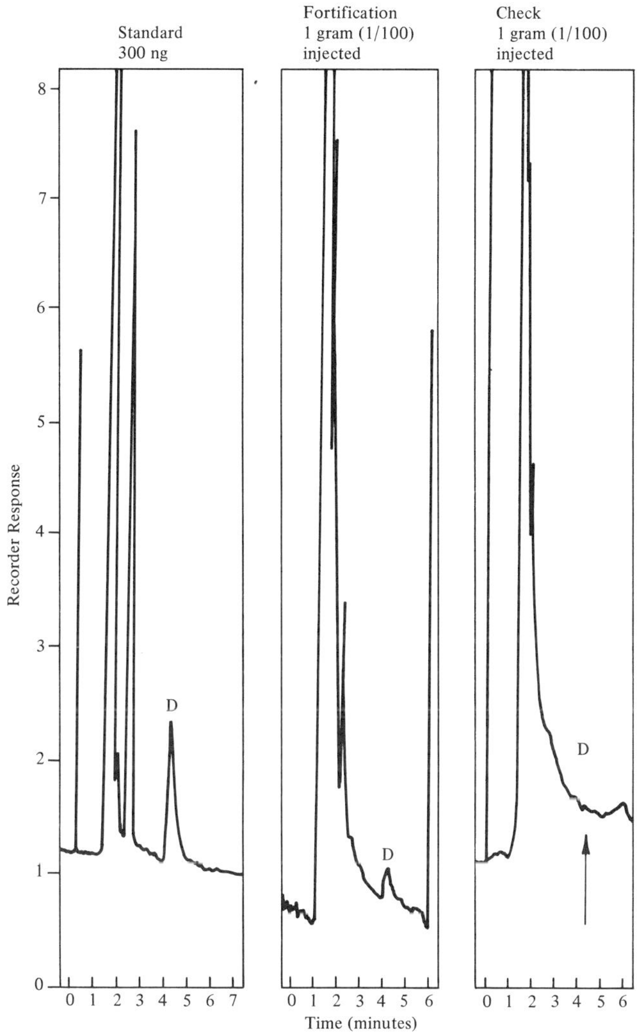

FIG. 3. Typical liquid chromatograms of soil extracts; (D) desmethyl karbutilate.

solutions be free of suspended solids to avoid deposits building up in the column and adversely affecting the columns performance.

ii. *Alternative Systems*

The soil and water residue methods described in the body of this report are the preferred procedures. However, alternative systems have also been successfully used. The following information may prove valuable to potential users who can not exactly duplicate the HPLC in-

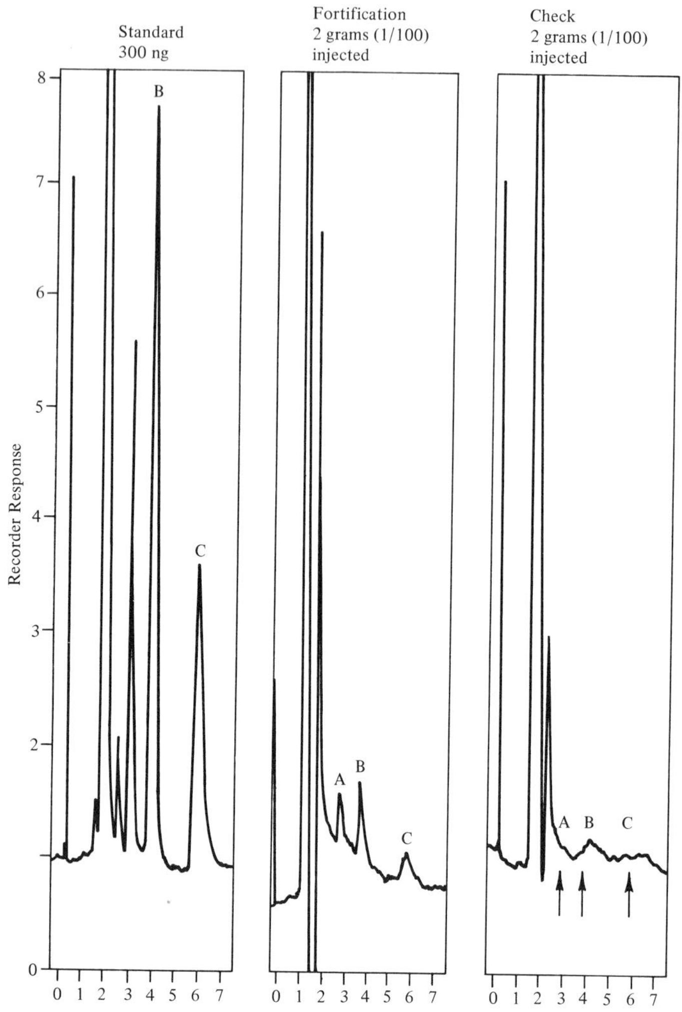

FIG. 4. Typical liquid chromatograms of water extracts; (A) karbutilate, (B) phenol of karbutilate, (C) monomethyl karbutilate.

strumentation, columns, or conditions specified in the preferred procedures.

(a) *Soil.* Prepare the samples as described in Sections d,ii,a and d,iii. Inject the samples for determination of karbutilate and monomethyl karbutilate using the following conditions:

Instrument: DuPont model 820 liquid chromatograph (or equivalent) equipped with a UV detector (254 nm) and a 10 mV recorder.

Column: 1 meter × 3.2 mm i.d. (6.4 mm o.d.), type 316 stainless steel, packed with 9.55 gm Corasil II (activated at 120°C for 24 hours).

Mobil phase: 7% (v/v) tetrahydrofuran in dichloromethane.

Temperature: Ambient.

Pressure: 425 psi.

Flow rate: 1 ml/minute.

Chart speed: 5 minutes/inch.

Detector: UV photometer fixed at 254 nm.

Absorbance: 2×10^{-2} AUFS.

Coarse zero: 13–15.

With the instrument equilibrated under these conditions, a retention time, i.e., from the point of injection to the peak center, of 3 minutes for karbutilate and 6 minutes for monomethyl karbutilate, should be obtained.

(b) *Column Cleanup* (*Soil*). Soil samples having karbutilate residue levels lower than 0.5 ppm are difficult to quantitate accurately because of early eluting interferences without the following cleanup. Determination of the need for this cleanup is based on the chromatographic data obtained in Section e,ii,a.

Thoroughly mix 2 gm neutral Norite with 5 gm Celite 545, and add the mixture to a 22 mm i.d. × 600 mm liquid chromatographic column containing a glass wool plug and 5 gm sodium sulfate in the bottom. Tap the sides of the column to allow the Norite–Celite mixture to pack. Place a glass wool plug on top of the packed absorbent mixture. Wet the column with 50 ml chloroform and drain the solvent to the top of the glass wool cap. Discard the collected washings. Add the concentrated analysis solution as used in Section e,ii,a to the Norite–Celite column. Rinse the column with 200 ml chloroform and discard the collected washings. Add 150 ml 40% (v/v) methanol in dichloromethane to the column and collect, by gravity flow, all the solvent possible.

Wash the methanol–dichloromethane column eluant with 50 ml distilled deionized water. Extract the water wash fraction with 150 ml dichloromethane, and combine the two dichloromethane fractions. Dry the combined fractions with sodium sulfate. Concentrate the dried sample on a rotary evaporator to about 10 ml and transfer to a graduated centrifuge tube.

Concentrate the solution in the graduated centrifuge tube to less than 1 ml using a gentle stream of nitrogen. Dilute the sample to exactly 1 ml with dichloromethane. Inject a suitable portion, e.g., 10 μl, into a liquid chromatograph operating under the conditions described in the Analysis Section e,ii,a.

(c) *Water.* Adjust 200 ml of subsampled water to pH 11 using a pH meter and 1 *N* sodium hydroxide. Extract the water twice with 200 ml dichloromethane. Readjust the aqueous phase to pH 4–4.5 using 1 *N* hydrochloric acid. Extract the acidified solution twice with 200 ml ethyl acetate. Concentrate the dichloromethane and ethyl acetate phases separately to about 5 ml on a rotary evaporator and transfer to separate graduated centrifuge tubes. Further concentrate the samples by blowing a stream of nitrogen over the surface to less than 1 ml and then dilute to exactly 1 ml with the corresponding solvent. Inject the dichloromethane solution to determine karbutilate and monomethyl karbutilate. Inject the ethyl acetate solution to determine the phenolic hydrolysis product of karbutilate. For both determinations, use the Corasil II column conditions listed in Section e,ii,a with the exception of changing the mobil phase to 1% (v/v) methanol in chloroform.

In the 1% (v/v) methanol–chloroform mobil phase, karbutilate and monomethyl karbutilate (methylene chloride concentrate) should have retention times of 4 minutes and 8 minutes, respectively. Karbutilate phenol (ethyl acetate concentrate) will have a retention time of 8 minutes in the methanol–chloroform mobil phase.

(d) *Notes.* Method sensitivity achieved with the Corasil II alternate system was equivalent to the Microporisil system. Recoveries for both systems were also very similar. Slight adjustment (1–2%) in the concentration of the polar component of the mobil phases may be necessary in order to compensate for variations between columns in the retentive nature of the Corasil II. Heating some batches of Corasil II up to 160°C may be necessary, on occasion, to obtain a suitably retentive support.

The Microporasil column is preferred because its greater resolution allows the method of analysis to be very direct. However, the Corasil II system is also a very practical system when the column resolution problem between monomethyl karbutilate and karbutilate phenol is overcome by chemical manipulations in the procedure.

Reference

Knaak, J. B., and Robinson, R. A. (1971). Agricultural Chemical Division, FMC Corporation, Middleport, New York (unpublished report).

26

MCPB

A. GUARDIGLI, J. VERNETTI, AND M. A. GALLO

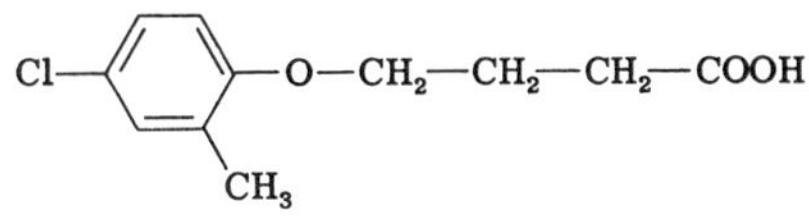

2-Methyl-4-chlorophenoxybutyric acid

I. GENERAL

A. Empirical Formula

$C_{11}H_{13}ClO_3$ (Mol. wt. 228.6).

B. Alternative Names

Can-Trol (MCPB sodium salt), trade name of Rhodia, Inc., Agricultural Division; Thistrol (MCPB sodium salt), trade name of Amchem Products, Inc.

C. Source of Analytical Standard

Agricultural Division of Rhodia, Inc., Somerset, New Jersey.

D. Biological Properties

The value of phenoxybutyric acids as selective herbicides in legumes was well known as early as 1955 (Wain, 1955; Shaw and Gentner, 1957; Fawcett *et al.*, 1958, 1959). The selectivity is based mainly on the differential rate of beta oxidation of MCPB by weeds and target crop (Kearney and Kaufman, 1969).

Production of such important crops as peas, alfalfa, soybeans, and other legumes have been adversely affected for lack of suitable methods of weed control. In peas (particularly green peas), berries of nightshade (*Solanum nigrum*), or the undeveloped flower head of Canada thistle (*Cirsium arvense*) are often difficult if not impossible to separate from the shelled pea. With the availability of 4-(2-methyl-4-chlorophenoxy)butyric acid (MCPB), it is possible to achieve effective weed control in legumes. The herbicide may be used safely at rates between 1 pound a.i./acre and 2 pounds a.i./acre.

The acute oral toxicity (LD_{50} values) of 4-(MCPB) sodium salt is approximately 700–1000 mg/kg for mice, rats, guinea pigs, and rabbits. Toxic symptoms, after administration, progress from ataxia to hyperreactivity to coma to death. Noncomatose animals generally survive. Spontaneous convulsions and tremors are seen in many animals during the hyperactive stage. Dacryorrhea and excessive salivation were evident in rats that died beyond 4–6 hours after dosing.

The results of eye irritation tests on New Zealand white rabbits showed that undiluted 4-(MCPB) sodium salt was minimally irritating if the eye was flushed within 4 seconds and was moderately irritating if the eye was not flushed. For inhalation studies, 4-(MCPB) sodium salt as a 2% solution was atomized into cages of guinea pigs for 1 hour per day for 1 week. No ill effects were noticed in any animals.

Daily oral administration of 700 mg/kg of 4-(MCPB) sodium salt to young rats for 24 days resulted in 100% mortality, while 135 mg/kg orally for the same duration had no effect on the experimental animals. Necropsy of the rats that died showed the gastric and superior intestinal mucosa to be eroded but not perforated with markedly constricted blood vessels. Examination of the lungs, heart, liver, spleen, and kidney revealed no abnormalities.

E. History

Synerholm and Zimmerman (1947) discovered that even-numbered long-chain alkanoic acids substituted (ω substituted) for acetic acid in 2,4-D produced active compounds when sprayed on tomato plants, whereas odd-number chains did not. This was attributed to breakdown of the chains to the active acetic moeity by the β-oxidase system in plants. Wain (1954) found similar results with the 2-methyl-4-chlorophenoxyalkanoic acid series. Meanwhile Carpenter and Soundy (1954) developed MCPB as a selective herbicide. In further work, Wain and Wightman (1954) discovered that some plants, such as peas and clover, do not possess an adequate β-oxidase system and are not affected by the longer-chain alkanoic acid derivatives even though susceptible to the acetic acid derivative. Thus the basis of selectivity of MCPB was revealed. May and Baker, Ltd. (1956) patented MCPB based on this selectivity for use in specific crops.

F. Physical Properties

MCPB is a white crystalline solid with a melting point of 99°–100°C. Water solubility is 45 ppm at 25°C. It is also soluble in polar solvents such as ketones, alcohols, and ethers. Solubility in other solvents at 25°C is shown in Table I.

TABLE I
SOLUBILITY OF MCPB

Solvent	Solubility (% wt/vol)
Dichloromethane	16.0
Carbon tetrachloride	2.9
1,2-Dichloroethane	7.0
Chlorobenzene	7.0
Hexane	6.5
Toluene	0.8

The infrared spectrum is given in Fig. 1.

G. Chemical Properties

1. METHOD OF SYNTHESIS

MCPB is synthesized by reacting the dehydrated sodium salt of ortho cresol, suspended in an inert solvent, with γ-butyrolactone. At the end of the reaction (3–5 hours), water is added and the solvent is removed by steam distillation. The solution is acidified, and the 2-methylphenoxybutyric acid is separated by filtration. The acid is chlorinated in a suitable solvent with gaseous chlorine to form MCPB. The MCPB is removed from the solvent by extraction with a sodium hydroxide solution and then precipitated with mineral acids, filtered, and dried.

2. CHEMICAL PROPERTIES

MCPB is a weaker acid than the 2,4-D and MCPA. The ionization constant (pK_a) is $1.4–1.5 \times 10^{-5}$ at 25°C, whereas that of 2,4-D is $1.7–1.8 \times 10^{-3}$, and that of MCPA is $8.1–8.8 \times 10^{-4}$ at the same conditions (Nelson and Faust, 1969). It undergoes the usual carboxylic acid reactions forming salts and esters readily. The chemical is stable under normal storage conditions.

H. Formulation

MCPB is produced for use as sodium salt solutions.

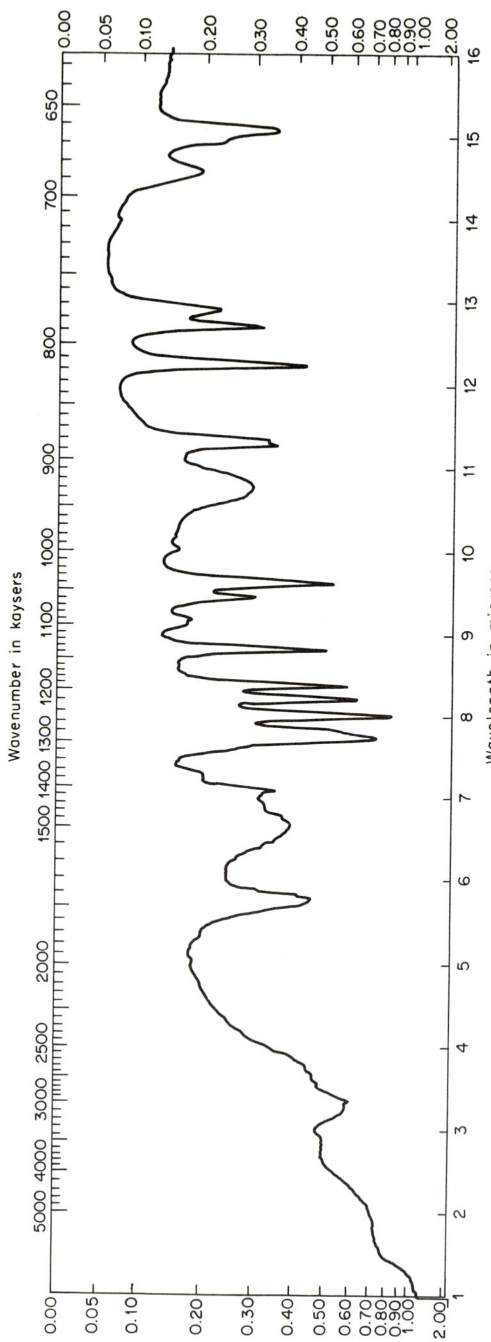

FIG. 1. Infrared spectrum of MCPB (2-methyl-4-chlorophenoxybutyric acid), double crystallized (Beckman spectrophotometer model IR 4).

II. ANALYSIS

A. Formulation Analysis

1. REVIEW OF METHODS

a. *Total Acidity Analysis*

The aqueous salt solutions are acidified and extracted with ethyl ether. The extract is evaporated and the solids are dissolved in a polar solvent with water (approximately 10%). The solids are titrated with standard sodium hydroxide and calculated as MCPB.

b. *Total Chlorine Analysis*

Aqueous salt solutions, which are not normally used as formulations, can be analyzed by the Parr bomb (Association of Official Analytical Chemists, 1970) or by equivalent procedures such as sodium biphenyl or sodium amide reduction methods (Rhodia, Inc., 1968). The latter two must be done with the dry extract. Correction for inorganic chloride is made by the normal Volhard method in any case.

c. *Infrared Spectrophotometric Method*

As an alternate method of analysis, Rhodia, Inc., uses the following differential infrared procedure. The difference reading of absorbance between a known standard in the reference beam and an unknown sample in the sample beam is determined at 800 cm^{-1} using 0.5-mm sodium chloride cells. Twenty micrograms MCPB over milliliter carbon disulfide is placed in both cells, and the cell difference is determined in absorbance units by baseline techniques. Expanded scale absorbance is helpful if available. The cell in the sample beam is removed and refilled with the unknown sample, and the difference is determined again. By correcting the sample–standard reading with the cell difference reading, the absorbance difference is obtained. Using a previously determined absorbance coefficient (from comparing different standard concentration solutions within the range of assay), the difference in concentration is calculated. The concentration difference is added or subtracted to the standard concentration to give the actual concentration of MCPB in the sample. A difference reading of 0.004 absorbance units is approximately equal to 1% concentration difference. Normally, the error is ±0.001 absorbance units so that the analysis error is ±0.25%. The advantage of the method is that an exact reference standard can be made from technical materials and the difference spectra obtained with a sample formulation, in addition to the analysis, can be examined for contaminants. It is imperative that the cell thickness remain constant and that cell calibrations be taken. Depending on cell usage, calibrations should be made every 1 to 3 months.

Extraction and preparation of formulations of aqueous salt solutions is best done by the method of St. John (1967). A more detailed description of this type of assay is shown by Pease and Deye (1967).

d. *Ultraviolet Analysis*

No specific method could be found for MCPB. It is assumed that the method of Hill (1952) for MCPA could be used with the appropriate absorption peak wavelengths.

e. *Gas Chromatography*

The normal internal standard procedure is best used for MCPB gas chromatography analysis. In the Rhodia, Inc., method (1970), the ether extracts of formulations, with MCPA added as the internal standard, is esterified to form the methyl esters with diazomethane or methanolic HCl. A $\frac{1}{4}$ inch × 4 foot stainless steel column packed with 10% OV-225 on acid washed and silanized Chromosorb W provides the separation of MCPA from MCPB and the isomers of MCPB. Temperature programming from 180–200°C is desirable for the best separation and peak area measurement. Thermal conductivity detectors are used which give a peak area ratio MCPA:MCPB of 0.95 (equal weights of acid equivalents). The regular internal standard calculation is used to obtain the percentage of MCPB. The isomers may be similarly determined if needed.

2. RECOMMENDED METHOD

a. *Principle*

The total organic acids as MCPB are obtained by acid base titration of the extracted acids from the aqueous salt formulation and are corrected for the isomeric acids using the normalized isomer content obtained by gas chromatography of the methyl esters.

b. *Reagents*

Isopropyl alcohol, Analytical Reagent Grade, diluted, 9 parts with 1 part water.

Sulfuric acid, Analytical Reagent Grade, 10% in water.

Sodium hydroxide, standard 0.10 *N,* standardized with potassium acid phthalate.

Diethyl ether, Analytical Reagent Grade.

Methanolic HCl Solution. Prepare by adding 5 ml of freshly distilled acetyl chloride to 100 ml of distilled methanol dropwise and swirling (use hood and safety glasses or face shield). Let stand 15 minutes and cool. Cap and keep in the refrigerator. The concentration is

3% w/w HCl solution. This solution should be prepared fresh every 2 weeks.

c. *Apparatus*

Separatory funnels, 300 ml.
pH meter, Beckman Zeromatic or equivalent.
Burette, 50 ml.
Tall beaker, 250 ml.
Beaker, 250 ml.

Gas chromatograph, Hewlett Packard 7620A or equivalent, with thermal conductivity detector. The $\frac{1}{4}$ inch × 4 foot stainless steel column is packed with 10% OV-225 on silanized Chromosorb W. The oven temperature is 190°C and the helium flow rate is 60 ml/minute.

d. *Procedure*

i. *Total Organic Acid*

Accurately weigh enough formulation into a weighing bottle to provide 1–1.2 gm MCPB for analysis. Transfer with distilled water into a separatory funnel containing 50 ml distilled water. Add 5 ml 10% H_2SO_4 and swirl to mix thoroughly. Add 50 ml ether and extract. Repeat with 30 and 20 ml ether and combine the extracts. Wash the extracts with two 10-ml portions of water to remove mineral acid and transfer the ether extract to a 250-ml tall beaker and evaporate to dryness on a steam bath. Transfer the contents to a 250 ml beaker with the isopropyl alcohol–water mixture using approximately 100–150 ml. Titrate the mixture to pH 9.0 with standard sodium hydroxide and calculate the total organic acid (or MCPB) as follows:

% total organic acid as MCPB

$$= \frac{\text{standard NaOH (ml)} \times \text{normality} \times 228.68}{\text{Weight of sample} \times 10}$$

ii. *Normalized Isomer Content*

Place 2–3 ml of the formulation into a separatory funnel containing 75 ml water, then add 5 ml 10% H_2SO_4, swirl, and extract with two 50-ml portions of ether. Combine the extracts and evaporate to approximately 10 ml on a steam bath. Transfer approximately 1 ml into a 15-ml conical centrifuge tube and add 2–3 ml methanolic HCl solution. Place the tube on a steam bath for 5–10 minutes until the HCl fumes cease to be evolved. Cool and inject 2–4 μl of the methyl esters into the gas chromatograph at the specified conditions. Sum the peak areas of the isomers and calculate the percentage of MCPB. The retention times in one system are shown in Table II.

TABLE II
GAS CHROMATOGRAPHIC DATA

Peak	Retention time (minutes)	Isomer
A	5.75	2-Methylphenoxybutyric acid
B	10.75	2-Methyl-6-chlorophenoxybutyric acid
C	14.85	2-Methyl-4-chlorophenoxybutytic acid
D	20.20	2-Methyl-4,6-dichlorophenoxybutyric acid

e. *GLC Parameters*

Column: 6-foot $\frac{1}{4}$-inch glass column packed with 10% OV 101 on Chromosorb W 80/100 mesh.

Column Temperature: 190°C (isothermal).

Injection Temperature: 240°C.

Carrier Gas: Helium, 60 cc/minute.

$$\% \text{ MCPB isomer content (normalized)} = \frac{C}{A + B + C + D} \times 100$$

i. *Calculation of Percentage of MCPB*

% MCPB in formulation.

$$= \frac{(\% \text{ total organic acid as MCPB}) \times (\% \text{ MCPB from Section ii})}{100}$$

ii. *Discussion*

The sensitivity of the instrument to detect the methyl esters of the acids is essentially equivalent (weight basis) by thermal conductivity detection. Since the bulk of the isomers of MCPB is usually the 6-chloro compound, no weight correction is necessary. The nonchlorinated and 4,6-dichloro compounds are approximately equivalent in concentration such that the molecular weight average is basically the same as MCPB. However, if either of the latter two compounds show high concentrations, then the molecular weights will have to be considered in the calculation.

B. Residue Analysis

1. REVIEW OF METHODS

Foliar broadcast applications of 4-(2-methyl-4-chlorophenoxy)butyric acid [4-(MCPB)], have been used to control broadleaf weeds in pea (*Pisum sativum* L.) fields since its discovery (Wain and Wightman,

1954). However, Glastonbury *et al.* (1959) of the Research Laboratory of May & Baker, Ltd., England, were the first to report on the persistence of MCPB residues in peas under field conditions. The crop samples sprayed at the three- to four-leaf stage with a commercial formulation of the sodium salt of MCPB were gathered immediately after spraying and thereafter at convenient intervals. The samples were dried and extracted with sodium hydroxide. The proteins were precipitated with phosphotungstic acid, and the free MCPB was isolated by solvent extraction and estimated by infrared spectrophotometry. Quantitative results were obtained from this nonquantitative procedure by isotope dilution analysis using MCPB labeled in the carboxyl group with carbon-14. The analytical method had a limiting sensitivity of 1 ppm calculated on a fresh weight basis.

Because of the continuing interest in the use of MCPB as a selective weed control in peas, Brannock and Freed (private communication, 1963) tested various analytical residue procedures for suitability of MCPB analysis on peas. Since the 2-methyl-4-chlorophenyl ring has an electron affinity up to a thousandfold less than that of the dichlorophenyl ring, the herbicide was determined by using the coulometric detector with the Dohrmann gas chromatograph.

Three different approaches to the extraction, purification, and detection of MCPB were attempted. These were devised in part to assess not only the amount of MCPB that may be present as a residue but also to determine whether methylchlorophenol and the methylchlorophenoxyacetic acid (MCPA) might be present as a residue. In the first procedure, the samples were extracted with alcohol, digested with base, acidified, and extracted into benzene. The benzene solution was then chromatographed on a basic alumina column and the organic acid eluted with sodium bicarbonate. The sample was then back-extracted into an organic solvent after acidification, evaporated to dryness, esterified, and injected into the gas chromatograph.

The second procedure followed essentially that described above, but instead of esterification, the sample was treated with pyridine hydrochloride after the method of Marquardt and Luce (1961). This treatment cleaves the ether linkage and converts the compound into the methylchlorophenol. The sample was then extracted into a minimum quantity of organic solvent and appropriate proportions equivalent to 1–5 gm of dry samples were injected into the gas chromatograph. The extraction and purification methods outlined above resulted in the extraction of substances giving strong signals with the coulometric detector and electron capture detector which obscured the precise identification of the compounds of interest.

The third approach to extraction and purification consisted of an extraction with a benzene and glacial acetic acid solvent. The authors found it necessary to use this system to determine if a residue of material remained in the plant tissues.

The methyl ester of MCPB was analyzed on a Dohrmann microcoulometric gas chromatograph with a gas chromatograph column packed with 10% polyglycol (Carbowax) coating for best resolution.

Bache *et al.* (1964a) showed that snap beans (*Phaseolus vulgaris*) were capable of converting MCPB into MCPA by β-oxidation of the side chain, as found previously for 4-(2,4-DB). In developing a method for determining residues of MCPA and MCPB by electron affinity, these authors found that because of their low response, the herbicides must first be nitrated to (presumably) the 5- and 6-nitro derivatives, and then methylated. (See also Volume V, p. 445, of this treatise for detailed discussion).

In the same year, Bache *et al.* (1964b) reported on the elimination of MCPA and MCPB in the urine from cows fed these herbicides. The residue analyses were also based on the electron capturing nitro and methyl ester derivatives.

Gutenman and Lisk (1963) used bromination to prepare the derivatives for gas chromatographic detection as mentioned in Volume VI, p. 33 of this treatise. Abbott *et al.* (1964), described a paper and thin-layer chromatographic procedure for detecting, separating, and identifying six herbicides: MCPA, MCPB, 2,4-D, 2,4-DB, 2,4,5-T in soil and water, and dalapon in water only. The extraction of these compounds from soil was found to be most difficult, but satisfactory recoveries were obtained by the extraction of soil slurry in dilute sulfuric acid with ether. The application of thin-layer chromatography to the identification of the components of herbicide mixtures was used by Abbott and Wagstaffe (1969) on mixed herbicide formulations including: phenoxyalkanoic acids, triazines, carbamates, and substituted ureas. They reported various separation systems and visualization agents.

2. RECOMMENDED METHOD

a. *Principle*

The GLC and TLC procedures described below are particularly designed to determine total residue resulting from the proposed use of MCPB as the sodium salt (Guardigli, 1974). The crop materials are subjected to alkaline hydrolysis–digestion prior to solvent extraction, insuring complete yield of residues present in all forms (Gunther, 1966; Chow *et al.*, 1971). The methods permit the detection of the intact herbicide including the formulated compound, free acids of MCPB or

MCPA, the possible chlorophenol metabolic products, and ester and/or glycoside complexes conjugated with plant proteins and glucose.

The parent compound MCPB and metabolite MCPA are determined by electron capture GLC after conversion to nitro and methyl ester derivatives (Bache *et al.*, 1964a). The electron capture detector and the principles of gas chromatography are described in Volumes I and VI of this treatise. The derivatives are readily separated by a polar column, and the peaks obtained are quantitated as acid equivalents. The nitro and methyl ester derivatives constitute an additional tool for precise peak identification and for quantitative determination of residues in crops treated with MCPB formulations. The total chlorophenols are determined by thin-layer chromatography (TLC) as free radicals by a very sensitive and specific color reaction (Nurnberg, 1961). The limits of detection, calculated as acid and phenol equivalents are about 0.04 ppm and 0.1 ppm, respectively. Recoveries exceeded 80%.

b. *Reagents*

Acetone, Nanograde, Mallinckrodt.

Chloroform, Nanograde, Mallinckrodt.

Hexane, Nanograde, Mallinckrodt.

Diethyl ether, anhydrous, Nanograde, Mallinckrodt.

Petroleum ether, Nanograde, Mallinckrodt.

Formic acid, Analytical Reagent Grade, Mallinckrodt.

Orthophosphoric acid, Analytical Reagent Grade, Mallinckrodt.

Boron trifluoride–methanol, 14% w/v esterification reagent, Applied Science Laboratories, Inc.

Methanol, Nanograde, Mallinckrodt.

Potassium phosphate dibasic, Reagent Grade.

Spray reagent, 1.0% 2,4-dichloroquinonechlorimide in methanol (Brinkmann Instruments Westbury, New York).

Sodium hydroxide, Reagent Grade.

Sodium sulfate, anhydrous, Reagent Grade.

Sodium bicarbonate, Reagent Grade.

Sodium nitrate, Reagent Grade.

Super Cell "Hy-flow," Johns Manville Corp.

Aluminum oxide, acidic, activity grade 1 Woelm, distributed by ICN Pharmaceuticals Inc. Cleveland, Ohio.

Florisil PR Grade, 60–100 mesh, used "as received without activation," Floridin Co., Berkeley Spring, West Virginia.

c. *Apparatus*

Assorted laboratory glassware.

Food cutter, Hobart model 84181-D with slicer attachment #12.

Rotary vacuum evaporators, Büchler or equivalent (all glass system).

Heating jacket, hot plates, baths.

Glass chromatography columns, Mini-columns, size 213, 10 mm i.d. × 300 mm. Chromaflex-Kontes, Vineland, New Jersey.

Gas chromatograph, Micro-Tek MT 220 equipped with electron capture detector. Ionization source ^{63}Ni.

Strip chart recorder, Westronics 1 mV.

Carrier gas cylinder, 5% methane in argon.

Microliter Hamilton syringes, 10 μl volume.

TLC plates, Camag, precoated silica gel glass plates 20 × 20 cm. Camag Inc., New Berlin, Wisconsin.

Developing tanks, Arthur H. Thomas Co.

UV viewer, Desaga U.V.I.S. for UV 254 mμ and 366 mμ with visible light illuminator. Brinkman Instruments Inc. Westbury, New York.

Photographic equipment, Polaroid MP-3 Land camera equipped with color correction filter Photar B-3.

d. *Reference Standard*

2-methyl-4-chlorophenoxybutyric acid (MCPB) and 2-methyl-4-chlorophenoxyacetic acid (MCPA), Rhodia, Inc., Agricultural Division, Somerset, New Jersey; 2-methyl-4-chlorophenol, Eastman Kodak Co., Rochester, N.Y.

e. *Experimental Procedure*

i. *Plant Material Preparation*

Process the entire representative field sample (i.e., vine and trash) with the Hobart food chopper model 84181-D. Grind-mix the legume seeds (i.e., peas) and maintain frozen for later analysis. Transfer 25 gm of the finely chopped sample into a 500-ml flask through a powder funnel. Add 100 ml deionized water and 50 ml 0.5 *N* sodium hydroxide.

ii. *Alkaline Hydrolysis—Digestion and Extraction*

Connect the flask to a water cooled condenser and let digest and reflux gently over a heating mantle for 1 hour to hydrolyze the phenoxys present in any form—sodium salt, ester, or "bound." Completely *cool* the hydrolyzed sample. Filter the aqueous mixture under reduced pressure through a Buchner funnel containing a no. 1 Whatman filter paper moistened with water and covered with a glass wool pad and a layer of Hy-flow Super Cell. Wash the filter cake with 100 ml deionized water. Transfer the combined alkaline aqueous filtrates to a 1-liter separatory funnel. Add 50 ml chloroform. Shake the mixture gently. Allow the solvent layer to separate. A quick separation may be obtained by centrifuging the solvent–water mixture.

iii. *Liquid–Liquid Partition Cleanup*

Collect the solvent layer into a second separatory funnel. Repeat the extraction two more times with 50-ml portions of chloroform each time. Backwash the combined chloroform extract twice with 50 ml of 0.5 *N* NaOH. Discard the solvent layer and add the caustic washings to the aqueous layer in the first separatory funnel. Acidify the aqueous layer with 100 ml 2 *N* H_2SO_4 to a pH of 2 or less, then add 50 ml chloroform. Shake the mixture well, allow the layers to separate, and filter the solvent phase through anhydrous sodium sulfate into a 250-ml evaporating flask. Repeat this operation four times with 50-ml portions of chloroform each time. Evaporate the combined organic extract to about 5 ml volume using a rotary vacuum evaporator with a water bath temperature not exceeding 50°C. The concentrated residue contains the phenoxy free acids and the phenol compounds.

iv. *Separation of Phenoxy Acids from Phenols by Acidic Alumina Column*

Place 5 gm Woelm acidic alumina, activity grade 1, into a 300 mm × 10 mm i.d. chromatographic tube having a coarse sintered glass disk. Wash the column with 50 ml chloroform. Transfer the sample residue to the column by washing the flask and eluting with 100 ml chloroform. Discard effluents containing pigments and impurities. Elute with 75 ml 5% methanol in chloroform. Collect this eluate into a 250-ml flask and set aside. This solution contains the compounds needed for the TLC determination of phenols.

Continue the elution from the alumina column with 100 ml 5% aqueous $NaHCO_3$ solution. Collect the eluates in a clean 250-ml filtering flask under reduced pressure. Acidify the solution with 3 ml 1 : 1 H_2SO_4–water. Transfer the acidified solution (pH 1–2) to a separatory funnel washing the flask three times with small portions (10 ml) of chloroform. Shake the separatory funnel vigorously. Allow the phases to separate and filter the chloroform, bottom layer, through a funnel containing Whatman no. 1 filter paper and anhydrous Na_2SO_4 into a 250-ml evaporating flask. Extract the aqueous phase three times with 50 ml chloroform each time. Shake, allow to separate, and collect the chloroform each time into the same 250-ml evaporating flask. Evaporate the solvent to near dryness in a rotary vacuum evaporator using a water bath at temperatures less than 50°C. Gently remove the last traces of solvent with a jet of dry clean air.

f. *Gas Chromatography*

i. *Conversion of MCPB and MCPA Acid to Nitro Derivatives*

The nitrating solution must be prepared immediately before use. Prepare the 2% sodium nitrate solution in concentrated orthophosphoric acid by heating the mixture to 85–90°C until the $NaNO_3$ is dissolved. Add 5 ml of the hot mixture to the residue into a 250-ml flask. Place the flask in a 85–90°C oil bath for 30 minutes with occasional swirling. Remove it and let cool. Transfer the contents to a 250-ml separatory funnel with 100 ml of 10 gm per liter sodium sulfate solution and 25 ml ethyl ether. Shake, allow to separate, and collect the ethyl ether into a 250-ml evaporating flask, by filtering the solvent through Whatman no. 1 filter paper and a small amount of anhydrous sodium sulfate. Repeat the extraction four times with 25 ml ethyl ether each time. Evaporate the combined ether extracts to near dryness by using a rotary vacuum evaporator. Water bath temperature is kept at not more than 30–35°C.

ii. *Esterification of the MCPB and MCPA Nitro Derivatives to Methyl Esters*

Add 3 ml of the esterification reagent boron trifluoride–methanol to the above residue. Gently reflux the solution on a heating mantle for 20 minutes, using a water-cooled condenser attached to the flask. Cool, rinse the condenser with a few milliliters of hexane, and transfer the methyl ester to a 250-ml separatory funnel by rinsing the esterification flask with 50 ml 0.4 *M* K_2HPO_4 and 30 ml hexane in small portions. Transfer all washings to a separatory funnel and shake gently. Allow to separate, and collect the aqueous phase in a separate 250-ml separatory funnel. Repeat the extraction of the aqueous phase three times with 25ml of hexane each time. Shake gently, let separate, and collect the hexane layer each time by filtering through anhydrous sodium sulfate into a 250 ml evaporating flask. Wash the funnel containing the filter paper and Na_2SO_4 with a few additional milliliters of hexane. Evaporate the hexane to near dryness with the aid of a rotary vacuum evaporator. Water bath temperature should be kept between 40 and 50°C.

iii. *Florisil Column Chromatography*

Set up the column by preparing a suspension of 10 gm Florisil in benzene. Insert a plug of glass wool at the bottom of a 300 mm × 10 mm i.d. chromatographic tube with a coarse sintered glass disk. Pour the suspension into the column in small fractions. Pack each fraction by suction and always insure that a layer of benzene is maintained above the adsorbent. Add anhydrous sodium sulfate (about $\frac{1}{4}$ inch) and a plug of glass wool to the column. Wash the column with 100 ml benzene. The column should not be allowed to go dry at any time.

The activity of the Florisil at hand should be predetermined by eluting a known amount of phenoxy derivative standard solution, and by injecting a known concentration (i.e., 0.5 μg/ml acid equivalent) into the gas chromatograph. The quantitative elution of the hexane semidry residue containing the phenoxy nitro and methyl ester derivatives can be accomplished by percolating 100 ml 1% ethyl acetate in benzene.

g. *Determination by Gas Liquid Chromatography*

i. *Operating Parameter*

Instrument: Micro-tek MT 220 equipped with electron capture detector. Ionization source ^{63}Ni or equivalent.

Column: 4 feet × 3.5 mm i.d. glass column packed with 5% w/w OV-210 on 80–100 mesh Gas Chrom Q.

Carrier gas flow rate: 80–85 cc/minute (5% methane in argon).

Temperatures

Column oven: 180°C.

Detector cell: 265°C.

Inlet: 220°C.

Electrometer range: Input × 10^2, output × 32.

Voltage output: 46 V.

Pulse rate: 270 μseconds.

Pulse width: 9 μseconds.

Recorder chart speed: $\frac{1}{4}$ inch/minute.

At the above conditions, the retention time of the MCPB derivative methyl ester is about 5 minutes and that of the MCPA nitro derivatives methyl esters are about 2.5 minutes and about 3.5 minutes, presumably for "isomer 6" and "5," respectively. The MCPA derivative peak emerging at the retention time of about 2.5 minutes was selected for quantitative purposes.

ii. *Operating Procedure*

Adjust the final volume of sample solution so that microliter injections of given concentrations of MCPB and MCPA acid equivalent correspond to amounts located within the linear range of instrument response. Usually, 1–2 μl of sample solutions are injected. The peak height H is measured in millimeters. Injections are made of the same volume (i.e., 1–2 μl) of standards corresponding to C nanograms per microliter of MCPB or MCPA acid equivalent. The peak height h is measured in millimeters.

With adequate instrument conditioning and detector consistency, the unknown quantity (nanograms/microliter) injected is:

$$\text{nanograms/microliter or microgram/milliliter} = \frac{H \times C}{h} = \mu\text{g/ml}$$

Thus

$$\text{ppm} = \frac{\mu\text{g/ml} \times \text{sample dilution (ml)}}{\text{sample weight (gm)}}$$

iii. *Preparation of Standard Curves*

Prepare the following stock solutions:

1. MCPB free acid in acetone nanograde containing 100 μg/ml of acid equivalent.

2. MCPA free acid in acetone nanograde containing 100 μg/ml of acid equivalent.

3. MCPB sodium salt in distilled water containing 100 μg/ml of acid equivalent.

Dilute the stock solutions to working standards and proceed from the alkaline hydrolysis–digestion through the nitration, methylation, and cleanup steps as described in this procedure. A series of solutions are diluted to correspond to 0.1–0.5 μg/ml of MCPB and MCPA acid equivalent. Equal volumes (microliters) of each solution are injected into the gas chromatograph, and the calibration curve is constructed by plotting the corresponding nanograms against the peak height.

The minimum detectable level is 0.1 μg/ml MCPB or MCPA acid equivalent. Therefore, by using a 25-gm sample and a final volume of 10 ml, the sensitivity of the method is 0.04 ppm acid equivalent.

h. *Thin-Layer Chromatography* (*TLC*)

i. *Determination of Phenolic Products*

In a rotary vacuum evaporator, partially concentrate the methanol–chloroform eluates containing the phenols that were set aside earlier. Transfer the concentrates (5–10 ml) to a Kontes tube heater apparatus provided with an evaporative concentrator, and concentrate the solution to exactly 1 ml or 0.5 ml volume, according to the desired sensitivity. The basic device and technique for the evaporator are described by Beroza and Bowman (1967).

ii. *TLC Spot Development*

With a pencil draw a line completely across the plate 14 cm from the imaginary spotting line that is about 2.5 cm from the bottom of the plate. For maximum sensitivity, spot the most concentrated sample possible (i.e., 200 μl of a 25 gm or 10 gm crop substrate in 1 ml or 0.5 ml of solvent) along with 0.5 μg, 1 μg, 2 μg, 3 μg, and 5 μg of phenolic compound reference standard.

Place about 200 ml of developing solvent (50 parts petroleum ether, 50 parts ethyl ether, 0.3 parts formic acid) into the bottom of a tank lined with blotting paper. The tank must be air tight. Allow the tank to saturate with solvent vapors. Position the plate in the tank, quickly cover the tank, and develop the plate until the solvent front reaches the scribed line. Let the plate air dry, and make preliminary identification under shortwave UV light.

iii. *Spot Visualization*

Spray the plate lightly with 1% sodium hydroxide in ethanol. Place it in the oven at 110°C for 30 minutes. Cool. Spray it uniformly and heavily with 1.0% w/v of 2,6-dichloroquinonechlorimide in methanol. Characteristic intense blue spots appear almost immediately. The maximum color intensity is reached in about 15 minutes.

This color reagent is specific for the detection of the hydroxylated benzene ring. The 2-methyl-4-chlorophenol migrates to an $R_f = \sim 0.45$. The size and intensity of the sample spots are visually compared with known amounts of standards migrating at the same R_f, on the same plate, and under the same conditions. The accuracy of this method is ±15%; the sensitivity is 0.5 μg/ml.

i. *Evaluation of the Extraction Procedure*

In order to determine the extraction efficiency of the method, the semidry filter cake substrate remaining after the hydrolysis–digestion

TABLE III
RECOVERIES OF MCPB, MCPA, AND 2-METHYL-4-CHLOROPHENOL ADDED TO UNTREATED CROP SAMPLES

Equivalents added (ppm)			Equivalents recovered (%)		
MCPA	MCPB	2-Methyl-4-chlorophenol	MCPA	MCPB	2-Methyl-4-chlorophenol
0.2	0.2	0.2	93.2	95.0	~80.0
0.2	0.2	0.5	89.5	81.2	~80.0
0.4	0.4	0.5	88.0	93.8	~80.0
0.4	0.4	1.0	91.0	92.0	~90.0
0.5	1.0	–	85.6	87.0	–
0	2.0	–	87.0	90.0	–
–	5.0	–	–	91.5	–
–	10.0	–	–	88.2	–
–	50.0	–	–	90.5	–

and filtration was placed in a thimble and subjected to exhaustive Soxhlet extraction with chloroform–methanol (1 : 1) for 12 hours as reported by Burke *et al.* (1971). The Soxhlet extract was evaporated and analyzed by following the described procedure. No significant additional residues could be detected, indicating the extracting procedure used in the method to be satisfactory.

j. *Applicability of Recommended Methods*

The gas–liquid and thin-layer chromatographic methods for MCPB and metabolites have been used on a variety of leguminous crop parts. Typical recovery data are shown in Table III.

References

Abbot, D. C., and Wagstaffe, P. S. (1969). *J. Chromatogr.* **43,** 361.

Abbot, D. C., Egan, H., Hammond, E. W., and Thompson, J. (1964). *Analyst* **89,** 480.

Association of Official Analytical Chemists. (1970). "Method of Analysis," 11th ed., p. 95. Ass. Offic. Anal. Chem., Washington, D.C.

Bache, C. A., Lisk, D. J., and Loos, M. A. (1964a). *J. Ass. Offic. Anal. Chem.* **47,** 348.

Bache, C. A., Lisk, D. J., Wagner, D. G., and Warner, R. G. (1964b). *J. Dairy Sci.* **47,** 93.

Beroza, M., and Bowman, M. C. (1967). *Anal. Chem.* **39.**

Burke, J. A., Porter, M. L., and Young, J. (1971). *J. Ass. Offic. Anal. Chem.* **54,** 142.

Carpenter, K., and Soundy, M. (1954). *Proc. Brit. Weed Contr. Conf., 2nd, 1954* p. 327.

Chow, C., Montgomery, M. L., and Yu, T. C. (1971). *Bull. Environ. Contam. Toxicol.* **6,** 576.

Fawcett, C. H., Taylor, H. F., Wain, R. L., and Wightman, F. (1958). *Proc. Roy. Soc., Ser. B* **148,** 543.

Fawcett, C. H., Pascal, R. M., Pybus, M. B., Taylor, H. F., Wain, R. L., and Wightman, F. (1959). *Proc. Roy. Soc. (London), Ser. B* **150,** 95.

Glastonbury, H. A., Stevenson, M. D., and Ball, R. W. E. (1959). *Weeds* **1,** 362.

Guardigli, A. (1974). Analytical Method 123. Agricultural Division of Rhodia, Inc., Somerset, New Jersey.

Gunther, F. A. (1966). *Residue Rev.* **14,** 83.

Gutenmann, W. H., and Lisk, D. J. (1963). *J. Ass. Offic. Agr. Chem.* **46,** 859.

Hill, R. (1952). *Analyst* **77,** 67.

Kearney, P. C., and Kaufman, D. D. (1969). "Degradation of Herbicides." p. 39. Dekker, Inc., New York.

Marquardt, R. P., and Luce, E. N. (1961). *J. Agr. Food Chem.* **9,** 266.

May & Baker, Ltd. (1956). British Patent 758,980.

Nelson, N. H., and Faust, S. O. (1969). *Environ. Sci. Technol.* **3,** 1186.

Nurnberg, E. (1961). *Deut. Apoth.-Ztg.* **101,** 268.

Pease, H. L., and Deye, J. F. (1967). *In* "Analytical Methods for Pesticides, Plant Growth Regulators and Food Additives" (G. Zweig, ed.), Vol. 5, p. 337. Academic Press, New York.

Rhodia, Inc. (1968). "Methods of Analysis H 197 and I 106." Agricultural Division of Rhodia, Inc., Somerset, New Jersey.

Rhodia, Inc. (1970). "Control Laboratory Alternative Procedure." Agricultural Division of Rhodia, Inc., Somerset, New Jersey.

St. John, L. E., Jr. (1967). *In* "Analytical Methods for Pesticides, Plant Growth Regulators, and Food Additives" (G. Zweig, ed.), Vol. 5, p. 439. Academic Press, New York.

Shaw, W. C., and Gentner, W. A. (1957). *Weeds* **5,** 75.

Synerholm, M. E., and Zimmerman, P. W. (1947). *Contrib. Boyce Thompson Inst.* **14,** 369.

Wain, R. L. (1954). *Proc. Brit. Weed Contr. Conf., 2nd, 1954* p. 311.

Wain, R. L. (1955). *J. Agr. Food Chem.* **3,** 128.

Wain, R. L., and Wightman, F. (1954). *Proc. Roy. Soc. (London), Ser. B,* **142,** 525–536.

27

Metoxuron

M. WISSON, C. VAN HOEK, AND H. H. SAUER

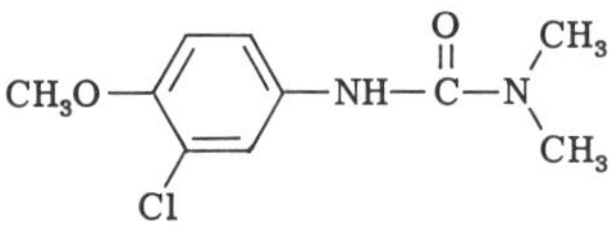

N′-(3-Chloro-4-methoxyphenyl)-*N*,*N*-dimethylurea

I. GENERAL

A. Empirical Formula

$C_{10}H_{13}ClN_2O_2$ (Mol. wt. 228.68).

B. Alternative Names

3-(3-Chloro-4-methoxyphenyl)-1,1-dimethylurea; Dosanex® is the registered trademark of Sandoz Ltd., Basel, Switzerland.

C. Source of Analytical Standard

Sandoz Ltd., Agrochemical Research Division, Basel, Switzerland.

D. Biological Properties

1. HERBICIDAL ACTIVITY

Dosanex® is a selective herbicide for postemergence use in winter-sown varieties of wheat and barley, as well as for pre- and post-emergence use in carrots. It is recommended for the control of annual grasses, blackgrass, loose silky bentgrass, wild oat, rye-grass, meadow grass, and for the control of the main broad-leaved weeds (Berg, 1968; Glenister and Griffiths, 1968; Griffiths and Ummel, 1971).

2. TOXICITY

The mammalian toxicity of this herbicide is low. The acute oral LD_{50} for rats is 3200 mg/kg, and the acute dermal LD_{50} is greater than 2000 mg/kg; the product shows neither dermal irritation nor allergic skin reaction. In a 90-day feeding study, rats and dogs received respective doses of up to 1250 and 2500 ppm administered daily with their food. No symptoms connected with this treatment were observed.

E. History

Metoxuron was developed by Sandoz Ltd., Basel (Swiss patents 472,180 and 484,609) and has been registered for use on carrots and cereals.

F. Physical Properties

Metoxuron is a colorless crystalline solid with a melting point of 126–127°C and a vapor pressure of 3.2×10^{-8} Torr at 20°C. Its water solubility is 678 ppm at 24°C, and it is soluble in acetone, cyclohexanone, and hot alcohol; moderately soluble in ether, benzene, and toluene; but insoluble in petroleum ether.

G. Chemical Properties

1. Method of Synthesis

Metoxuron is prepared commercially by the following reaction:

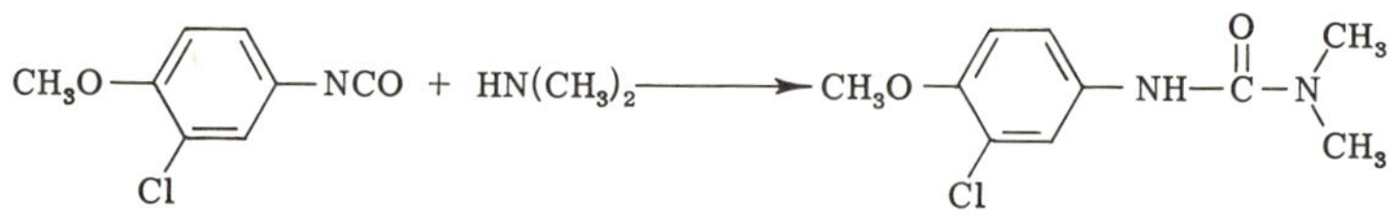

2. Chemical Stability

Metoxuron is thermally stable up to 70°C. It is stable in aqueous solutions at different pH values under normal conditions (for half-life periods, see Table I), but is hydrolyzed by strong acids or alkalis at elevated temperature. Metoxuron is degraded quickly by sunlight and

TABLE I
HALF-LIFE PERIODS OF METOXURON (5 PPM LEVEL) IN AQUEOUS BUFFER SOLUTION

pH	At 50°C (days)	At 20°C[a] (months)
3	18	11
5	21	13
7	24	15
9	>30	>20
11	26	16

[a] Calculated from values at 50°C.

UV light when in solution or when applied on glass and thin-layer chromatography (TLC) plates, as well as on soil surfaces.

H. Formulation

This herbicide is available as a wettable powder containing 80% active ingredient (w/w) and as flowable containing 45% active ingredient (w/w).

II. ANALYSIS

A. Formulations

1. REVIEW OF METHODS

Formulations of metoxuron can be assayed mainly by two methods; first, by quantitative hydrolysis according to the following reaction:

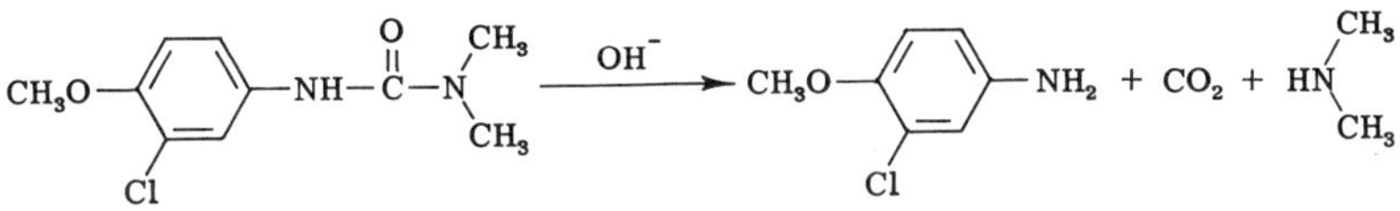

The dimethylamine, thus formed in quantitative yield, is distilled into a standard trap and determined by titration with acid (Lowen and Baker, 1952). The second method is based on quantitation of the active ingredient by UV spectrometry, after separation from by-products on TLC.

2. RECOMMENDED METHOD—HYDROLYSIS PROCEDURE

a. *Principle*

Metoxuron is separated from the formulation with methylene chloride; this solution is extracted with hydrochloric acid to remove interfering amines. After evaporation of the solvent, the residue is hydrolyzed with potassium hydroxide–propylene glycol. The liberated dimethylamine is distilled, absorbed into a boric acid solution, and titrated with standard hydrochloric acid.

The possible by-products *N'*-(3,4-dichlorophenyl)-*N,N*-dimethylurea, *N'*-(3-chloro-4-hydroxyphenyl)-*N,N*-dimethylurea, and *N'*-(4-methoxyphenyl)-*N,N*-dimethylurea are determined by a semi-quantitative TLC method and subtracted from the volumetric value.

b. *Reagents*

Methylene chloride.

Hydrochloric acid, 1 *N*. For preparation and standardization, see CIPAC-Handbook (Raw, 1970).

Propylene glycol, Reagent Grade.

Potassium hydroxide, Reagent Grade.

Silicone grease.

Boric acid, Reagent Grade.

Tashiro indicator. Dissolve 40 mg methylene blue and 60 mg methyl red in 100 ml ethanol (95%).

Reference sample of N'-(3,4-dichlorophenyl)-N,N-dimethylurea.

Reference sample of N'-(3-chloro-4-hydroxyphenyl)-N,N-dimethylurea.

Reference sample of N'-(4-methoxyphenyl)-N,N-dimethylurea.

Tetrahydrofuran, Analytical Grade.

TLC precoated glass plates, silica gel 60 F_{254}, layer thickness 0.25 mm, 20 × 20 cm, Merck, Darmstadt, Germany; catalog no. 5715, or equivalent.

Chloroform, Reagent Grade.

Ethylacetate, Reagent Grade.

c. *Apparatus*

Distillation apparatus with ground joints, illustrated in Fig. 1.

Sintered glass crucible 3G4 and adapter to fit conical suction.

Micropipettes, 5 μl, Drummond Microcaps (Drummond Scientific Co., Broomall, Pennsylvania) or equivalent.

UV lamp, 254 nm.

d. *Experimental Procedure*

Weigh accurately 3.5 gm of sample into a 200-ml beaker, add 100 ml methylene chloride, and agitate on a magnetic stirrer for 5 minutes. Filter through the sintered glass crucible, and rinse the beaker and the crucible with portions of methylene chloride so that the total of extract is about 200 ml. Use only a slight vacuum to prevent the crystallizing of metoxuron on the wall of the crucible.

Transfer the extract quantitatively into a 500-ml separatory funnel, add 50 ml 1 *N* hydrochloric acid, shake vigorously for 1 minute, and drain the methylene chloride layer into a second separatory funnel. Add 50 ml 1 *N* hydrochloric acid, shake for 30 seconds, and run off the organic layer into a 500-ml round-bottom flask. Wash the two aqueous extracts separately and successively twice with each 100 ml methylene chloride, and combine the extracts in the 500-ml round-bottom flask. Discard the aqueous layers. Evaporate the methylene chloride to dryness on a rotatory evaporator under reduced pressure at a maximum temperature of 40°C.

Add 100 ml propylene glycol, 40 gm potassium hydroxide, and some distillation stones to the residue, and immediately connect the

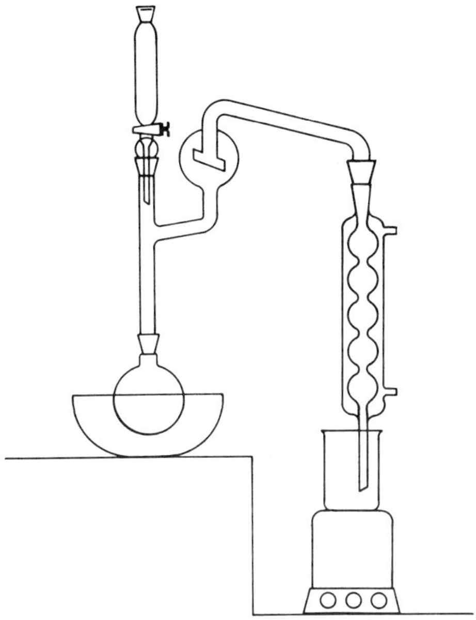

FIG. 1. Distillation apparatus for hydrolyzing metoxuron formulations.

flask tightly with the distillation apparatus. All joints should be lubricated with a thin film of silicone grease. The end of the delivery tube from the condenser is placed into a 400-ml beaker so that it is submerged in a solution consisting of 150 ml redistilled water, 0.2 gm boric acid and 1.0 ml Tashiro indicator solution.

Warm the contents of the flask gently until all particles of the metoxuron have dissolved, then boil for 10 minutes so that propylene glycol distills into the condenser. The dimethylamine is titrated continuously with 1 *N* hydrochloric acid. The distillation of dimethylamine is completed by carefully adding water drop by drop from the dropping funnel to the boiling content of the round-bottom flask (1 drop/second). The titration is continued until the color of the indicator has not further changed for 2 minutes. For every series of analyses, a blank determination, omitting the sample, is carried out.

e. *Calculation*

$$\% \text{ Metoxuron} = \frac{(a - b) \times N \times 22.87}{W} - (\% \text{ by-products})$$

where a = milliliters hydrochloric acid required for sample determination

b = milliliters hydrochloric acid required for blank determination

N = normality of hydrochloric acid
W = weight of sample in grams.

f. *Determination of By-Products*

Weigh accurately 1 gm of sample into a 20-ml volumetric flask, dissolve and fill up to volume with tetrahydrofuran. Prepare the by-product reference solution by dissolving 100 mg of each of the by-products in the same 50-ml volumetric flask with tetrahydrofuran. Pipette from this stock solution the following portions into 20-ml volumetric flasks and dilute with tetrahydrofuran:

1.0 ml	corresponding to 0.2%	of each by-product
2.0 ml	corresponding to 0.4%	of each by-product
3.0 ml	corresponding to 0.6%	of each by-product
4.0 ml	corresponding to 0.8%	of each by-product
5.0 ml	corresponding to 1.0%	of each by-product

Spot portions of 5 μl of each sample, as well as the reference solutions, side by side on the TLC plate. After evaporation of the solvent, place the plate in a TLC developing tank and develop by the ascending technique using chloroform–ethyl acetate (80:20) as a solvent. Remove the plate, and after evaporation of the developing solvent, evaluate for the contents of by-products by visual assessment under the UV light (Table II).

g. *Discussion*

The method was tested in a CIPAC collaborative study in 1974. The results of eighteen collaborators had shown that the method is well reproducible and rather accurate. The results were presented at the eighteenth CIPAC meeting in London of June 10, 1974.

TABLE II
R_f VALUES OF METOXURON AND BY-PRODUCTS ON TLC GLASS PLATES (MERCK)[a]

Compound	R_f
Metoxuron	0.25
N'-(3,4-Dichlorophenyl)-N,N-dimethylurea	0.34
N'-(3-Chloro-4-hydroxyphenyl)-N,N-dimethylurea	0.08
N'-(4-methoxyphenyl)-N,N-dimethylurea	0.13

[a] TLC developing tank presaturated for 30 minutes (without filter paper insert); 13 cm solvent migration above spotting line; chloroform–ethyl acetate (80:20) solvent.

3. SUPPLEMENTARY METHOD—SPECTROPHOTOMETRIC METHOD

a. *Principle*

The active ingredient, dissolved in methanol, is separated from by-products and formulation additives by TLC, extracted from the silica gel layer with ethanol and assayed by scanning the UV spectrum.

b. *Reagents*

Methanol, Reagent Grade.

TLC precoated plastic sheets, Polygram® Sil G/UV_{254}, layer thickness 0.25 mm, 20 × 20 cm, Macherey-Nagel & Co., Düren, Germany, or equivalent.

Chloroform, Reagent Grade.

Acetone, Reagent Grade.

Ethyl acetate, purified over alumina Woelm basic, activity grade I (M. Woelm, Eschwege, Germany) or equivalent.

Ethanol, 95%, Uvasol® (Merck, Darmstadt, Germany) or equivalent.

Metoxuron, pure analytical standard.

c. *Apparatus*

Micropipettes, 5 μl.

Centrifuge.

UV lamp, 254 nm.

Double-beam spectrophotometer, Perkin-Elmer model 124 or equivalent.

Cells of 1-cm path length.

Recorder, Perkin-Elmer model 165 or equivalent.

d. *Experimental Procedure*

i. *Preparation of Sample*

Weigh accurately an amount of sample equivalent to 1.0 gm of metoxuron and transfer into a 50-ml volumetric flask. Dissolve with methanol, shake vigorously for 1 minute and allow the suspended material to settle.

Spot twice a 5-μl portion of this solution on the TLC plate leaving space for the blank sample. Develop the plate by the ascending technique using mixtures of chloroform–ethyl acetate or chloroform–acetone–ethyl acetate as solvents (see Table III). Allow the solvent to evaporate for about 20 minutes, and observe the spots under the UV light as quickly as possible to prevent photodecomposition. Mark a 2.8-cm square around the spots and cut out using a pair of scissors. Transfer the square into a centrifuge tube, and add exactly 10 ml of ethanol by

TABLE III
SOLVENTS AND R_f VALUES OF METOXURON AND BY-PRODUCTS ON TLC PLASTIC SHEETS (POLYGRAM)[a]

Compound	R_f	
	Chloroform–ethyl acetate (80:20) developed twice	Chloroform–acetone–ethyl acetate (100:100:1)
Metoxuron	0.28	0.47
N′-(3,4-Dichlorophenyl)-*N*,*N*-dimethylurea	0.39	0.53
N′-(3-Chloro-4-hydroxyphenyl)-*N*,*N*-dimethylurea	0.11	0.41
N′-(4-Methoxyphenyl)-*N*,*N*-dimethylurea	0.21	0.43
N,*N*′-Bis(3-chloro-4-methoxyphenyl)urea	0.51	0.57

[a] TLC developing tank with filter paper insert, presaturated for 30 minutes; 16 cm solvent migration above spotting line.

means of a pipette. Immediately stopper the tube, shake vigorously for 1 minute and then centrifuge at 4000 rpm. Transfer cautiously the supernatant solution into the cell, and scan in the spectrophotometer from 350 nm to 220 nm against the blank sample.

ii. *Preparation of Blank Sample*

Prepare a blank sample for each TLC plate by cutting out a 2.8-cm blank square of the R_f value of the sample and treat, as described for the sample.

iii. *Calibration*

Weigh accurately 1.0 gm of the pure standard metoxuron, and treat as it is outlined for the sample. Scan in the spectrophotometer from 350 nm to 220 nm against the blank sample.

e. *Calculation*

Determine the absorbance at 245 nm corrected for baseline at 350 nm.

$$\% \text{ metoxuron} = \frac{a \times K \times 100}{A \times k}$$

where a = absorbance of the sample
K = weight of the pure standard in grams per 100 ml
A = absorbance of the pure standard
k = weight of the sample in grams per 100 ml

f. *Discussion*

The by-products mentioned in Section 2,f, as well as the *N*,*N*′-bis(3-chloro-4-methoxyphenyl)urea do not interfere in this procedure. The reproducibility is satisfactory with a relative standard deviation of less than 1.5%.

B. Residues

1. REVIEW OF METHODS

Residues of metoxuron in crop and soil are determined by acid hydrolysis to *o*-chloro-*p*-anisidine and quantitated colorimetrically using a modified procedure of the methods described by Lowen and Baker (1952) and Katz (1966, 1967). The values obtained include the residues of the monomethyl-metabolite of metoxuron, *N*-(3-chloro-4-methoxyphenyl)-*N*′-methylurea. The second method permits the separate identification and semiquantitative estimation of metoxuron and its monomethyl metabolite by means of TLC. This procedure may serve as a helpful confirmatory method.

2. RECOMMENDED METHOD—COLORIMETRIC PROCEDURE

a. *Principle*

The residues of metoxuron and its metabolite are extracted with acetonitrile, hydrolyzed to *o*-chloro-*p*-anisidine by 1 *N* phosphoric acid, and cleaned up by partition between water and methylene chloride. The anisidine is diazotized, coupled with *N*-1-naphthylethylenediamine to produce a colored compound, which is separated from the remaining plant material by column chromatography. It is then determined spectrophotometrically at 572 nm.

b. *Reagents*

Sodium sulfate, pure, anhydrous.

Acetonitrile, technical, redistilled from P_2O_5–H_3PO_4 (Food and Drug Administration, 1971).

Petroleum ether, redistilled and purified over alumina Woelm basic, activity grade I (M. Woelm, Eschwege, Germany) or equivalent.

Sodium chloride, Reagent Grade.

Methylene chloride, as described for petroleum ether.

Phosphoric acid, Reagent Grade, 1 *N*.

Nitrogen, pure.

Hydrochloric acid, Reagent Grade, 37%, 1 *N* and 2 *N*.

Sodium nitrite, Reagent Grade; 2% aqueous solution fresh prepared daily.

Sulfamic acid, Reagent Grade; 10% aqueous solution fresh prepared daily.

N-(1-Naphthyl)ethylenediamine dihydrochloride, Reagent Grade, 2% in 2 *N* hydrochloric acid.

Cellulose powder, microcrystalline, Merck or equivalent.

Acetic acid, glacial, Reagent Grade.

o-Chloro-p-anisidine, Reagent Grade.

Chromatographic column. Prepare a suspension of 2 gm of the cellulose powder in 10–20 ml of 1 *N* HCl for each column. Pour this solution in a column having at the lower end a plug of wadding. Open the stopcock and allow the solution to drain just to the top of the cellulose powder.

c. *Apparatus*

Grinder.

Blender, Waring or equivalent, with 1-liter glass jars.

Shaking machine.

Food chopper, Anliker, Bassersdorf, Switzerland, or equivalent.

Two-neck round-bottom flasks, 50 and 100 ml sizes.

Heating mantels, 50 and 100 ml sizes.

Double-beam spectrophotometer, Perkin-Elmer model 124 or equivalent.

Cells of 1-cm path length.

Chromatographic column, with stopcock, 20 cm × 1 cm i.d.

d. *Experimental Procedure*

i. *Extraction*

(a) *Extraction of Grain.* Grind the entire sample and mix it thoroughly. Place 20–200 gm of material in a blender jar, add 50 gm sodium sulfate and 200 ml acetonitrile, and blend for 3 minutes. Transfer to a 1-liter screw-cap bottle and shake it mechanically for 30 minutes. Filter with suction through a filter paper and collect the filtrate. Return the filter cake into the blender jar with the aid of 150 ml acetonitrile, and mix for 2 minutes. Shake the slurry again for 30 minutes and filter into the same filter flask. Evaporate at a maximum temperature of 40°C until complete removal of acetonitrile is achieved.

(b) *Extraction of Vegetables.* Grind the entire sample in a food chopper with an equal weight of dry ice. Place the sample in frozen storage overnight to allow the dry ice to sublime. Place 20–200 gm of material in a blender jar, add 50–200 gm sodium sulfate and 150–200 ml acetonitrile. Blend for about 3 minutes and proceed as under (a).

(c) *Extraction of Soil.* Mix the entire sample thoroughly and eliminate stones larger than 1-cm diameter. Take the subsamples from these undried samples for analysis.

A 100–200-gm portion of the mixed sample is airdried overnight to determine the water content of the soil sample. The parts per million of the analytical sample has to be corrected by multiplying with the factor (see Section vii).

$$f = \frac{\text{soil undried (gm)}}{\text{soil airdried (gm)}}$$

Shake 100–200 gm of the undried sample mechanically in a 1-liter screw-cap bottle with 200 ml acetonitrile for 30 minutes. Filter with suction through a filter paper and collect the filtrate. Return the filter cake into the bottle with the aid of 200 ml acetonitrile. Shake the slurry again for 30 minutes and filter into the same filter flask. Evaporate at a maximum temperature of 40°C until complete removal of acetonitrile.

ii. *Cleanup Procedure*

Transfer the crude extract of plant material or soil to a 250-ml separatory funnel. Rinse the filter flask successively with 20 ml redistilled water and 40 ml petroleum ether, and add the rinsing each time to the separatory funnel. Also add 5 ml of a saturated aqueous sodium chloride solution.

Shake the funnel for 30 seconds and drain the lower phase through a filter paper into a second separatory funnel. Extract the petroleum ether phase twice more successively with 25 ml water and 5 ml saturated sodium chloride solution.

Re-extract the combined aqueous phases with five portions of each 80 ml methylene chloride. Run off the organic layers through a funnel containing a plug of wadding and sodium sulfate into a 500-ml round-bottom flask. Evaporate the combined extracts to dryness in a rotary evaporator under reduced pressure at 40°C.

iii. *Hydrolysis Procedure*

Transfer the remainder or a portion with methylene chloride into a two-neck round-bottom flask. Evaporate the solvent completely by means of a rotatory vacuum evaporator at 40°C. Add 30 ml 1 *N* phosphoric acid and a few boiling chips. Connect the flask to a condenser, pass a slow stream of nitrogen through the solution and reflux gently for 6 hours.

Allow to cool and transfer the solution to a 250-ml separatory funnel, add 2 ml concentrated hydrochloric acid, and shake. Extract the hydrolysis solution twice with each 20 ml methylene chloride. Discard the organic phases and filter the hydrolysis solution through a wet filter paper.

iv. *Diazotization and Coupling*

Transfer the hydrolysis solution or a portion into a 100-ml flask and cool to 0–5°C in an ice bath. Add 1 ml sodium nitrite solution and keep at about 5°C for 20 minutes. Add 2 ml sulfamic acid to destroy the excess nitrite, shake the flask vigorously, and set aside until effervescence subsides (about 30 minutes). Pipette 4 ml *N*-(1-naphthyl)ethylenediamine dihydrochloride solution into the flask and mix thoroughly. Allow to stand at room temperature for 90 minutes.

v. *Chromatographic Separation*

Transfer the dye solution quantitatively onto the cellulose column by careful pipetting against the inside wall. Rinse the flask and the wall of the column with a small portion of 1 *N* hydrochloric acid. Develop the column with 20 ml of a mixture of 1 *N* hydrochloric acid–glacial acetic acid (9 : 1) (v/v). Discard these solutions. Elute the azo dyestuff of the anisidine derivative with a mixture of each 10 ml 1 *N* hydrochloric acid and glacial acetic acid (v/v) and collect the eluate in a 20-ml volumetric flask.

Transfer a portion of the dye solution to a 1-cm cell and scan in the spectrophotometer from 700 nm to 370 nm against 1 *N* hydrochloric acid–glacial acetic acid (1 : 1). Read the absorbance at 572 nm corrected for the baseline at 700 nm. The dyestuff is stable for at least 5 hours.

vi. *Calibration and Standard Curve*

Prepare standard solutions containing known amount of *o*-chloro-*p*-anisidine ranging from 2–50 μg and carry through the procedure of diazotization, coupling, chromatographic separation, and spectrophotometric quantitation. Plot absorbance versus *o*-chloro-*p*-anisidine concentration expressed as micrograms per 20 ml.

vii. *Calculation*

$$\text{ppm Metoxuron} = \frac{A \times 1.45}{W} \times D \times f$$

where A = amount of *o*-chloro-*p*-anisidine in micrograms per 20 ml
W = weight of sample in gm
D = dilution factor
f = correction factor (see Section i)

e. *Discussion*

The procedure described above has been applied successfully to grain and straw of wheat and barley, carrots, alfalfa, potatoes, sugar beets, sunflower seeds, and various soil types. Recoveries at 0.1–0.5 ppm levels generally have been in the range between 70% and 90%. Due to absorbance values of various blank samples, the limit of detection has ranged from 0.02 to 0.05 ppm (except for straw, 0.03–0.12 ppm). The residue values obtained include the residues present as metoxuron and its metabolite *N*-(3-chloro-4-methoxyphenyl)-*N'*-methylurea.

3. CONFIRMATORY METHOD—TLC METHOD

a. *Principles*

The residues of metoxuron and its metabolite are extracted with acetonitrile. Interfering substances are removed from the extracts by exploiting the difference in the distribution coefficients in two-phase systems (petroleum ether–water and water–chloroform) and then by adsorption on a column. The subsequent separation and estimation is carried out by visual assessment on TLC plates.

b. *Reagents*

Chloroform, redistilled and purified over alumina Woelm basic, activity grade I (M. Woelm, Eschwege, Germany) or equivalent.

Toluene, Reagent Grade.

Alumina, Woelm, neutral.

Adsorption mixture. 5 gm activated charcoal (Norit A), 5 gm magnesium oxide, 5 gm Celite 545, and 85 gm alumina neutral are homogeneously mixed.

Column I. A glass tube (2 cm × 30 cm) with a ground-glass cock is charged with 3 gm alumina suspended in a solution of chloroform and acetonitrile (9 : 1).

Column II. A similar tube is charged with 1 gm alumina suspended in acetonitrile; 0.5 gm of adsorption mixture is charged above this, and then 2 gm alumina above that.

TLC precoated plastic sheets, Polygram® Sil G/UV$_{254}$, layer thickness 0.25 mm, 20 × 20 cm (Macherey-Nagel & Co., Düren, Germany) or equivalent.

Reference sample of metoxuron.

Reference sample of N-(3-chloro-4-methoxyphenyl)-N'-methylurea.

Acetone, Reagent Grade.

Ammonia solution, pure, 25%.

Methanol, Reagent Grade.

Silver nitrite, Reagent Grade.

Silver nitrite 1% stock solution. 1 gm silver nitrite dissolved in 50 ml acetonitrile, 25 ml each of acetone and water.

Ammoniacal silver nitrite solution. 5 parts 1% stock solution, 25 parts ammonia solution, 20 parts acetonitrile, and 50 parts methanol.

c. *Apparatus*

UV lamp, 254 nm.

d. *Experimental Procedure*

i. *Preparation of Sample*

Extraction and cleanup procedure by partition between two solvents are carried out as described for recommended method.

ii. *Cleanup Procedure*

Transfer the residue quantitatively by means of two portions each of 20 ml of a chloroform–acetonitrile solution (9:1) to column I, or, if the extract is still heavily contaminated, to column II. Elute the active ingredient slowly with three lots of 80 ml of the solution of chloroform and acetonitrile (9:1). Evaporate the combined eluates to dryness under reduced pressure at 40°C. Dissolve the residue in toluene, transfer to a 20-ml volumetric flask and dilute to the volume with toluene.

iii *Determination*

Spot a portion of the extract on a TLC plate side by side with various quantities of metoxuron and its metabolite. Develop with mixed solvents chloroform–acetone (8:2). Expose the plate to UV light for 15 minutes and then spray with ammoniacal silver nitrite solution. Place the damp plate under the UV lamp for a further 15 minutes. If necessary, repeat the spraying and irradiation procedure until the optimal degree of staining is attained.

e. *Discussion*

If this method is used for semiquantitative estimation of metoxuron and *N*-(3-chloro-4-methoxyphenyl)-*N'*-methylurea, it is desirable that the amounts range between 0.1 μg and 0.6 μg. Due to the very sensitive staining that render visible amounts even as small as 0.05 μg, the limit of detection may range between 0.05 ppm and 0.15 ppm.

The accuracy of the assay by comparison of the size and intensity of the spots obtained from the samples with those from standard substances is about ±25%.

REFERENCES

Berg, W. (1968). *Z. Pflanzenkr. (Pflanzenpathol.) Pflanzenschutz* Sonderheft IV, 233.

Food and Drug Administration. (1971). "Pesticide Analytical Manual," Vol. 1, p. 120. FDA, Washington, D.C.

Glenister, M., and Griffiths, G. P. (1968). *Proc. Brit. Weed Contr. Conf., 9th, 1968,* Vol. 1, p. 46.

Griffiths, G. P., and Ummel, E. (1971). *Proc. Brit. Weed Contr. Conf., 10th, 1970,* Vol. 1, pp. 77 and 186; Vol. 2, p. 849.

Katz, S. E. (1966). *J. Ass. Offic. Anal. Chem.* **49,** 452.

Katz, S. E. (1967). *J. Ass. Offic. Anal. Chem.* **50,** 911.

Lowen, W. K., and Baker, H. M. (1952). *Anal. Chem.* **24,** 1475.

Raw, G. R. (1970). "CIPAC Handbook," Vol. I, p. 752.

28

Oryzalin

O. D. Decker and W. S. Johnson

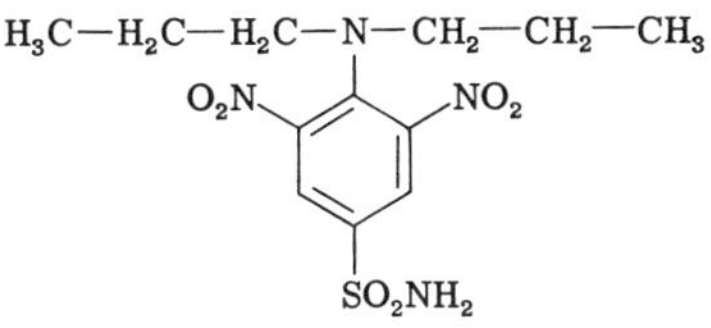

3,5-Dinitro-N^4, N^4-dipropylsulfanilamide

I. GENERAL

A. Empirical Formula

$C_{12}H_{18}N_4O_6S$ (Mol. wt. 346.36).

B. Alternative Names

Dirimal and Surflan®. Surflan® is a registered trademark of Elanco Products Company, Division of Eli Lilly and Company.

C. Source of Analytical Standard

Elanco Products Company, 640 South Alabama Street, Indianapolis, Indiana.

D. Biological Properties

Oryzalin is a selective, preemergence herbicide for the control of annual grasses and certain broadleaf weeds in soybeans and other selected crops.

The acute oral LD_{50} of technical oryzalin for adult male and female rats and gerbils is greater than 10 gm/kg. Exposure of rats for 1 hour to a mist of formulated oryzalin, 3.56 mg/liter, showed no adverse effects. Technical oryzalin at 2 gm/kg on rabbit skin caused only slight irritation. When oryzalin was placed in the rabbit eye, the irritation was no greater than would be expected from any inert particulate material. Dietary administration of technical oryzalin to rats and oral administration to dogs for 3 months at a level of 750 ppm caused no significant changes. Baby chicks fed for 3 weeks at 500 ppm were normal.

E. History

The herbicidal properties of oryzalin were described by Gramlich, *et al.* (1969).

The use of oryzalin as a herbicide is covered by U.S. patent 3,367,949 issued to Eli Lilly and Company, February 6, 1969.

F. Physical Properties

Pure oryzalin is a yellow–orange crystalline solid that melts at 141–142°C and is soluble in water to the extent of 2.5 ppm at 25°C. It is readily soluble in polar organic solvents such as acetone, methanol, ethanol, and acetonitrile. It is only slightly soluble in benzene and xylene and is insoluble in hexane. The vapor pressure is less than 1×10^{-7} mm Hg at 30°C.

The infrared spectrum of an oryzalin mull exhibits strong absorption maxima at 3370, 3280, 1608, 1536, 1348, 1187, 1098, 980, 905, and 705 cm^{-1}. The ultraviolet absorption spectrum of a dilute methanol solution exhibits maxima at 283 and 383 nm, the latter giving a molar absorptivity of 2780.

G. Chemical Properties

1. METHOD OF SYNTHESIS

Oryzalin can be synthesized from chlorobenzene by a series of reactions described in *Organic Reactions,* Coll. Vol. IV, p. 364, *Chemical Abstracts, 79,* P78390C (1973), and U.S. Patent 3,367,949.

2. CHEMICAL REACTIONS

Oryzalin exhibits acidic properties, being highly soluble in alkaline solutions and titratable with a strong base in a nonaqueous media.

3. CHEMICAL STABILITY

Oryzalin is subject to decomposition by ultraviolet irradiation.

H. Formulations

Oryzalin is formulated as a 75% wettable powder.

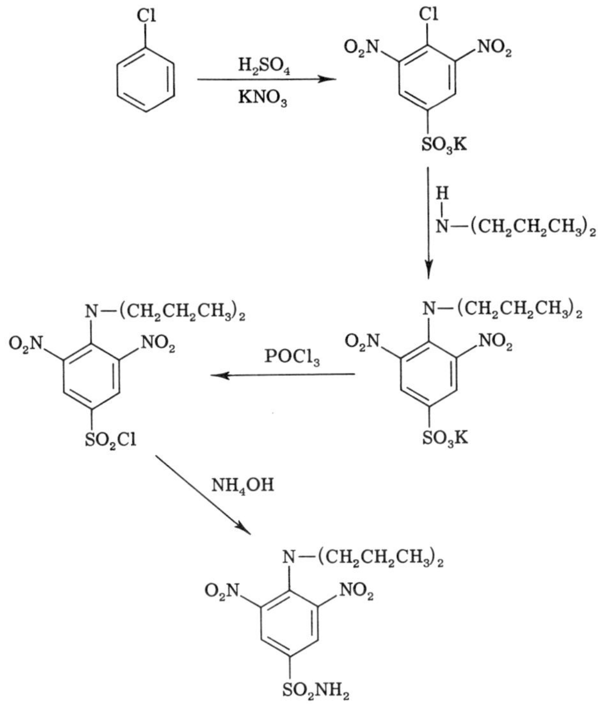

II. ANALYSIS

A. Formulation Analysis

1. REVIEW OF METHODS

Oryzalin can be titrated in acetonitrile with tetrabutyl ammonium hydroxide in benzene–methanol (9 : 1). However, some impurities in technical material are also titratable, so this method is not generally recommended. The spectrophotometric measurement of oryzalin following the chromatographic removal of colored impurities is the preferred procedure.

2. RECOMMENDED METHOD

a. *Principle*

Oryzalin is removed from solid formulations with methanolic HCl, solubilized in benzene, and purified on a Florisil column. Measurement is achieved spectrophotometrically at 383 nm.

b. *Reagents*

Sodium sulfate, anhydrous.

Benzene, analytical reagent.

Ethyl acetate, analytical reagent.

Methanolic hydrochloric acid, 1%. Mix 30 ml concentrated hydrochloric acid with 970 ml methanol.

Oryzalin standard, 1.0 mg/ml in benzene. Dissolve the sample in about 10 drops of ethyl acetate, then dilute to volume with benzene.

Florisil, deactivated 100–200 mesh. Determine the apparent moisture of the Florisil (Floridin Co., Berkeley Springs, West Virginia) by heating at 600°C for 4 hours. Add deionized water to the Florisil "as received" to yield a moisture content of 8–8.5%. Mix for an hour, then allow it to stand in the sealed container overnight before using.

c. *Apparatus*

Glass columns, 14 × 240 mm, fitted with glass or Teflon stopcocks and 250 ml reservoir.

Rotary vacuum evaporator, Rinco or equivalent.

Spectrophotometer capable of making measurements at 383 nm.

d. *Experimental Procedure*

i. *Florisil Column Preparation*

Insert a glass wool pledget into the bottom of a chromatographic column and, with gentle tapping, add, in order, 3.5 gm anhydrous sodium sulfate, 5.0 gm deactivated Florisil, and 3.5 gm anhydrous sodium sulfate. With the stopcock open, add 20 ml benzene and allow it to drain to the level of the sodium sulfate.

ii. *Sample Preparation*

(a) *Technical Material.* Accurately weigh a sample equivalent to 100 mg of oryzalin into a 100-ml volumetric flask. Dissolve the sample in about 10 drops of ethyl acetate, then dilute to volume with benzene.

(b) *Dry Formulations.* Weigh a sample equivalent to 100 mg of oryzalin into a 250-ml glass-stoppered Erlenmeyer flask. Add 100 ml of the methanolic HCl solution and stir for 30 minutes using a magnetic stirring bar and stirrer. Allow the solution to clear, then transfer 10.0 ml into a 125-ml round-bottom flask. Take to dryness using a flash evaporator and 50°C water bath. Pipette 10.0 ml benzene into the flask and swirl to dissolve the residue. Add 2 gm anhydrous sodium sulfate to absorb any residual water, and clarify the solution. Transfer a 4.0-ml portion of the benzene solution to the prepared Florisil column and wash the sample onto the column with benzene. Wash the column with 20 ml benzene. Elute the column with benzene–ethyl acetate (95:5) until the main yellow band is 1 inch from the bottom of the Florisil, discarding the eluate to this point. Continue eluting with the same solvent system and collect the yellow band in a 100-ml volumetric flask. Dilute the flask to volume with the benzene–ethyl acetate mixture. Repeat the column procedure with a 4.0-ml portion of the standard oryzalin solution. Measure the absorbance of the sample and standard at 383 nm using the benzene–ethyl acetate solution as the reference standard. The standard at this concentration will contain 0.040 mg/ml.

iii. *Calculations*

$$\% \text{ oryzalin} = \frac{A_{383}\text{ (sample)}}{A_{383}\text{ (standard)}} \times 0.040 \text{ mg/ml} \times \frac{100 \text{ ml}}{\text{sample weight (mg)}} \times \frac{100 \text{ ml}}{4 \text{ ml}} \times \text{purity of standard}$$

iv. *Interferences*

The Florisil cleanup adequately removes substances in technical grade material and the commercial formulation that would interfere with the analysis.

v. *Sensitivity*

Precision of this analytical method is approximately ±1.5% at 95% confidence level.

vi. *Discussion of Method*

Properly deactivated Florisil should allow for complete elution of the oryzalin from the column with 70–90 ml of the benzene–ethyl acetate mixture after collection begins. If not properly deactivated, much larger volumes are required to elute the oryzalin from the column. It is recommended that each newly deactivated batch of Florisil be checked by eluting a 4.0-ml portion of the standard oryzalin from a column.

If comparison with a standard curve is made, a standard need not be run with each set of samples. If a direct standard is used, the column chromatographic step is not necessary unless the purity is less than 99%. Such reference standards should be chromatographed along with the samples to be analyzed. Oryzalin has acidic properties and will bind itself with basic clays. Therefore, it is necessary to use the acidified methanol as the extraction solvent for the formulated material.

B. Residue Analysis

1. RECOMMENDED METHOD

a. *Principle*

Oryzalin is quantitatively extracted from plant tissue and soil with methanol. The extract is purified by liquid–liquid partitioning and column chromatography over alumina. Oryzalin is converted to the N^1,N^1-dimethyl derivative (N^1,N^1-dimethyl-3,5-dinitro-N^4,N^4-dipropylsulfanilamide) for detection and measurement by gas–liquid chromatography using an electron capture detector.

b. *Reagents*

Stock Solution. Accurately weigh 10 mg of analytical standard, transfer quantitatively to a 200-ml volumetric flask with benzene, and dilute to volume (50 μg/ml). Store in a refrigerator and prepare fresh every 3 months. Dilutions should be protected from light and prepared fresh weekly.

Methanol, analytical reagent.
Sodium chloride, analytical reagent, 5% w/v in deionized water.
n-Hexane, analytical reagent, redistilled before use.
Carbon tetrachloride, analytical reagent.
Sodium hydroxide, analytical reagent, 1.25 *N*, in deionized water.
Dichloromethane, analytical reagent, redistilled before use.
Sodium sulfate, anhydrous, analytical reagent.
Methyl iodide, analytical reagent, redistilled before use.
Alumina, Alcoa F-20, use as received and standardize as follows:

Place a glass wool plug in the bottom of a 14 mm × 250 mm chromatographic column equipped with a Teflon stopcock and 250-ml reservoir. Add 30 ml benzene to the column. Add 20 gm alumina, and rinse down with benzene. Remove entrapped air by stirring, and allow the alumina to settle. Gently add 5 ml anhydrous sodium sulfate to the top of the alumina and drain the benzene to the surface of the sodium sulfate. Add 250 μg of the dimethyl derivative of oryzalin to a flask containing 5 ml of benzene. Transfer to the column and start to drain. Rinse the flask with two successive 5-ml portions of benzene, allowing each to go into the column before the next addition. Add an additional 50 ml benzene to the column, allow to drain at a flow rate of 3–5 ml/minute and discard. Add 85 ml of 95:5 benzene–ethyl acetate to the column and observe the movement of the yellow band as the column drains. Note the volume of solvent required to move the leading edge of the band to within 1 inch of the bottom of the column and discard this amount of eluate as a forerun. Continue the elution of the yellow band and record the volume required to completely remove the yellow color from the column. This volume, plus an additional 20 ml, will be collected in the column chromatographic procedure to insure complete removal of the compound. Normally, 60–65 ml of 95:5 benzene–ethyl acetate should be collected.

Benzene, analytical reagent, redistilled before use.
Ethyl acetate, analytical reagent.

All reagents should be checked for impurities by electron capture gas chromatography before use in the assay.

c. *Apparatus*

Blender, Omni-mixer (Ivan Sorvall Co.) or equivalent.

Gas chromatograph, Hewlett-Packard model 5713A, with a model 18731A ^{63}Ni linear EC detector, and a model 7123A 1-mV recorder. The chromatographic column is 120 cm × 3 mm i.d. glass tubing packed with 5% XE-60 on 80–100 mesh Chromosorb W HP (Applied Science Laboratories, State College, Pennsylvania). The column is conditioned at 250°C for 16 hours prior to use.

Chromatographic columns, 14 mm i.d. × 250 mm, equipped with a 250-ml reservoir top, and Teflon stopcock.

Rotary vacuum evaporator, Rinco or equivalent.

d. *Experimental Procedure*

i. *Sample Preparation*

(a) *Crops.* Weigh a 25-gm sample of ground and mixed crop tissue into a quart Mason jar. Add sufficient methanol to make 200 ml total liquid with allowance for the water content of the sample. Blend the sample on an Omni-mixer at moderate speed for 5 minutes. Allow the insolubles to settle, and transfer 20 ml of the supernatant liquid to a 250-ml separatory funnel containing 80 ml 5% sodium chloride solution. Add 20 ml additional methanol to the separatory funnel. Wash the aqueous methanol mixture twice with 30-ml portions of hexane and discard the hexane washes. Add 10 ml 1.25 *N* sodium hydroxide to the aqueous methanol to raise the pH to >12. Wash the aqueous methanol with 30 ml of carbon tetrachloride and discard the wash. Add 20 ml or more of saturated boric acid to the aqueous methanol to give a pH of 8–9. Extract the aqueous methanol twice with 20 ml dichloromethane passing each extract through anhydrous sodium sulfate into a 125-ml boiling flask. Rinse the sodium sulfate with 10 ml dichloromethane. Remove the dichloromethane by rotary vacuum evaporation using a 50°C water bath. Add 20 ml methanol to the flask and swirl to dissolve. Proceed with the derivatization.

(b) *Soils*. Weigh a 25-gm sample of well-mixed soil into a quart Mason jar. Add 200 ml methanol and blend the sample on an Omni-mixer at moderate speed for 5 minutes. Allow the insolubles to settle, and transfer 20 ml of the supernatant liquid to a 125-ml boiling flask. Proceed with the derivatization.

ii. *Derivatization*

Add 0.5 gm anhydrous sodium carbonate and 3 ml methyl iodide to the methanol solution of crop or soil extract. Stopper the flask lightly and incubate overnight at 50°C. Following incubation, add 40 ml distilled water to the flask, and transfer the mixture to a 125-ml separatory funnel. Rinse the flask with 20 ml dichloromethane and add to the separatory funnel. Shake the funnel and pass the lower dichloromethane layer through sodium sulfate into a 125-ml boiling flask. Repeat the extraction with an additional 20-ml portion of dichloromethane and combine the extracts. Rinse the sodium sulfate with 10 ml dichloromethane. Remove the solvent by rotary vacuum evaporation using a 50°C water bath.

iii. *Column Chromatography*

Prepare an alumina column using Alcoa F-20 alumina that has been previously standardized. Dissolve the residue in the flask in 5 ml benzene, transfer to the column and start to drain. Rinse the flask with two successive 5 ml portions of benzene, allowing each to go into the column before the next addition. Add an additional 50 ml benzene to the column, allow to drain at a flow rate of 3–5 ml/minute and discard. Add 85 ml 95:5 benzene–ethyl acetate to the column and collect the volume of eluate as determined by the alumina standardization. Remove the solvent by rotary vacuum evaporation using a 50°C water bath.

iv. *Gas Chromatography*

Column temperature: 230°C.
Detector temperature: 300°C.
Injection temperature: 280°C.

Electrometer attenuation: Setting to provide 30–40% full scale recorder deflection on injection of 0.3–0.4 ng of the dimethyl derivative of oryzalin. For the gas chromatograph described, the cell current is constant at 10^{-9} A, and the signal measured is the pulse frequency required to maintain the standing current.

Carrier gas: Argon (90): Methane (10).
Carrier gas flow rate: 75 ml/minute.

A series of concentrations of standard solutions of dimethyl oryzalin (e.g., 0.025–0.25 μg/ml) are prepared to determine the response curve for the electron capture detector. To facilitate standard curve interpolations, the GC standards may be prepared in terms of oryzalin equivalents (1 μg oryzalin = 1.08 μg dimethyl derivative). Otherwise, the mass of dimethyl derivative corresponding to a particular point on the curve must be multiplied by 0.925 to correct for differences in molecular weight of the oryzalin and its dimethyl derivative. Inject 3–3.5 μl of each standard solution into the gas chromatograph. Plot peak response versus nanograms injected to obtain the standard response curve. Check the response with each set of samples assayed, and frequently inject a standard dilution (e.g., 0.1 μg/ml) during the course of the assay to determine whether detector response has changed.

Dissolve crop extracts in 1.0 ml benzene and soil extracts in 2.0 ml benzene. Inject the same volume of sample extract as used for the response curve. One microliter of crop extract is equivalent to 2.5 mg of crop tissue, and 1 μl of soil extract is equivalent to 1.25 mg of soil.

v. *Recovery*

Standard recovery samples are assayed with each set of experimental samples. Recovery samples are prepared by adding a methanol solution of oryzalin to 25 gm of control crop tissue or soil to provide a fortification level of 0.05 or 0.10 ppm, respectively. The efficiency of the derivatization of oryzalin is monitored by reacting an amount of oryzalin standard equivalent to the amount present in the recovery sample. The methylated oryzalin standard should also be passed over an alumina column prior to gas chromatography.

Standard recovery samples have given values of 70–75% for crops such as soybeans, peaches, plums, prunes, citrus fruits, and nut meats. Recovery from soil is about 92%. The within-day coefficient of variation for the methylation of oryzalin is about 6%.

vi. *Sensitivity*

The test sensitivity is about 0.01 ppm as determined by the recovery of oryzalin from soybean seed.

vii. *Sample Calculation*

The number of nanograms of oryzalin present in the number of milligrams of plant tissue or soil injected into the gas chromatograph may be obtained from the standard curve. The value obtained may be corrected to reflect assay efficiency as determined from the recovery samples.

e. *Discussion*

Interferences seldom appeared in the gas chromatograms. A variety of other pesticides have been checked for interference with the detection and measurement of oryzalin and none was found.

Reference

Gramlich, J. V., Golab, T., Van der Schans, C. D., and Steenwyk, P. L. (1969). *Abstracts, Weed Science Society of America* (Abstract 22).

29

Pronamide

I. L. ADLER, L. D. HAINES, AND J. P. WARGO, JR.

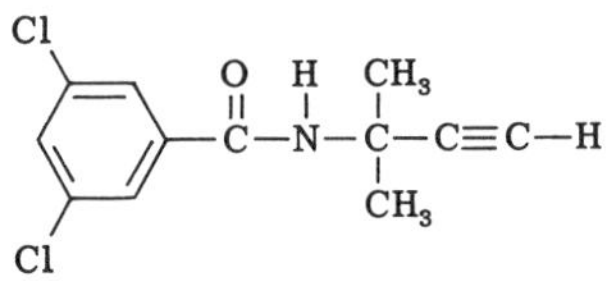

3,5-Dichloro-*N*-(1,1-dimethyl-2-propynyl)benzamide

I. GENERAL

A. Empirical Formula

$C_{12}H_{11}NOCl_2$ (Mol. wt. 256.13).

B. Alternative Names

3,5-Dichloro-*N*-(1,1-dimethyl-2-propynyl)benzamide, sometimes named *N*-(1,1-dimethylpropynyl)-3,5-dichlorobenzamide, was offered experimentally under the code number RH-315. It was introduced commercially as Kerb® herbicide. The Weed Society of America has adopted the common name pronamide.

C. Source of Analytical Standard

Agricultural and Sanitary Chemicals Deptartment, Rohm and Haas Company, Independence Mall West, Philadelphia, Pennsylvania.

D. Biological Properties

1. HERBICIDAL APPLICATIONS

Pronamide is a selective pre- and postemergence herbicide for use to control certain annual and perennial grasses, volunteer grains, and annual broadleaf weeds. The actual and potential fields of selective application are in lettuce, alfalfa, clover, birdsfoot trefoil, crown vetch, cane fruit, blueberries, tree fruits, woody ornamentals, Christmas trees, bermudagrass turf, sugar beets, peanuts, and beans.

2. TOXICITY

The acute oral LD_{50} for 3,5-dichloro-*N*-(1,1-dimethyl-2-propynyl)benzamide to male and female albino rats is 8350 mg/kg and 5620 mg/kg, respectively. The acute oral LD_{50} to mongrel dogs is 10,000 mg/kg.

E. History

3,5-Duchloro-*N*-(1,1-dimethyl-2-propynyl)benzamide is covered as a novel compound by U.S. patent 3,534,098 and its herbicide composition and use by U.S. patent 3,640,699 issued to Rohm and Haas Company.

F. Physical Properties

3,5-Dichloro-*N*-(1,1-dimethyl-2-propynyl)benzamide is an odorless, white, crystalline solid at ambient temperatures, having a melting point of 155–156°C. The solid is readily soluble in most organic solvents, but is only soluble to the extent of less than 25 ppm in water. The vapor pressure of 3,5-dichloro-*N*-(1,1-dimethyl-2-propynyl)benzamide is 8.5×10^{-5} Torr at 25°C.

G. Chemical Properties

1. METHOD OF SYNTHESIS

3,5-Dichloro-*N*-(1,1-dimethyl-2-propynyl)benzamide is prepared by the reaction of 3,5-dichlorobenzoyl chloride with 2-amino-2-methyl-3-butyne.

2. CHEMICAL REACTIONS

3,5-Dichloro-*N*-(1,1-dimethyl-2-propynyl)benzamide may be chemically hydrolyzed in the presence of sulfuric acid. This reaction is used in the analytical method for the determination of pronamide residues. Other reactions of the compound have been reported by Yih and Swithenbank (1971a,b). The metabolism of pronamide in soil has been reported by Fisher (1974).

H. Formulations

Pronamide is commercially available (as Kerb® herbicide) as a wettable powder containing 50% active ingredient.

II. ANALYSIS

A. Formulation Analysis

1. REVIEW OF METHOD

The analytical method for measuring the active ingredient in Kerb® formulations involves Soxhlet extraction of the wettable powder to remove 3,5-dichloro-*N*-(1,1-dimethyl-2-propynyl)benzamide followed by gas chromatography of the extract.

2. RECOMMENDED METHOD

a. *Principle*

A sample of Kerb® wettable powder is Soxhlet extracted with benzene to remove the active ingredient. A portion of the extract is then analyzed by gas chromatography using a flame ionization detector.

b. *Reagents*

Dioxane, Reagent Grade.
Stilbene, Eastman White Label.
Gas Chrom Q, 80–100 mesh.
QF-1, GLC liquid phase.
Benzene, Reagent Grade.

c. *Apparatus*

Gas chromatograph having a flame ionization detector.
Recorder, 1 mV.
Glass columns. A 5-foot column was used having a 3 mm o.d. and a 1.6–1.8 mm i.d.
Microliter syringe, 5 or 10 μl.
Extraction thimbles, 10 mm × 50 mm fat extracted seamless Whatman thimble.
Micro-Soxhlet extraction apparatus.

d. *Experimental Procedure*

i. *Instrumental*

Pack a 5-foot glass column with a packing of 5% QF-1 on 80–100 mesh Gas Chrom Q. Install the column for on-column injections. Condition column for at least 4 hours at 225°C. Set the gas chromatograph for the following operating conditions:

Helium gas flow: 30 ml/minute

Temperature
Column: 160°C.
Injector: 190°C.
Detector: 250°C.

ii. *Internal Standard Dilution*

Prepare a 0.4% (w/v) solution of stilbene in dioxane. A 50-ml volume is usually sufficient.

iii. *Calibration*

Weigh exactly 150 mg of a 3,5-dichloro-*N*-(1,1-dimethyl-2-propynyl)benzamide analytical standard into a 25-ml volumetric flask, and add to the mark with dioxane. Make the following dilutions: Pipette 5-ml, 4-ml, 3-ml, 2-ml, and 1-ml portions into separate 10-ml volumetric flasks. Pipette 2 ml of the internal standard solution into each flask. Add to the mark of each flask with dioxane. These dilutions correspond to the 100%, 80%, 60%, 40% and 20% calibration points.

Measure the peak height of each 3,5-dichloro-*N*-(1,1-dimethyl-2-propynyl)benzamide and internal standard peak. Calculate the peak height ratio for each calibration point.

(a) *3,5-dichloro-N-(1,1-dimethyl-2-propynyl)benzamide (internal standard)*. Plot the above ratio versus percentage of 3,5-dichloro-*N*-(1,1-dimethyl-2-propynyl)benzamide. It is recommended that two or three injections be made for each calibration point and sample.

iv. *Sample Analysis*

Weigh accurately 600 mg of the wettable powder sample into an extraction thimble. Extract in a micro-Soxhlet apparatus for 18 hours using 15 ml benzene. Transfer the benzene solution of the extract to a 100-ml volumetric flask. Rinse the apparatus with dioxane and add the rinse solution to the 100-ml flask. Add to the mark with dioxane. Pipette 5 ml of the above solution into a 10-ml volumetric flask. Pipette 2 ml of the internal standard solution into the flask. Add to the mark with dioxane.

Inject $\frac{1}{2}$–1-μl portions onto the column. Measure the peak heights of the 3,5-dichloro-*N*-(1,1-dimethyl-2-propynyl)benzamide and the internal standard.

e. *Calculations*

Measure the peak heights of the 3,5-dichloro-*N*-(1,1-dimethyl-2-propynyl)benzamide and the internal standard. Calculate the peak height ratio:

$$\frac{\text{3,5-dichloro-}N\text{-(1,1-dimethyl-2-propynyl)benzamide peak height}}{\text{internal standard peak height}}$$

Read percentage of 3,5-dichloro-*N*-(1,1-dimethyl-2-propynyl)benzamide from the calibration curve prepared previously.

f. *Discussion*

Under the stated conditions, 3,5-dichloro-*N*-(1,1-dimethyl-2-propynyl)benzamide elutes in 3.7 minutes, and the stilbene internal standard elutes in 1.8 minutes. Both peaks are measured from the solvent peak.

B. Residue Analysis

1. REVIEW OF METHODS

The analytical method for the determination of pronamide residues has been described by Adler *et al.* (1972a).

2. RECOMMENDED METHOD

a. *Principle*

Terminal residues of pronamide and/or its metabolites are determined by digesting the sample with sulfuric acid and methanol to convert residues to methyl-3,5-dichlorobenzoate, hereinafter designated as MDCB. This compound is codistilled from the reaction medium, purified by chromatography on Florisil, and measured by electron capture gas–liquid chromatography. The method is sensitive to 0.01 ppm.

b. *Reagents*

Florisil, 100–200 mesh (Floridin Co.); heat 24 hours at 120°C, deactive by adding 5% water, and equilibrate by tumbling ≥ 2 hours.

Methanol, Reagent Grade (Baker analyzed); use without further purification.

Solvents, Petroleum ether, benzene, and hexane, all technical grade; distill in glass before use.

c. *Apparatus*

Gas chromatograph with a ^{63}Ni electron capture detector.

Glass chromatographic tubes, 1 × 40 cm with Teflon stopcocks.

Traps with Teflon stopcocks, Dean-Stark (Kontes Glass Co., no. K-751250).

Recorder, 1 mV.

Glass coiled gas chromatograph columns 8 feet × 3 mm i.d.

d. *Experimental Procedure*

i. *Sample Preparation and Digestion*

Homogenize the sample in Hobart food chopper or Wiley mill (alfalfa hay). Weigh a 10-gm homogeneous sample into a 500-ml single neck, round-bottom flask. Add 100 ml 25% H_2SO_4, 50 ml methanol, and several boiling granules to the flask. Connect a 500-ml flask to the Dean-Stark trap equipped with a water-cooled condenser, and reflux gently for about 16 hours (overnight is convenient). After the reflux period, open the stopcock on the trap and collect the distillate.

ii. *Extraction and Separation*

Pour the distillate into a 500-ml separatory funnel and rinse the distillation receiver with a total of 75 ml water. Add the washings and 25 ml petroleum ether to the separatory funnel and vigorously shake contents for 2 minutes.

After extraction, remove and discard the lower, aqueous layer. Pour the upper, organic layer from the top of the separatory funnel directly into the chromatographic column containing 7 cm of specially deactivated Florisil and topped with 1–1.5 cm anhydrous Na_2SO_4.

Rinse the separatory funnel twice with 5-ml portions of petroleum ether, and add rinsings to the column. Let all of the solvent percolate through the column. Elute methyl-3,5-dichlorobenzoate (MDCB) from the Florisil column with 25 ml 20% benzene in hexane.

iii. *Gas Chromatography*

(a) *Operating Conditions.* Column packing 10% OV-17 on Gas Chrom Q; 90% argon–10% methane (P-10, Matheson) at flow rate of 60 ml/minute at exit port (scrub gas through Drierite and 4A molecular sieves before it reaches the gas chromatograph); temperatures (°C); column, 180; detector, 260; and injection block, 260; electrometer setting 3.2×10^{-10} A; chart speed 0.5 inches/minute. Under these conditions the retention time of MDCB is 2 minutes. Before initial use, condition the column 1 week at 225°C. Thereafter, keep column conditioned by injecting 10 μl Silyl 8 (Pierce Chemical Co.) daily.

(b) *Analysis.* Inject 10 μl eluant (or appropriate dilution) from the Florisil column into the gas chromatograph and measure MDCB. (**Note:** Do not attempt to improve sensitivity by concentrating Florisil column eluant, since the methyl ester is lost in concentration step.)

Inject 10-μl portions of solutions containing known amounts of MDCB daily, and determine the slope of the linear function of weight injected (nanograms MDCB) versus peak height. Use a factor of 1.25 to convert weight of MDCB to weight of pronamide.

iv. *Calculations*

Calculate concentration of residue (expressed as pronamide) in the sample as follows:

ng pronamide/10 μl injected = peak height (inches) × slope (ng MDCB/inch) × (1.25 ng pronamide/ng MDCB)

μg pronamide (total) = (ng pronamide/10 μl) × [(1000 μl/ml)/(1000 ng/μg)] × total volume (ml)

ppm pronamide in sample (corrected for recovery) = (μg pronamide (total)/gm sample) × (1/recovery factor)

= peak height (inches) × slope (ng MDCB/inch) × 1.25/[total volume (ml)/10 μl] × gm sample × recovery factor

v. *Sensitivity and Recovery*

The GLC analytical procedure described above is sensitive to 0.01 ppm pronamide. The method has been validated by recovering methyl-3,5-dichlorobenzoate-^{14}C from ^{14}C-labeled pronamide, which was added to samples before the acid digestion step. Recoveries were determined by measuring the peak due to MDCB on the gas chromatogram, as well as by radioassay of the final column eluant. The results (Table I) show excellent agreement between these two methods of determining recoveries.

Amounts of pronamide ranging from 0.42 to 22.6 μg have been recovered as MDCB from reagents (no sample used), with an average recovery of 79% (standard deviation = 6.6). Six known metabolites (as well as a mixture containing equal weights of each of the metabolites) and pronamide have also been converted to MDCB in reagents with about a 70% yield. Methyl-3,5-dichlorobenzoate was recovered in about 70% yield when this compound was subjected to the procedure, and this, therefore, appears to be the factor which limits recovery (Table II).

TABLE I

RECOVERY OF PRONAMIDE ADDED TO SAMPLE BEFORE DIGESTION STEP

	Range of fortifications		No. of fortifications	Average recoveries (%) ± St. D.)	
Sample	μg	ppm		GLC	Radioassay
Reagents	0.42–22.6	–	8	79.1 ± 6.6	–
Alfalfa hay	10.7–107.0	2.1–21	33	70.9 ± 7.4	70.0 ± 6.5
Lettuce	0.21–1.05	0.01–0.04	11	88.2 ± 10.4	90.6 ± 5.7
Milk	0.10–0.84	0.005–0.04	28	76.0 ± 7.3	74.0 ± 5.0
Eggs	0.12–1.00	0.01–0.10	35	68.8 ± 6.6	69.5 ± 5.6

TABLE II

RECOVERY OF METABOLITES AS MDCB

Compound[a]	No. of fortifi-cations	Range of fortifi-cations (μg)	Recovery (%)
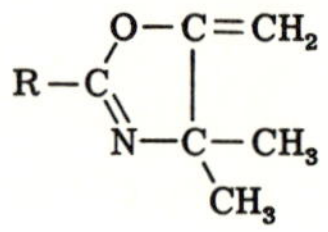	3	1.18–11.8	68.7
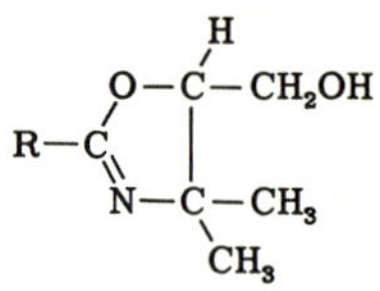	3	1–10	69.0
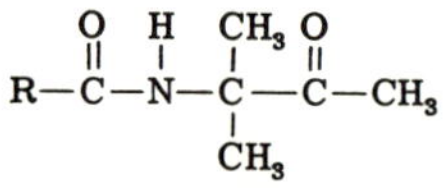	3	1–10	68.9
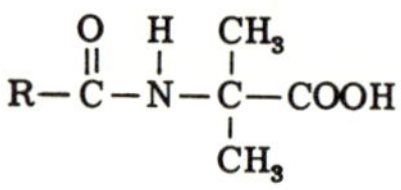	3	1–10	67.6
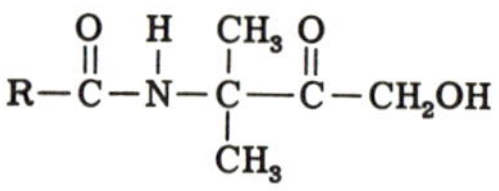	3	1.51–15.1	65.1
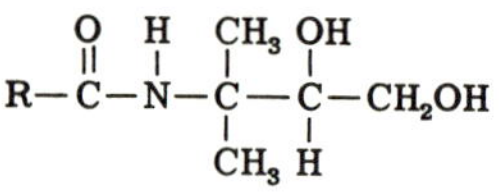	3	1.01–10.1	66.1
Mixture of Pronamide and above compounds	6	15.5–155	69.9

[a]

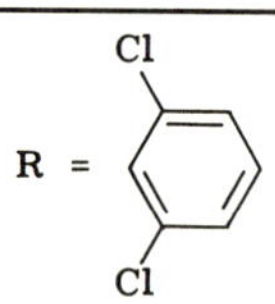

REFERENCES

Adler, I. L., Gordon, C. F., Haines, L. D., and Wargo, J. P., Jr., (1972a). *J. Ass. Offic. Anal. Chem.* **55,** 802–805.

Adler, I. L., Haines, L. D., and Wargo, J. P. (1972b). *J. Agr. Food Chem.* **20,** 1233–1235.

Fisher, J. D. (1974). *J. Agr. Food Chem.* **22,** 606–610.

Viste, L. L., Crovetti, A. J., and Horrom, B. W. (1970). *Science* **167,** 280.

Yih, R. Y., Swithenbank, C., and McRae, D. H. (1970). *Weed Sci.* **18,** 604–607.

Yih, R. Y., and Swithenbank, C. (1971a). *J. Agr. Food Chem.* **19,** 314–319.

Yih, R. Y., and Swithenbank, C. (1971b). *J. Agr. Food Chem.* **19,** 320–324.

30

Sencor®

C. A. Anderson

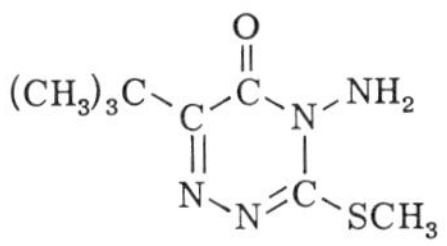

4-Amino-6-(1,1-dimethylethyl)-3-(methylthio)-1,2,4-triazin-5(4*H*)-one

I. GENERAL

A. Empirical Formula

$C_8H_{14}ON_4S$ (Mol. wt. 214.3)

B. Alternate Names

Bay 94337, metribuzin. Sencor® is a trade mark of the parent company of Farbenfabriken Bayer GmbH, Leverkusen, West Germany.

C. Source of Analytical Standard

Chemagro Agricultural Division of Mobay Corporation, Kansas City, Missouri, and the parent company of Leverkusen, West Germany.

D. Biological Properties

The acute oral LD_{50} of Sencor® technical to rats is 1100–1200 mg/kg. No mortality resulted when rats and rabbits were dermally exposed to 20,000 mg/kg Sencor® technical or 50% wettable powder. Neither the technical nor the 50% wettable powder are eye or skin irritants. No mortality resulted when rats were exposed for 1 hour to concentrations up to 20,000 μg per liter of air of Sencor® technical or Sencor® 50% wettable powder.

E. History

Sencor® herbicide was originated by the parent company of Leverkusen, West Germany and is being developed by Chemagro in the United States and Canada under license.

F. Physical Properties

Sencor® is a white to tan crystalline solid with melting point, 125°C; density, 1.28; vapor pressure, $< 10^{-5}$ mm Hg at 20°C, 2×10^{-4} mm Hg at 60°C; solubility, 1200 ppm in water, soluble in lower alcohols, insoluble in aromatics and aliphatics.

G. Chemical Properties

1. METHOD OF SYNTHESIS

4-Amino-6-(1,1-dimethylethyl)-3-(methylthio)-1,2,4-triazin-5(4*H*)-one is prepared by the methylation of 4-amino-6-(1,1-dimethylethyl)-3,4-dihydro-3-thioxo-1,2,4-triazin-5(2*H*)-one.

2. CHEMICAL REACTIONS

4-Amino-6-(1,1-dimethylethyl)-3-(methylthio)-1,2,4-triazin-5(4*H*)-one is stable in dilute acids and alkalies up to approximately pH 12 at 20°C. It reacts with ketones.

H. Formulations

The only formulation available is a 50% wettable powder at this time.

II. ANALYSIS

A. Formulation Analysis

1. REVIEW OF METHODS

Sencor® can be determined by infrared spectrophotometry, but Betker (1974) found that negative interference from the *N*-methyl substituted isomer, a synthesis by-product, causes low results. A correction can be applied for the interfering compound, but to do so requires determining its concentration by gas chromatography (GLC). This combined IR–GLC procedure is obviously more complicated and subject to more measurement errors than the direct GLC determination of Sencor® described below. Liquid chromatography is also applicable but is relatively time-consuming and not as reproducible as GLC.

2. RECOMMENDED METHODS

a. *Principle*

The preferred method for analyzing technical and formulated Sencor® is gas chromatography (GLC) using methylene chloride solvent (Betker, 1974). The advantages are speed, specificity, and relatively high precision and accuracy.

The active ingredient content of an unknown sample is determined by comparison with a standard Sencor® of known purity utilizing an internal standard added to the solutions of the sample and of the analytical standard. Solutions of the sample and standard are injected into the chromatograph, and the appropriate peak areas or heights of the eluted components, expressed as the Sencor® internal standard ratio, are compared.

b. *Reagents*

Column packing, 2.5% w XE-60 on 80–100 mesh Chromosorb G-AW-DMCS. Available from Applied Science Laboratories, Incorporated, Box 440, State College, Pennsylvania.

Di-n-butylphthalate (DNBP) solution, internal standard, 4% v/v solution of DNBP (Fisher Scientific No. D-30 or equivalent) in Reagent Grade methylene chloride.

Methylene chloride, Reagent Grade.

Sencor®, analytical standard of known purity (see Section C, above).

c. *Apparatus*

Column, gas chromatograph, 3 × $\frac{1}{4}$-inch stainless steel, packed by vibrating until full with 2.5% XE-60 on 80–100 mesh Chromosorb G-AW-DMCS. Condition the column overnight at 240°C, then immediately before use by making repeated injections of Sencor® solution under the operating conditions given below. Use this column for Sencor® determinations only.

Gas chromatograph, F & M 5750 or equivalent, equipped with a thermal conductivity detector.

Integrator, electronic (optional).

Shaker, mechanical.

Syringe, micro, 10-μl, Hamilton #701N or equivalent.

d. *Experimental Procedure*

i. *Operating Conditions*

Clean the injection port with water and acetone before setting the following conditions on the instrument. These conditions are typical for the F & M 5750; other makes or models of gas chromatographs may require slightly different conditions.

Attenuation (maximum): ×1.
Bridge current: 180–200 mA.
Carrier gas: helium, 60 ml/minute.
Program selector: isothermal.

Recorder range: 0.1 mV.
Sample size: variable; see procedure, below.
Temperature
Detector: 230–240°C
Injection Port: 230–240°C
Oven: 220°C

Adjust the oven temperature as necessary (but not over 250°C) and the carrier gas flow rate ±10 ml/minute to yield retention times and resolution similar to those shown in Fig. 1.

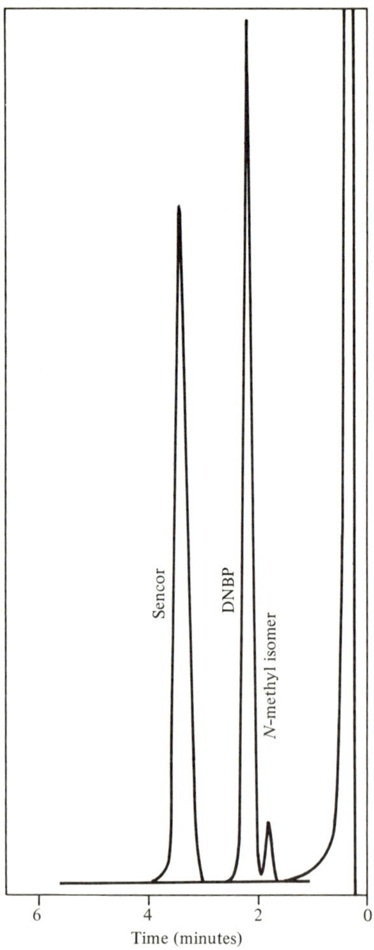

FIG. 1. Typical scan of technical material.

TABLE I

Material	Weight range (gm)
Sencor analytical standard	0.29–0.30
Sencor® Technical	0.31–0.33
Sencor® 50% wettable powder	0.56–0.58

ii. *Procedure*

Weigh the amount of sample and analytical standard shown in Table I to the nearest 0.1 mg into separate 25-ml volumetric flasks. Add 15 ml methylene chloride to each flask. Stopper the flasks and shake on a mechanical shaker for 15 minutes. Pipette 5 ml DNBP (internal standard) solution into each flask, then dilute to the mark with methylene chloride. Stopper and mix thoroughly. Allow particles large enough to clog the syringe to settle (it is not necessary for the supernatant solution to be completely clear).

(a) *Technique.* Inject 3 μl of the standard solution into the gas chromatograph at attenuation $\times 1$. Adjust the cell current as necessary within the limits shown above to give a DNBP peak height of 70–90% of scale. If the retention times and resolution are not like those shown in Fig. 1, check the flow rate and oven temperature. If the noise level is more than 0.5% of full scale, switch the attenuation to $\times 2$ and inject 6 μl in this and all subsequent injections.

If measurements are to be based on peak height, draw a baseline under the peaks as shown in Fig. 1. (Electronically measured areas may be used in lieu of peak heights in the following operation.) Measure the peak heights ± 0.5 mm and calculate the ratio of the Sencor® peak height to the DNBP peak height. Perform replicate injections of the standard until the Sencor®–DNBP peak height ratios for two successive injections agree within 2% (± 1% of their average). (If a suitable average ratio has not been obtained after a maximum of six injections, or if the peak resolution is not equivalent to that shown in Fig. 1, repack the column.) Inject the sample solution twice. If more than one sample is to be analyzed, inject each in duplicate. Inject the standard again after every six to eight sample injections, and also at the end of a run.

If the internal standard peak is significantly higher in the sample than it is in the standard, weigh another portion of the sample as in Table I above and label it "blank." Dissolve or extract this blank as above, but *do not* add internal standard to this solution at any point. Dilute the blank solution to volume with methylene chloride. Stopper, mix thoroughly and inject into the gas chromatograph, using the same injection volume, attenuation and temperatures as used for the sample. Measure the peak height of any peak that occurs *at the same retention time* as the maximum for DNBP (interfering peak). Subtract the height of this interfering peak from the height of the DNBP peak for the corresponding sample, and use the difference as the DNBP peak height for this sample.

Calculate the Sencor®–DNBP peak height ratio for each standard injection. The ratios for all standards immediately preceding and following a given series of samples must agree within 2.0% (±1.0% of the average ratio). If they do, use the average ratio in calculating results for the intervening samples; if they do not, repeat the injections of the standard and of all intervening samples.

Calculate the Sencor®–DNBP peak height ratios for all sample injections and average the ratios for the duplicate injections of a sample. If any duplicate ratios differ by more than 2.0% (±1.0% of their average), reject them and repeat the injections of that sample, preceding and following it with standard injections as before.

e. *Calculations*

The ratio of Sencor®–DNBP (internal standard) peak heights produced by an unknown sample is compared to the corresponding ratio produced under identical conditions by a Sencor® standard of known purity.

$$\%\ \text{Sencor}^{\circledR} = \frac{(\text{Sencor}^{\circledR}\ \text{peak height})_u\ (\text{DNBP peak height})_s\ (\text{weight})_s\ (\%\ \text{purity})_s}{(\text{DNBP peak height})_u\ (\text{weight})_u\ (\text{Sencor}^{\circledR}\ \text{peak height})_s}$$

where u = unknown Sencor® sample and s = standard Sencor® sample.

f. *Discussion*

The conditions of this method provide for the separation of the Sencor® active ingredient from other components, including manufacturing impurities and formulation inerts. The use of an internal standard (di-*n*-butyl phthalate in this case) greatly increases precision by minimizing volumetric errors; the effects of inaccurate dilutions and variation in injection volume are eliminated. Accurate weighings and measurements of internal standard volumes are, of course, critical. Peak height reproducibility criteria are stipulated as a precaution against instrumental drift or malfunction.

B. Residue Analysis

1. REVIEW OF METHODS

Residues of Sencor® were first analyzed by Stanley and Schumann (1969). The method employed a single blender extraction followed by Florisil column cleanup and electron capture gas chromatographic analysis using a 5% OV-101 column. A modification of that method reported by von Stryk (1971) employed flame photometric detection in the sulfur mode. Later, Thornton and Schumann (1971) reported an electron capture gas chromatographic method for Sencor® and its deaminated diketo metabolite in soybeans. The most recently developed method (Thornton, 1974) employs a reflux extraction to remove conjugated residues and will detect Sencor® and the three metabolites that have been identified using electron capture gas chromatography.

2. RECOMMENDED METHODS—VARIOUS CROPS

a. *Principle*

Metabolism studies (Church and Flint, 1972) have shown the metabolism of Sencor® in crops following a stepwise pathway as shown in Fig. 2. Any of the compounds could conceivably be found as residues in crops. However, metabolism studies show it is unlikely that the DK(IV) or DA(II) metabolite would ever be present in mature crops grown above ground. This is in contrast to root crops, which have shown residues of the DK metabolite and, to a limited extent, DA. The procedure described here is primarily for measurement of Sencor® and the major metabolite, DADK(III). However, the method will also detect and measure DK and DA but with less precision.

Briefly, the procedure involves a refluxing step to release Sencor® and any metabolite that may be conjugated. Sencor® and metabolite residues are separated from the solids by filtration and the filtrate solution evaporated until only water remains. Following this, Sencor® and metabolites are separated from each other by liquid–liquid partition prior to individual gas chromatographic analysis employing electron capture detection. The entire analysis scheme is outlined in Fig. 3.

b. *Reagents*

Acetonitrile, Nanograde or equivalent.
Benzene, Nanograde or equivalent.
Benzene–acetonitrile mixture, 99:1 and 90:10.
Chloroform, Nanograde or equivalent.
Decyl alcohol, Eastman.
Florisil, PR grade, 60–100 mesh, 2.5% water deactivated. Heat in

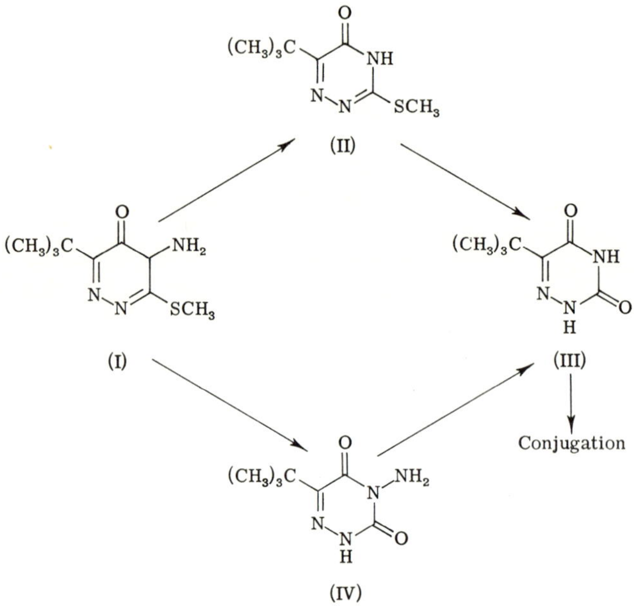

FIG. 2. Sencor® metabolic pathway. (I) Sencor®; (II) deaminated Sencor® metabolite (DA) [6-(1,1-dimethylethyl)-3-methylthio-1,2,4-triazine-5(4*H*)-one]; (III) deaminated diketo Sencor® metabolite (DADK) [6-(1,1-dimethylethyl)-1,2,4-triazine-3,5(2*H*,4*H*)-dione]; (IV) diketo Sencor® metabolite (DK) [4-amino-6-(1,1-dimethylethyl)-1,2,4-triazine-3,5(2*H*,4*H*)-dione].

an oven at 130°C for at least 24 hours and allow to cool in a tightly stoppered bottle. Add 2.5% water (2.5 ml water + 97.5 gm dried Florisil) and allow to equilibrate at room temperature for 24 hours in a tightly stoppered bottle before use.

Glass Beads, P series, 100–140 mesh. Potters Industries, Inc., Carlstadt, N. J.

Hexane, Nanograde or equivalent.

Hydrochloric acid, 4 *N*.

Silica gel, Fisher S-679, 7% water deactivated. Heat overnight at 130°C and allow to cool in a tightly stoppered bottle. Add 7 ml water to 93 gm of activated silica gel and mix thoroughly. Allow to equilibrate in a tightly stoppered bottle for 24 hours before use.

Sodium hydroxide, 0.1 *N*

Sodium sulfate, anhydrous, granular, ACS reagent.

Sencor® standard solution. Weigh 0.05 gm Sencor® into a 100-ml volumetric flask and dilute to volume with benzene. Make appropriate dilutions to obtain a standard solution of 0.125 μg/ml in benzene.

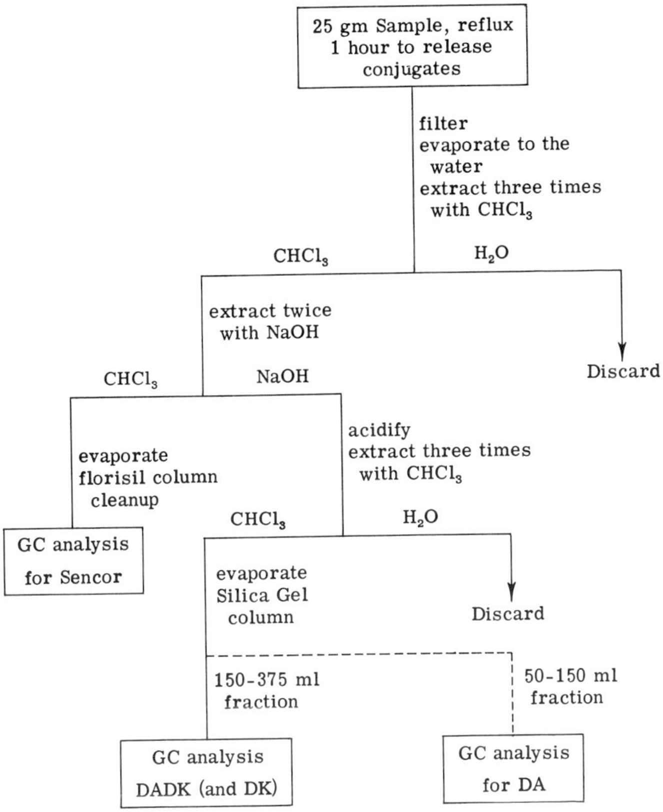

FIG. 3. Schematic of Sencor® and metabolite analysis in crops.

Metabolite standard solution. Weigh 0.05 gm DADK metabolite into a 100-ml volumetric flask and dilute to volume with benzene. DADK is not readily soluble in benzene; therefore, addition of acetonitrile helps to dissolve the initial weighed amount. Make all subsequent dilutions with benzene. Make appropriate dilutions to obtain a standard of 0.5 μg/ml DADK. The final dilution of the DADK standard should also contain 0.02 ml decyl alcohol per milliliter of standard solution.

c. *Apparatus*

Assorted laboratory glassware.

Chromatographic tubes, 20 × 400 mm with 300 ml integral reservoir.

Filter paper, Whatman GF/A glass filter paper, 9.0 cm.

Food chopper, Hobart or equivalent.

Gas chromatograph, Hewlett-Packard model 5750, equipped with high temperature ^{63}Ni electron capture detector or equivalent.

Rotary vacuum evaporator, Rotavapor model VE50 or equivalent.

Water bath, 35–40°C.

d. *Extraction and Preliminary Separation*

Grind the entire sample in a food chopper with dry ice. Mix thoroughly and place the sample in frozen storage overnight to allow the dry ice to sublime. Weigh 25 gm of the ground sample into a 1000-ml flat-bottom flask. Add 50 ml water and 200 ml acetonitrile (for bulky samples such as hay or straw, use 75 ml water and 300 ml acetonitrile; for extremely bulky samples, reduce sample size to 10 gm). Attach the flask to a reflux condenser and heat the sample at reflux for 1 hour. Allow the sample to cool, and rinse down the condenser with 3–5 ml acetonitrile. Filter the samples through a 9-cm Whatman GF/A filter paper in a size 2A Büchner funnel. Wash the flask and the solids with 50 ml acetonitrile (use 100 ml acetonitrile for bulky samples). For samples of high chlorophyll or wax content, it may be helpful in reducing emulsions if the sample is transferred to a separatory funnel at this point and washed with 200 ml hexane. Discard the hexane (upper phase).

Transfer the filtrate to a 1000-ml round-bottom flask and evaporate the sample on a rotary vacuum evaporator until only water remains (some residual acetonitrile can be tolerated). Transfer the water remaining after evaporation to a 500-ml separatory funnel. Rinse the flask with 75 ml chloroform and add the rinse to the separatory funnel. Shake the separatory funnel for 30 seconds. Drain the lower (chloroform) phase into a second 500-ml separatory funnel. Repeat the partition with two additional 75-ml portions of chloroform. Discard the water phase. Extract the combined chloroform phases with successive 50- and 25-ml portions of 0.1 *N* NaOH. Drain the chloroform into a 500-ml flask (if chloroform phase is cloudy, drain through a plug of granular sodium sulfate and rinse with a few milliliters of fresh chloroform). Evaporate the chloroform to dryness on a rotary vacuum evaporator. Remove the last traces of solvent with a stream of nitrogen. Reserve the residue for Sencor® cleanup (Section e, below).

Combine the two NaOH washes and acidify them with 3 ml 4 *N* HCl. Extract the aqueous solution with three 75-ml portions of chloroform and combine the extracts in a 500-ml flask. Add 0.5 ml decyl alcohol to the flask and evaporate the chloroform just to dryness. Remove the last traces of solvent with a stream of nitrogen. Reserve the sample for metabolite cleanup (Section f, below).

e. *Sencor® Cleanup*

Tamp a plug of glass wool into the bottom of a chromatographic tube. Cover the glass wool with about 1 cm of glass beads. Fill the chromatographic tube with benzene to the bottom of the reservoir, allowing a small amount of benzene to run out of the bottom of the tube. Tap the tube to dislodge trapped air. Add 10 gm Florisil (deactivated with 2.5% water) to the chromatographic tube and allow the Florisil to settle. Tap the tube to level the top of the column and dislodge any Florisil from the walls of the tube. Add about 5 gm granular sodium sulfate to the top of the column and drain the benzene to the top of the sodium sulfate. Add 5–10 ml benzene to the flask containing the Sencor® residue (Section d, above). Transfer the benzene to the chromatographic column and rinse the flask with two additional 5-ml portions of benzene. Pass the benzene solution through the column at the rate of 2 drops per second. Rinse the walls of the column with two 5-ml washes of benzene. Allow the benzene to drain to the top of the sodium sulfate. Discard the benzene wash.

Elute the column with 175 ml benzene–acetonitrile (99:1) at the rate of 2 drops per second. Discard the first 35 ml and collect the remaining 140 ml in a 250-ml round-bottom flask. Evaporate the eluant just to dryness on a rotary vacuum evaporator in a 40°C water bath. Remove the last traces of solvent with a stream of nitrogen. Dissolve the residue in 20.0 ml benzene. Stopper and hold for gas chromatographic analysis (Section g, below).

f. *Metabolite Cleanup*

Tamp a plug of glass wool into the bottom of a chromatographic tube. Cover the glass wool with approximately 1 cm of glass beads. Fill the chromatographic tube with benzene–acetonitrile (90:10) to the bottom of the reservoir, allowing a small amount to run out of the bottom of the tube. Tap the tube to dislodge trapped air. Add 10 gm silica gel (deactivated with 7% water) to the chromatographic tube. Allow the silica gel to settle. Tap the tube to level the top of the column and to dislodge any silica gel from the walls of the tube. Add about 5 gm granular Na_2SO_4 to the column and drain the solvent to the top of the Na_2SO_4. Transfer the metabolite residue in the flask described in Section d, above, to the column with four washes of 5–10 ml benzene–acetonitrile (90:10). Wash the column with additional benzene–acetonitrile to make 150 ml. Discard this wash.

Elute the DADK from the column with an additional 225 ml benzene–acetonitrile (90:10). Add 0.5 ml decyl alcohol to the flask and evaporate the benzene–acetonitrile just to dryness. Remove the last traces of solvent with a stream of nitrogen. This will not remove the decyl alcohol keeper in the flask. Transfer the residue in the flask to a 13-ml graduated tube with benzene, diluting the sample to 5.0 ml. Reserve the sample for gas chromatographic analysis (Section g, below).

g. *Gas Chromatographic Analysis*

Column, 1 m × 3 mm o.d. glass column packed with 5% OV-225 solution (Applied Science Labs, Inc., 1967) coated on 80–100 mesh Chromosorb W (H.P.).

Carrier gas, 5% methane in argon, 60 ml/minute, 40 psig.

Column oven temperature, Sencor® (and DA) 200°C (see Section i, below).

Column oven temperature, DADK (and DK) 180°C (see Section i, below).

Injection port temperature, 250°C.

Detector cell temperature, 270°C.

Pulse interval, 15 μseconds.

Chart speed, ½ inch/minute.

Inject 4 μl of sample or standard solution. Sencor® fractions should be injected along with a 0.125 μg/ml Sencor® standard. Metabolite fractions should be injected along with a 0.5 μg/ml metabolite standard. Identify the Sencor® or DADK by retention times and measure the area produced on the recorder strip chart with a polar planimeter. At the operating conditions stated, the retention times for Sencor® and DADK are approximately 4.0 minutes and 3.0 minutes, respectively.

h. *Calculations*

Calculation of Sencor® or metabolites in a sample is accomplished by use of the following equation where response for the unknown is compared to the response obtained for a known quantity in a standard solution.

$$\text{ppm} = \frac{\text{sample area}}{\text{standard area}} \times \frac{\text{standard injection (ng)}}{\text{weight of sample (gm)}} \times \frac{\text{final volume (ml)}}{\text{sample injection } (\mu\text{l})}$$

Any residues of DADK result from conversion of the active ingredient Sencor® by metabolic action. Thus, these residues should be reported in Sencor® equivalents, taking into account the molecular weight difference. Molecular weight conversion factors for converting from apparent DADK to Sencor® equivalents is 214/169 = 1.266. Factors for DK and DA are 1.163 and 1.075, respectively (see Section i, below).

i. *Discussion*

The method described in this manuscript measures Sencor® and its DADK metabolite in a wide variety of crops. A procedure was developed earlier for the analysis of Sencor® and metabolites in potatoes (Thornton and Stanley, 1972), but it did not give sufficient cleanup for many other crops. The method was modified by changing the gas chromatographic cleanup for Sencor® (Stanley and Thornton, 1972) and also by changing from 10% water-deactivated silica gel to 7% water-deactivated silica gel for cleaning up the metabolite fraction. With the 7% water-deactivated silica gel, most of the interfering crop extractives elute in the first 150 ml and are discarded.

As stated earlier, the DA and DK compounds are also possible conversion products of Sencor®. This procedure can be used for analysis of these compounds, if desired. DK and DA behave almost identically to DADK in the various partitions in the procedure. DK elutes from 7% water-deactivated silica gel with practically the same pattern as DADK. Therefore, any DK present in the sample will be seen in the DADK fraction with a retention time about twice that of DADK (approximately 6.7 minutes). In some dry samples, such as straw, DK recovery may be somewhat less than is found for Sencor® or DADK.

The DA compound elutes from the silica gel column in the fraction just prior to DADK–DK. In the event analysis for DA is desired, the eluate from 50 to 150 ml of benzene–acetonitrile (90:10) should be collected and analyzed separately for DA. Although it would be more practical to collect all three metabolites in a common fraction, it is not possible because of interference peaks for DADK, which occur in the first 150-ml of eluate. The elution pattern for all three metabolites from silica gel is shown in Fig. 4. DA is analyzed at the same gas chromatographic conditions as Sencor® and has a retention time of 7.5 minutes.

For plant metabolism studies utilizing ^{14}C-labeled Sencor®, conjugated metabolites in solids or water fractions were released by acid hydrolysis (Church and Flint, 1972). In the residue method for potatoes, a steaming procedure was found to be as effective as acid hydrolysis in releasing conjugated residues, and it did not degrade Sencor® (Thornton and Stanley, 1972). When the steaming procedure was tried with sugarcane, it gave more crop extractives, appeared to cause more decomposition of Sencor® to a nondetectable product and gave lower recoveries of the metabolites than did a reflux extraction procedure. One hour reflux with acetonitrile–water (4:1) released maximum residues from field-treated sugarcane samples and caused the least amount of decomposition of added metabolites in laboratory-fortified samples. Additional studies with alfalfa that had been treated with ^{14}C-labeled Sencor® and

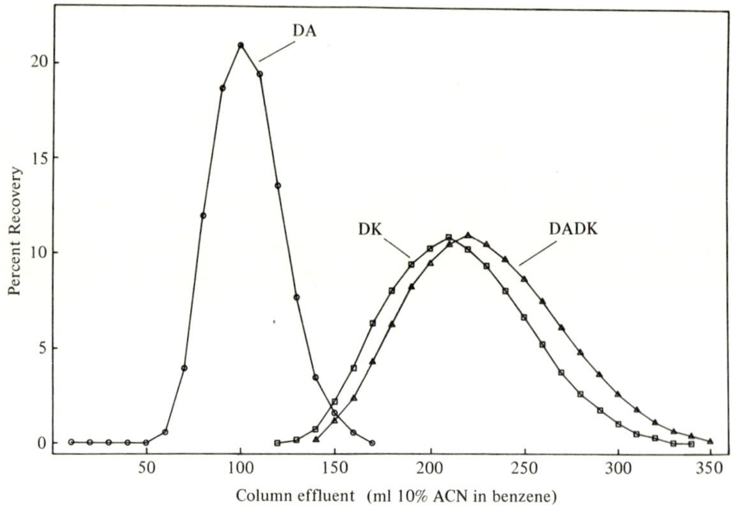

FIG. 4. Elution pattern of DA, DK, and DADK on 7% water-deactivated silica gel.

sampled 49 days later showed the 1-hour acetonitrile–water reflux extraction to be as effective in solubilizing activity as repeated reflux extractions or even acid hydrolysis using 2 *N* HCl (Stanley, 1974). Thus, a 1-hour reflux was adopted for general use.

Sencor®, DADK, DK, and DA in a common solution can be resolved from each other by gas chromatography using a column of 5% OV-225 on Chromosorb W (H.P.). However, because of the great difference in chemical characteristics between Sencor® and its metabolites, it is preferable to separate them into fractions for cleanup. This is accomplished by a chloroform–aqueous NaOH partition. Thus Sencor®, which is quantitatively removed by the chloroform, is analyzed separately after being cleaned up by Florisil column chromatography. DADK, DK, and DA are removed from the aqueous solution after acidification and further cleaned up by silica gel column chromatography prior to gas chromatography. Interferences from coextractives carried through the cleanup procedures are minimized by analyzing the various fractions at different gas chromatographic temperatures.

Recovery experiments were run on many different crops and products by adding Sencor® and metabolites before the reflux extraction step. Recovery of Sencor®, DADK, and DA were generally in the 75–100% range, while DK recoveries were somewhat less. Low recoveries for DK are not a major problem, since DK is considered a minor metabolite. Some of the crops that have been analyzed by this procedure are alfalfa, asparagus, barley, corn (cob, kernel, forage), lentils, peas, potatoes, sainfoin, soybeans, sugar cane (and products), tomatoes, and wheat.

Standard curves have been run to determine linearity of response for Sencor® and the three metabolites. Response for Sencor® was linear over a thirtyfold range up to 1.5 ng injected. Response for the three metabolites was linear over a thirtyfold range up to approximately 7.5 ng. In all cases, the upper limit amounts are equivalent to about 0.15 ppm if the sample size and portions described in the procedure are used. Samples containing residues in excess of this amount should be diluted and reinjected to make sure response is linear.

If 0.1 square inch is considered the smallest area that can be accurately measured with a polar planimeter, the level of detectability for the method is determined by the amount of Sencor® or metabolite necessary to produce this area. In general, 0.1 ppm of each compound produced an area of 1 square inch or better, indicating the level of detectability to be approximately 0.01 ppm. Due to changes in sensitivity from day to day and from one instrument to another, this limit will vary. Also, the size of any control peak will vary with the procedure, the crop and the instrument sensitivity. Electron capture response for the metabolites is less than that for Sencor®. This difference in response is compensated for by dissolving final metabolite residues in smaller volumes.

3. RECOMMENDED METHOD—SOIL

a. *Principle*

The procedure described here will extract and characterize residues of Sencor® and all three of its metabolites in soil. Briefly, the procedure involves a refluxing step to release Sencor® and any of its metabolites that may be conjugated or adsorbed to the solids. Free residues are separated from the solids by filtration. Following this, Sencor® may be cleaned up on a Florisil column for rapid GC analysis. Where the metabolite analysis is desired, Sencor® and metabolites are first separated on a silica gel column with the Sencor® fraction further cleaned up on a Florisil column. Residues are ultimately measured by electron capture gas chromatography.

b. *Reagents*

See Section 2,b, above.

c. *Apparatus*

See Section 2,c, above.

d. *Extraction*

Mix soil samples so that the entire sample is homogeneous. Allow wet samples to dry to the point where they can be sufficiently mixed. Weigh 25 gm of mixed soil into a 1-liter flat-bottom boiling flask. Weigh a separate 100-gm portion of the same soil into a 250-ml beaker, and place in an oven at 130°C overnight for moisture determination. To the 25-gm portion in the boiling flask, add 30 ml water and 270 ml acetonitrile. Attach the flask to a reflux condenser and heat the sample at reflux for 1 hour. Cool the sample sufficiently so it may be handled and filter with vacuum through a 9-cm Whatman no. 541 filter paper in a size 2A Büchner funnel. Rinse the flask with 100 ml fresh acetonitrile and use this to wash the filter cake.

Transfer the filtrate to a 1000-ml round-bottom flask and evaporate on a rotary vacuum evaporator until only water remains (5–10 ml). Transfer the water remaining after evaporation to a 250-ml separatory funnel with sufficient water to make the total water 50 ml. Rinse the flask with 75 ml chloroform and add the rinse to the separatory funnel. Shake the separatory funnel for 30 seconds. Drain the lower (chloroform) phase into a 300-ml round-bottom flask. Repeat the partition with two additional 75-ml portions of chloroform. Add 0.5 ml decyl alcohol to the flask and evaporate the combined chloroform extracts to dryness on a rotary vacuum evaporator at 40°C. Remove the last traces of solvent with a stream of nitrogen. If analysis for Sencor® only is desired, proceed from this point to Section f, below.

e. *Silica Gel Column Separation*

Prepare the silica gel column exactly as described in the previous method for Sencor® (Section 2). Place a clean receiver under the column and transfer the residue in the flask (Section d, above) to the column with four rinses of 5 ml benzene–acetonitrile (90:10), allowing each rinse to drain into the column before adding the next. Add 375 ml benzene–acetonitrile (90:10) to the column reservoir. Adjust the drip rate to about 2–3 drops per second.

Collect the first 60 ml that elutes. This fraction contains any Sencor® residue present in the sample. Reserve this fraction for Florisil column cleanup. Change receivers and collect the balance of the eluate in a 500-ml round-bottom flask. This fraction contains the metabolites DA, DADK, and DK. Add 0.5 ml decyl alcohol to the metabolite fraction and evaporate just to dryness on a rotary vacuum evaporator at 40°C. Remove the last traces of solvent with a stream of nitrogen. This will not remove the decyl alcohol keeper in the flask. Transfer the residue in the flask to a 13-ml graduated tube with benzene, diluting the sample to 5.0 ml. Stopper and hold for gas chromatographic analysis (Section g, below).

f. *Florisil Column Cleanup for Sencor®*

Prepare the Florisil column exactly as described in Section 2. Evaporate the 60 ml of eluate from the silica gel column just to dryness or use the initial extract solution if analyzing for Sencor® only. Transfer the residue in the flask to the column, using four 5 ml rinses, allowing each rinse to drain into the column before adding the next. Discard the benzene.

Add 175 ml benzene–acetonitrile (99 : 1) to the column reservoir and adjust the elution rate to 2–3 drops per second. Discard the first 25 ml benzene–acetonitrile (99 : 1) that elutes. Change receivers and collect the balance of the eluate in a 250-ml round-bottom flask. Evaporate the eluate just to dryness on a rotary vacuum evaporator at 40°C. Remove the last traces of solvent with a stream of nitrogen. Dissolve the residue in 20 ml benzene. Stopper and hold for gas chromatographic analysis (Section g, below).

g. *Gas Chromatographic Analysis*

Column, 2 feet × $\frac{1}{8}$ inch o.d. glass column packed with 5% Reoplex 400 solution coated (Applied Science, 1967) on 80–100 mesh Gas Chrom Q.

Carrier gas, 5% methane in argon, 70 ml/minute, 40 psig.
Column oven temperature, 200°C.
Injection port temperature, 250°C.
Detector cell temperature, 270°C.
Pulse interval, 15 μseconds.
Chart speed, $\frac{1}{2}$ inch/minute.

Inject 4 μl of sample or standard solution. Sencor® fractions should be injected along with a 0.125 μg/ml Sencor® standard in benzene solution. Metabolite fractions should be injected along with a standard containing 2.5 μg each of DA, DADK, and DK, plus 0.5 ml decyl alcohol diluted to 5 ml in benzene (0.5 μg/ml each metabolite).

Identify Sencor® or metabolites by retention times and measure the area produced on a recorder strip chart with a polar planimeter. At the operating conditions stated, the retention times for the various compounds are as follows: Sencor®, 3.0 minutes; DADK, 2.5 minutes; DK, 3.4 minutes; DA, 6.2 minutes.

h. *Calculations*

Calculation of the parts per million of Sencor® or metabolites in a sample is accomplished by use of the following equation where response for the unknown is compared to the response obtained for a known quantity in a standard solution:

$$\text{ppm} = \frac{\text{sample area}}{\text{standard area}} \times \frac{\text{standard injection (ng)}}{\text{sample weight (gm)}} \times \frac{\text{final volume (ml)}}{\text{sample injection } (\mu\text{l})} \times \frac{100}{100 - \text{moisture } (\%)}$$

Any residues of DADK, DK, or DA apparently derive from the conversion of Sencor® by metabolic or environmental action. Thus, these residues should be reported in Sencor® equivalents, taking into account the molecular weight difference.

When analyzing for Sencor® and all metabolites, total Sencor® equivalents should first be calculated before converting residues to dry weight basis.

i. *Discussion*

The method described in this chapter measures Sencor® and also its three possible metabolites in soil. The procedure described here offers the flexibility of being able to screen rapidly for Sencor® residues or use the entire procedure to characterize all of the possible Sencor® conversion compounds.

Initial extraction of the soil is done using a mixture of 10% water in acetonitrile and refluxing for 1 hour to free residues that may be adsorbed to soil particles or bound to the organic matter. Such a procedure has been shown in the laboratory to be more effective than simple shaking techniques or even Soxhlet extraction. Similar results were also obtained by Mattson *et al.* (1970) for symmetrical triazine herbicides in soil.

Soil samples may be quickly analyzed for Sencor® alone by cleaning up the initial extract on a Florisil column, followed by GC analysis. For samples where complete characterization of the metabolite residues is desirable, it is necessary to separate Sencor® from the metabolites using the silica gel column. The Sencor® elutes in the first 60 ml, but must then be further cleaned up by using a Florisil column before GC analysis. The metabolite fraction from the silica gel column may be evaporated and analyzed by GC for DADK, DK, and DA metabolites. Simpler cleanup schemes using only Florisil were not possible because the DADK metabolite could not be recovered from Florisil.

Recovery experiments were run with Sencor® and all three metabolites on various types of soil by fortifying the dry soil before reflux extraction. Soils were generally fortified at the 0.05, 0.1, and 0.5 ppm levels with the various compounds. Recoveries of Sencor®, DADK, and DA were routinely in the 75–100% range. DK recoveries were gener-

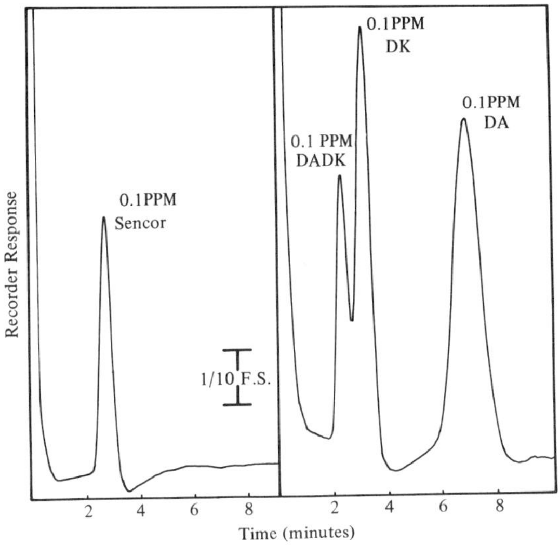

FIG. 5. Chromatograms of Sencor® and metabolite standards.

ally less, averaging about 65%. The lower recovery of DK is not of major importance since it is an insignificant metabolite in soil, occurring in only minor amounts (Khasawinah, 1973).

Chromatograms for Sencor® and metabolites are shown in Fig. 5. When portions and sample sizes are taken according to the procedure, the peaks shown in the figure represent 0.1 ppm of each compound. The level of detectability for Sencor® or metabolites in soil is established at approximately 0.01 ppm as discussed in the previous method for crops.

REFERENCES

Applied Science Labs, Inc. (1967). Technical Bulletin 2A, p. 5. State College, Pennsylvania.
Betker, W. R. (1974). CHEMAGRO Report No. 39988.
Church, D. D., and Flint, D. R. (1972). CHEMAGRO Report No. 32047.
Khasawinah, A. M. (1973). CHEMAGRO Report No. 39121.
Mattson, A. M., Kahrs, R. A., and Murphy, R. T. (1970). *Residue Rev.* **32,** 371.
Stanley, C. W. (1974). CHEMAGRO Report No. 40977.
Stanley, C. W., and Schumann, S. A. (1969). CHEMAGRO Report No. 25838.
Stanley, C. W., and Thornton, J. S. (1972). CHEMAGRO Report No. 35115.
Thornton, J. S. (1974). CHEMAGRO Report No. 40901.
Thornton, J. S., and Schumann, S. A. (1971). CHEMAGRO Report No. 30387.
Thornton, J. S., and Stanley, C. W. (1972). CHEMAGRO Report No. 33005.
von Stryk, F. G. (1971). *J. Chromatogr.* **56,** 348.

Part V
Growth Regulators and Miscellaneous Pesticides

31

Ancymidol

RICHARD FRANK AND EDGAR W. DAY

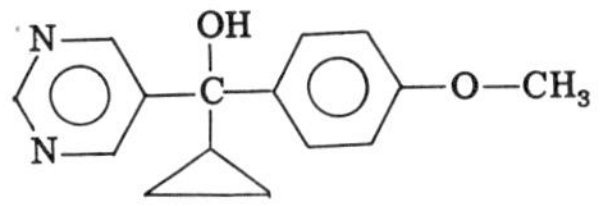

I. GENERAL

A. Empirical Formula

$C_{13}H_{16}N_2O_2$ (Mol. wt. 256.28).

B. Alternative Name

A-Rest® is a registered trademark of Elanco Products Co., a division of Eli Lilly and Company.

C. Source of Analytical Standard

Elanco Products Co., 640 South Alabama Street, Indianapolis, Indiana.

D. Biological Properties

Ancymidol is a plant growth regulator that reduces internode elongation. It has shown activity on a wide range of plant species and is effective on most species when applied to either foliage or soil. Ancymidol has been used extensively with excellent results on chrysanthemums, poinsettias, and Easter lilies producing compact, sturdy plants.

Technical ancymidol has a very low mammalian toxicity. The acute oral toxicity (LD_{50}) to the adult rat and mouse is 4500 and 5000 mg/kg, respectively. When rats are fed 8000 ppm ancymidol in their feed for 3 months, all survived with no changes attributed to the treatment. Dogs were dosed daily for 90 days with 200 mg/kg ancymidol (equivalent to 8000 ppm) and all survived with no significant changes which could be attributed to the treatment. The material caused moderate irritation to the eyes of rabbits.

E. Physical Properties

Ancymidol is an odorless, white crystalline solid that melts at 110°–111°C. The chemical is weakly basic and unstable in the presence of acid (pH 4 or less). The vapor pressure of the material is less than 1×10^{-6} mm Hg at 50°C. It is thermally stable up to 200°C, where vaporization begins to occur. Ancymidol is freely soluble in solvents such as acetone, ethyl acetate, chloroform, and methylcellosolve, moderately soluble in aromatic hydrocarbons, and only slightly soluble in saturated hydrocarbons. Water solubility is approximately 650 ppm at 25°C.

F. Chemical Properties

Ancymidol is prepared by the following reactions.

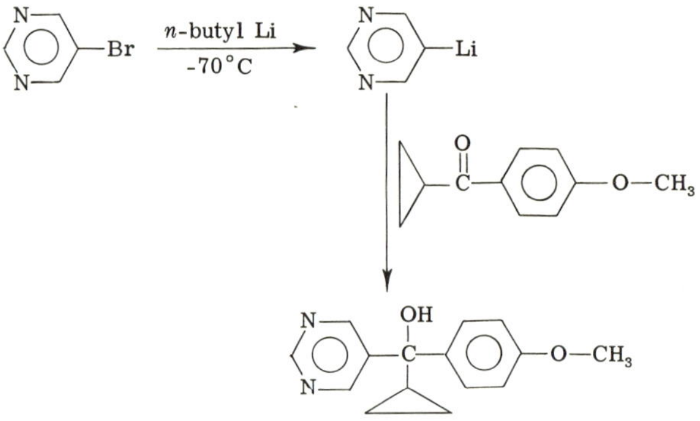

Under acidic conditions (pH 4 or less) ancymidol degrades rapidly; however, when the pH is maintained above 6, the compound is stable even at temperatures up to 50°C.

G. Formulations

Ancymidol is formulated as a 0.026% aqueous concentrate and a 4.5% emulsifiable concentrate. These are marketed under the trademark A-Rest®.

II. ANALYSIS

A. Formulation Analysis

1. RECOMMENDED METHOD

a. *Principle*

A weighed sample of the 4.5% emulsifiable concentrate is diluted to the proper concentration with chloroform, and the ancymidol is measured by gas chromatography using flame ionization detection. A portion of the 0.026% aqueous concentrate is extracted with ethyl acetate. The extract is purified over a Florisil chromatographic column, and the ancymidol is measured as above.

b. *Equipment*

Glass chromatographic columns, 14 × 240 mm equipped with glass or Teflon stopcock.

Vacuum evaporator, Rinco evaporator or equivalent.

Gas chromatograph. Hewlett-Packard model 402 (Hewlett-Packard, Rt. 41, Avondale, Pennsylvania) equipped with a flame ionization detector, appropriate recorder, and accessories.

Operating conditions
Temperatures (°C)
Flash heater: 290.
Column oven: 215.
Detector: 280.
Gas flows (ml/minute)
Nitrogen carrier: 35.
Hydrogen: 35.
Oxygen: 300.
Electrometer range: 10.
Attenuation: 8 (AFS 3.2×10^{-10})

Gas chromatographic tube, 4 ft × 3.0 mm i.d. borosilicate glass packed with 2% OV-17 on 80–100 mesh Chromosorb W HP. Condition newly prepared column 16 hours at 275°C while purging with carrier gas.

Florisil column preparation. Insert glass wool pledget into the bottom of the column, and add with constant tapping in order: 1 gm sodium sulfate, 5 gm deactivated Florisil, and finally, 1 gm sodium sulfate. With stopcock open, add 20 ml chloroform and allow to drain to top of sodium sulfate; the column is now ready for use.

c. *Reagents*

Ancymidol standard solution. Weigh 50 mg ancymidol reference standard into a 50-ml volumetric flask and dilute to volume with chloroform. Dilute this solution 1 to 10 with chloroform for the working standard solution.

Diheptyl phthalate internal standard solution. Weigh 40 mg diheptyl phthalate (Eastman Organic Chemicals, Rochester, New York) into a 250-ml volumetric flask and dilute to volume with chloroform.

Ethyl acetate, analytical reagent.

Chloroform, analytical reagent.

Methanol, analytical reagent.

Florisil, 100–200 mesh (Floridin Co., Berkeley Springs, West Virginia). Determine the moisture of the "as received" material by heating at 110°C for 6 hours. Add enough water to give 7.5–8% total moisture.

Sodium sulfate, anhydrous granules, analytical reagent.

d. *Sample Preparation*

4.5% Emulsifiable concentrate. Weigh a sample equivalent to 10 mg ancymidol into a 100-ml volumetric flask and dilute to volume with chloroform.

0.026% Aqueous concentrate. Transfer by pipette 4.0 ml of the sample into a 125-ml separatory funnel containing 30 ml water and 10 ml saturated sodium sulfate solution. Extract this mixture with three 20-ml portions of ethyl acetate, passing each portion of ethyl acetate through a layer of anhydrous sodium sulfate, combining the portions in a 125-ml boiling flask. Evaporate the ethyl acetate using a rotary flash evaporator and 50°C water bath. Carefully transfer the residue in the flask to the prepared Florisil column using small portions of chloroform (15 ml total), and allow to drain to the top of the sodium sulfate. Wash the walls with 4 ml chloroform and again drain to the top of the sodium sulfate. Discard all eluate to this point. Elute column with 45 ml of 2% methanol in chloroform (v/v) and collect the total eluate. Remove the solvent collected using a flash evaporator and 50°C water bath. Dissolve the residue in 10.0 ml chloroform.

e. *Measurement*

By pipette, transfer 1.0 ml of the sample solution to a small vial and add exactly 1.0 ml of the internal standard solution, seal, and mix well. Inject 2 or 3 μl of the mixture into the gas chromatograph using any acceptable technique. Repeat all of the above for the ancymidol working standard solution. Measure the peak response (height or area) resulting from the ancymidol and internal standard and calculate the ratio R.

$$R = \frac{\text{peak response ancymidol}}{\text{peak response internal standard}}$$

f. *Calculations*

$$\%\ \text{ancymidol} = [R_{\text{sample}}/R_{\text{standard}}] \times \text{concentration of standard (mg/ml)} \times [\text{volume of } CHCl_3 \text{ to dissolve sample (ml)/sample weight (mg)}] \times \text{purity standard (\%)}.$$

For the 0.026% aqueous concentrate, assume sample weight of 4000 mg.

B. Soil Residue Analysis

1. RECOMMENDED METHOD

a. *Principle*

The ancymidol is extracted from the soil with an acetone–ethanol solvent mixture. The sample extract is purified by treating with charcoal absorbent and by alumina column chromatography if the charcoal treatment is not adequate for a particular soil. Final measurement is made by gas chromatography using flame ionization detection.

b. *Equipment*

Glass chromatographic columns, same as used in formulations section.

Vacuum evaporator, Rinco evaporator or equivalent.

Gas chromatograph, same equipment as given in formulations section. Electrometer settings may have to be adjusted to increase the sensitivity of detection.

Gas chromatographic tube, same as given in formulations section. A 4 foot × 4.0 mm i.d. borosilicate glass column packed with 5% W-98 on Chromosorb W HP, 80–100 mesh has also been used successfully for measurements of the ancymidol.

Omni-mixer or equivalent.

Filter paper, Whatman no. 1.

c. *Reagents*

Acetone, analytical reagent.

Ethanol, denatured 3A.

Charcoal, Darco G-60.

Hyflo Supercel filter aid, Johns Manville Co.

Sodium sulfate, anhydrous analytical reagent.

Methylene chloride, analytical reagent.

Benzene, analytical reagent.

2% methanol, analytical grade, in benzene.

5% sodium chloride solution, analytical reagent.

Chloroform, analytical reagent.

Alumina, Alcoa F-20.

5% ethyl acetate, analytical reagent, in benzene.

Ancymidol standard solution A, 1.00 mg/ml. Accurately weigh 50.0 ancymidol analytical standard into a 50-ml volumetric flask and dilute to the mark with chloroform.

Ancymidol standard solution B, 20 μg/ml (direct standard). Quantitatively dilute a portion of solution A 1 to 50 with chloroform.

d. *Experimental Procedure*

i. *Sample Preparation*

The sample is thoroughly blended and screened to achieve a homogeneous mixture.

ii. *Extraction and Cleanup*

Weigh a 100-gm portion of the sample into a 1 quart Mason jar, add 200 ml acetone–ethanol mixture (5 : 1) and blend the sample for 5 minutes on an Omni-mixer. Add 5 gm Darco G-60 charcoal, 10 gm Hyflo filter aid, and mix thoroughly. Allow the mixture to stand 5 minutes with occasional agitation. Filter the contents of the jar through Whatman no. 1 filter paper with suction into a 500-ml filter flask, washing the jar with two 100-ml portions of acetone–ethanol mixture and passing each portion through the filter cake into the sample flask. Transfer contents (original extract plus washings) of the sample flask to a 2000-ml separatory funnel containing 800 ml 5% sodium chloride solution. Extract the aqueous mixture with two 80-ml portions of methylene chloride, passing each portion through a funnel containing about 75 gm anhydrous sodium sulfate and into a 500-ml boiling flask. Rinse the sodium sulfate with 30 ml methylene chloride after the second extraction volume has drained. Concentrate the methylene chloride solution to a volume of about 30 ml on a rotary vacuum evaporator, using a 50°C water bath. Add 5 or 6 gm sodium sulfate to the flask, and swirl to absorb any water that might be present. Transfer the liquid remaining to a 125-ml boiling flask, and rinse the 500-ml flask with four 10-ml portions of methylene chloride, retaining the sodium sulfate in the larger flask and adding the rinses to the 125-ml flask. Evaporate the methylene chloride solution to dryness on a rotary vacuum evaporator and 50°C bath. Retain the residue for measurement or further cleanup.

A standard recovery of 0.2 ppm and a control sample are assayed with each set of experimental samples. The standard recovery sample is prepared by fortifying 100 gm of control soil with 20 μg ancymidol (1 ml of standard solution B). The control sample is assayed exactly as an experimental sample. Experimental samples may require dilution or concentration to arrive at a final concentration of approximately 20 μg/ml.

e. *Measurement*

Dissolve the residue in 1.0 ml chloroform. Inject 3.0 μl each of ancymidol standard solution B (direct standard), control sample, fortified control, and experimental samples into the gas chromatograph. If the chromatograms show inadequate cleanup with the charcoal treatment, the sample extract may be further purified by alumina column chromatography (see Section f, below).

f. *Supplemental Cleanup*

Into a 14 × 240 mm glass chromatographic column, insert a glass wool pledget and add standardized alumina with constant tapping until a 7.5 cm column has been formed. Add enough sodium sulfate to give a 1-cm layer at the top of the column. (See Section h, below.) Wash the column with 20 ml benzene and allow to drain to top of sodium sulfate. Dissolve the sample obtained at the end of Section d,ii, above, in 5 ml benzene and transfer to the alumina column. Allow the solution to flow into the column at a rate of about 3 ml/minute. Rinse the flask with a second 5-ml portion of benzene and add to the column. When the second portion of benzene has reached the top of the sodium sulfate, wash the column with 40 ml 5% ethyl acetate in benzene, and discard all the eluate to that point. Elute the column with approximately 70 ml of 2% methanol in benzene, discarding the first 20 ml of eluate and collect the remaining 50 ml (or fraction indicated in the standardization procedure) in a 125-ml boiling flask. Evaporate the column eluate to dryness on a rotary vacuum evaporator. Proceed to Section e, above.

g. *Calculations*

Measure the peak response (peak area or height) for each standard and sample. Percent recovery from fortified samples is calculated from the following equation.

$$\%\ \text{recovery} = \frac{\text{peak response recovery sample}}{\text{peak response direct standard}} \times 100$$

Ancymidol content of the experimental samples is calculated by the following equation.

$$\text{ppm ancymidol} = \frac{\text{peak response sample}}{\text{peak response of recovery sample}} \times 0.20\ \mu\text{g/gm} \times \text{dilution factor}$$

Dilution factor applies only if volume used to dissolve sample is different from volume used for recovery samples.

h. *Alumina Standardization*

The alumina, as received, must be deactivated prior to use to a known moisture content of 3.5–3.8%, as determined by weight loss at 100°C for 4 hours. This is accomplished by determining moisture on "as received" material and then adding water to bring the moisture content to the desired level. The alumina–water mixture is then prepared, allowed to stand overnight, and then mixed again before use. Standardization of the fraction of solvent to collect during the elution is done by placing 5 ml benzene containing 20 μg of ancymidol onto a prepared column and then washing and eluting the column as described in Section f, above. Evaporate the eluate from the fractions collected and dissolve in 1.0 ml chloroform. Compare the fractions by the gas chromatographic method to the 20 μg/ml direct standard.

i. *Discussion*

The sensitivity of the method is estimated to be about 0.02–0.05 ppm, and fortified recovery studies at 0.2–1.0 ppm have shown an average recovery of 85% and range from 76 to 93%.

32

PROTECT®

J. R. Riden

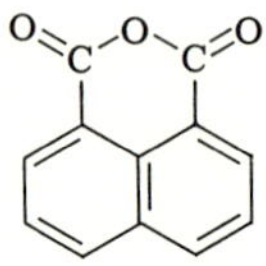

1,8-Naphthalic anhydride

I. GENERAL

A. Empirical Formula

$C_{12}H_6O_3$ (Mol. wt. 198.2).

B. Alternate Name

S-15718. PROTECT® is a registered trademark of Gulf Oil Corporation.

C. Source of Analytical Standard

Chemical Research Department, Gulf Oil Chemicals Company, Industrial & Specialty Chemicals Division, Crop Protection Products, 9009 West 67 Street, Merriam, Kansas.

D. Biological Properties

PROTECT® is used as a corn seed treatment to protect the corn plant from herbicidal injury by thiolcarbamate herbicides.

E. History

The concept of chemical seed treatments as herbicidal antidotes was first suggested by Hoffmann (1962). The use of 1,8-naphthalic anhydride as an antidote is covered by U.S. patent 3,564,768.

F. Physical Properties

1,8-Naphthalic anhydride is a light tan solid with a melting point of 270°–274°C. It has a low (<2 gm/100 ml) solubility in organic solvents and in water.

G. Chemical Properties

1. METHOD OF SYNTHESIS

1,8-Naphthalic anhydride is prepared by the oxidation of acenaphthene.

2. CHEMICAL REACTIONS

1,8-Naphthalic anhydride exists in equilibrium with 1,8-naphthalic acid in aqueous systems.

H. Formulations

PROTECT® is applied to the seed as the active (> 95% pure) ingredient.

II. ANALYSIS

A. Formulation Analysis

1. RECOMMENDED ANALYSIS

a. *Principle*

Infrared spectroscopy is used to analyze solutions of 1,8-naphthalic anhydride by investigating the 2100–1400 cm^{-1} region of the spectrum. In this region, the anhydride carbonyl absorption occurs as a doublet centered at 1750 cm^{-1}. One band of the doublet is at 1770 and the other at 1737 cm^{-1}. The 1770 band is intense and thus offers the sensitivity for the measurement.

b. *Reagents*

Hexafluoroisopropanol (HFIP), technical.

1,8-Naphthalic anhydride standard, obtained by recrystallization of technical material from concentrated HNO_3, washing with water, and vacuum-oven drying. Melting point should be 273°–274°C.

c. *Experimental Procedure*

i. *Sample Preparation*

Weigh approximately 200 mg of sample to the nearest 0.1 mg in a

10-ml volumetric flask. Dilute to volume with HFIP and dissolve by shaking or ultrasonic vibration.

ii. *Measurement*

Place the sample in a 0.05-mm-thick infrared sample cell. Obtain the spectrum on an infrared spectrometer over the 2100–1400 cm^{-1} region. The cell should be cleaned with HFIP, rinsed with $CHCl_3$, and then dried after each determination. Calculation of absorbance at 1700 cm^{-1} is made by using the baseline technique.

iii. *Calibration and Calculations*

Accurately weigh 1,8-naphthalic anhydride standard into six 10-ml volumetric flasks so the final concentrations will range from 20 to 400 mg per flask. Measure the absorbance and plot a concentration versus absorbance curve.

iv. *Discussion*

It has been found that the HFIP solvent or impurities in 1,8-naphthalic anhydride do not interfere with the measurement. An accurate absorptivity coefficient for the anhydride carbonyl cannot be calculated from the standard plot, since it will not be completely linear over the range of interest. For this reason, a new standard plot should be made whenever a different instrument is used or after extended periods of time. The accuracy of the method is therefore a function of the accuracy of concentration versus absorbance plot. Precision of the method is $\pm 0.5\%$ or better.

B. Residue Analysis

1. RECOMMENDED METHOD

a. *Principle*

1,8-Naphthalic anhydride is extracted from crop samples with methanol and cleaned up on a column of anion exchange resin. It is eluted from this column, extracted into toluene, and determined by gas chromatography using an electron capture detector.

b. *Reagents*

Methanol, Reagent Grade, redistilled.

Toluene, Reagent Grade, redistilled.

25% phosphoric acid solution, Reagent Grade phosphoric acid, 85%–distilled water (1 : 3 v/v).

10% ammonium hydroxide solution, Reagent Grade ammonium hydroxide–distilled water (1 : 9 v/v).

Bio-Rex 5 ion exchange resin (Bio-Rad Laboratories no. 1407831).
Sodium sulfate, anhydrous, Reagent Grade.

c. *Apparatus*

Waring Blendor, 1 quart capacity.

Centrifuge, International model 450 (or equivalent) with 300 ml capacity perforated stainless steel basket (International model 1341) and stainless steel draining chamber (International model 1320).

Rotary evaporator.

Centrifuge tubes, 12 or 15 ml graduated and stoppered.

Gas chromatograph equipped with pulsed ^{63}Ni electron capture detector.

Centrifuge filter paper, Whatman grade 54SFC, catalog no. 5 CMSX 100 m.

Chromatographic columns, 20 mm i.d. × 150 mm with a coarse fritted disc, Teflon stopcock, and having a 45 mm i.d. × 70 mm reservoir.

Glass wool.

Sea sand.

Filter discs, Whatman 2.0 cm GF/A glass fiber.

Separatory funnels, 250 ml with Teflon stopcock.

d. *Experimental Procedure*

i. *Extraction*

Green crop samples are chopped or shredded until the pieces are small enough to be easily handled by a Waring Blendor. The chopped samples are mixed by tumbling in a large plastic bag.

A 100-gm portion of corn grain or green crop sample is ground with 400 ml freshly distilled methanol for 3–5 minutes in a Waring Blendor. The brei is placed into a basket centrifuge and the filtrate collected in a 1-liter graduated cylinder. It may be necessary to repass the initial filtrate through the centrifuge to obtain a clear filtrate. The Waring Blendor is rinsed with 50–100-ml portions of fresh methanol into the centrifuge until about 400 ml fresh methanol has been used in rinsing the Blendor and filter cake.

The volume of the combined filtrates is measured, and a portion equivalent to 10 gm of crop sample is transferred to a 250-ml Erlenmeyer flask and diluted with half its volume of water. Sometimes, after standing a short time, a precipitate forms in the filtrate. This precipitate should be removed by filtration through a funnel plugged with glass wool before the portion is removed.

ii. *Cleanup*

The aqueous methanolic extract is passed over an ion exchange resin column previously prepared as follows: The chromatographic column is charged with 10 gm Bio-Rex 5 ion exchange resin. Fifty milliliters of the ammonium hydroxide solution is added, and the resin bed is stirred gently with a glass rod. The column is allowed to drain, and 50 ml water is added with gentle stirring. While the column is draining and after the resin particles have settled, a 20-mm filter disc is placed on the column to settle over the resin bed. After the column has drained, about $\frac{1}{2}$ inch of sea sand is added gently so as not to disturb the resin bed. The column is now washed with successive 50-ml portions of water until the column eluent is neutral to pH paper.

The aqueous methanol extracts are placed on the column. If necessary, nitrogen pressure is applied to the columns to provide a flow through the column of about 100 ml/hour. Then 2 successive 50-ml washes of a 1 : 1 (v/v) solution of methanol and water are passed through the column followed by 50 ml of water.

The column is now swept with 200 ml phosphoric acid solution. The phosphoric acid eluent is collected in a 250-ml Erlenmeyer flask. The column may be allowed to drain overnight if necessary. The collected phosphoric acid eluent is heated to about 70°C and cooled in a water bath or an icewater bath to ambient temperature. The cooled acid solution is now transferred to a 250-ml separatory funnel fitted with a Teflon stopcock and extracted three times with 10-ml portions of freshly distilled toluene.

The toluene extracts are combined in a 125-ml Erlenmeyer flask and dried over anhydrous sodium sulfate for about 30 minutes. The toluene extracts are filtered into a 100-ml round-bottom, single-neck flask through a funnel and glass wool plug. The Erlenmeyer flask and sodium sulfate are rinsed with two 10-ml portions of toluene through the funnel into the flask. The combined filtrates are reduced to about 5 ml volume on a rotary evaporator under water aspirator pressure at about 50°C. The concentrated toluene extracts are transferred to the graduated centrifuge tube and brought to 10.0 ml volume by successive small rinses of the flask with fresh toluene.

iii. *Gas Chromatography*

A 4-μl sample of the toluene extract is injected into a Hewlett-Packard 5750 gas chromatograph equipped with a pulsed ^{63}Ni detector. A 6 foot × $\frac{1}{4}$ inch glass column packed with 3% PPE20 on 80–100 mesh Supelcoport support is used to resolve 1,8-naphthalic anhydride. Helium at 60 psi is used as a carrier gas with a flow rate of 60 ml/minute. A purge gas mixture of 10% methane and 90% argon is used at 60 psi with the flow rate adjusted for maximum response, usually 65–80 ml/minute.

Temperatures are maintained at 265°C for the column, 275°C at the injection port, and 290°C at the detector. Detector pulse interval is 150 μseconds. Electrometer sensitivity is set to range 10 and attenuation 8. Under these conditions, the retention time for 1,8-naphthalic anhydride is 8 minutes measured from the leading edge of the solvent peak.

iv. *Preparation of Standard Curve*

A series of standard solutions containing from 10 to 100 ng 1,8-naphthalic anhydride per milliliter of toluene is prepared. A 4-μl injection of these solutions into the gas chromatograph under the conditions given in the previous section gives the peak height responses linear to concentration at this range and reproducible although nonlinear at higher levels. The crop samples are injected with the same syringe. Standards are injected frequently between crop samples, as standard response is affected by changing cylinders or injection port septums or by overnight shutdown. A gradual increase in response may also be observed during the course of a series of sample injections; therefore, it is recommended that standard sample injections be included after every two or three crop samples.

The 4-μl injection quantities described in the above paragraph are identical and reproducible for both preparation of a standard curve and obtaining crop sample responses. In absolute measurements, these syringes actually delivered 4.2 μl when the barrel contained 4.0 μl. The injection of 4.2 μl of the 100-ng standard (420 pg 1,8-naphthalic anhydride) gives a peak response of 98–100 mm at a range of 10 and an attenuation of 8. This recorder response corresponds to an electrometer output of 3.9×10^{-11} A.

e. *Calculations*

In this procedure, the 1,8-naphthalic anhydride in a 10-gm crop sample is extracted, cleaned up, and brought to a 10-ml volume in toluene prior to gas chromatography. The gas chromatographic peaks found for crop samples are converted into the corresponding nanograms of 1,8-naphthalic anhydride by using an external standard of 1,8-naphthalic anhydride at comparable levels and referring to the standard curve. As each milliliter of this final plant extract is equivalent to 1 gm of crop material, the nanograms of 1,8-naphthalic anhydride found correspond to parts per billion of 1,8-naphthalic anhydride in the crop sample.

f. *Discussion*

The method for the determination of 1,8-naphthalic anhydride in corn is straightforward, quick, easy, and requires no exotic reagents or apparatus. 1,8-Naphthalic anhydride is determined per se by gas chromatography after cleanup and extraction. As 1,8-naphthalic anhydride exists in equilibrium with 1,8-naphthalic acid under certain conditions, this property was considered in developing the method. 1,8-Naphthalic anhydride may be eluted from the ion exchange column in this equilibrium state; therefore, the phosphoric acid eluent is heated to convert any 1,8-naphthalic acid to the more stable anhydride. Quantitative extraction into toluene is accomplished, and negligible conversion back to the acid form occurs in or beyond this step.

The electron-capturing ability of 1,8-naphthalic anhydride is very high. The ECD/FID sensitivity ratio is about 4000. A 40-pg injection of 1,8-naphthalic anhydride is quantitatively measured above an instrument background noise level of one-fifth the response peak. With the detection of such low levels of 1,8-naphthalic anhydride, it would be expected that the analytical procedure for 1,8-naphthalic anhydride would be extremely sensitive; however, backgrounds from untreated samples are usually present. These are normally of the order of 0–0.01 ppm for untreated samples of corn grain and silage. Therefore, a sensitivity of 0.02 ppm may be assigned for the determination of 1,8-naphthalic anhydride in corn grain and silage. Recoveries of 70–110% were found in samples fortified at this level. This analytical procedure is considered to be specific for 1,8-naphthalic anhydride, as no known pesticides interfere.

Reference

Hoffmann, O. L. (1962). *Weeds* **10,** 322.

33

Succinic Acid 2,2-dimethylhydrazide

V. P. LYNCH

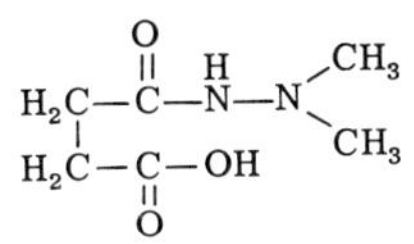

I. GENERAL*

A. Alternative Name

Daminozide is the common name proposed by ISO for succinic acid 2,2-dimethylhydrazide.

II. RESIDUE ANALYSIS

See Volume V, pp. 502–505 for description of a spectrophotometric method based on the reaction of unsymmetrical dimethylhydrazine with trisodium pentacyanoammino ferroate. The following method has been previously described by Lynch (1969).

1. PRINCIPLE

Succinic acid 2,2-dimethylhydrazide is hydrolyzed in strong aqueous alkali to yield unsymmetrical dimethylhydrazine (UDMH). The latter is distilled out and oxidized by selenium dioxide to yield formnaldehyde, which is determined spectrophotometrically using 2-hydrazinobenzothiazole.

2. APPARATUS

The apparatus consists of two distillation units (A and B).

Distillation unit A (hydrolysis stage): A 700-ml round-bottom flask fitted with a splash head (Quickfit SM7/13) to the other end of which is attached a double-surface condenser (effective length 17 cm) and a delivery tube.

Distillation unit B (distillation of formaldehyde): A 50-ml round-

* See Volume V, pp. 499–501 for general information.

bottom flask fitted with a U adaptor (Quickfit MF 22/1) to the other end of which is attached a Liebig condenser (effective length 15 cm) and a delivery tube.

3. REAGENTS

2-Hydrazinobenzothiazole (British Drug Houses Ltd.)., 0.5% solution in concentrated hydrochloric acid–water (1:9 v/v).

Potassium ferricyanide, 1% aqueous solution.

Potassium hydroxide, 10% aqueous solution.

Dimethylformamide, Reagent Grade.

Sodium hydroxide, pellets and 50% aqueous solution.

Selenium dioxide, 5% aqueous solution.

Sulfuric acid, 7.5 *N* and 10.0 *N*.

Formaldehyde solution in water, 4 μg/ml.

Daminozide solution in water, 20 μg/ml.

4. PROCEDURE

a. *Preparation of Standard Curve*

Portions of the standard formaldehyde solution in the 0–20-μg range were pipetted into 1 × 6 inch test tubes. The volume of the solution in the tubes was adjusted to 6 ml with distilled water. 1.25 ml 0.5% 2-hydrazinobenzothiazole solution were added to each of the tubes, and the solution was allowed to stand for 18 minutes with intermittent shaking. 1.25 ml 1% potassium ferricyanide solution was then added and the solution was allowed to stand for 25 minutes with intermittent shaking. 2.5 ml dimethylformamide was added (to dissolve the red precipitate) followed by 2.5 ml 10% potassium hydroxide solution. The contents of the tubes were transferred to 25-ml volumetric flasks and made up to volume with water. The absorbance of the solutions was read at 582 nm in 1-cm cells against a reagent blank. The absorbance was plotted against the concentration of formaldehyde.

b. *Application to Apples and Potatoes*

Fifty grams of finely shredded sample is placed in a 700-ml round-bottom flask, 50 ml water is added, the flask is fitted to the distillation apparatus (A), and 30 ml is distilled off. The distillate is discarded. 250 ml 50% sodium hydroxide solution plus 40 gm sodium hydroxide pellets are added to the flask, and another 30 ml is distilled off. This distillate is acidified by adding dropwise 10 *N* sulfuric acid; usually 3–4 drops are sufficient. Depending on the expected residue, all or part of the distillate is transferred to a 50-ml round-bottom flask and concentrated to 4 ml on a rotary film evaporator. One ml of the 5% selenium dioxide solution is

TABLE I

RECOVERY OF FORMALDEHYDE FROM THE SELENIUM DIOXIDE STAGE

Formaldehyde added (μg)	Recovery (%)
15.5	103, 98, 104
7.75	96, 103

TABLE II

RECOVERY OF UNSYMMETRICAL DIMETHYLHYDRAZINE (UDMH) THROUGH ENTIRE PROCEDURE

UDMH added (μg)	Recovery (%)
11.5	110
23.0	104
34.5	96, 99, 103

TABLE III

RECOVERY OF DAMINOZIDE[a]

Daminozide added (ppm)	Recovery (%)
From Apples	
0.5	106, 94, 88
1.0	90, 93, 91
2.0	83, 88, 97
5.0	93, 100
20.0	92, 97
From Potatoes	
0.5	85, 90
1.0	95
2.5	94
5.0	96, 98

[a] The control values for apples and potatoes were <0.1 ppm.

added, and the solution is allowed to stand in the flask in a water bath for 30 minutes at 50°C. The flask is transferred to the distillation apparatus (B), 10 ml 7.5 *N* sulfuric acid is added, and 6 ml is distilled off into a test tube containing 1.25 ml 0.5% 2-hydrazinobenzothiazole solution. The same procedure as for the standard curve was then followed. For conversion of formaldehyde to daminozide, the readings from the standard curve were multiplied by 160/30.

5. RESULTS

The results obtained are shown in Tables I, II, and III.

6. DISCUSSION

Under the conditions described, formaldehyde is a stable oxidation product of unsymmetrical dimethylhydrazine. The oxidation is quantitative, 1 mole of UDMH yielding 1 mole of formaldehyde.

REFERENCE

Lynch, V. P. (1969). *J. Sci. Food Agric.* **20,** 13–14.

Subject Index

N

O

U

Errata for Volume VII

On p. 399, 5,6-Dimethyl-2-*N*-dimethylamino-dimethylcarbamate-pyridin-4-yl, which appears under the structure, *should read:* 2-Dimethylamino-5,6-dimethylpyrimidin-4-yl dimethylcarbamate

On, p. 403, last paragraph, lines 3 and 4,5,6-dimethyl-2-*N*-dimethyl-2-*N*-methylformamido-dimethylcarbamate pyridin-4-yl (II) *should read:* 5,6-dimethyl-2-*N*-methylformamidopyrimidin-4-yl dimethylcarbamate (II)

On p. 403, last paragraph, lines 4 and 5, 5,6-dimethyl-2-*N*-methylamino-dimethylcarbamate pyridin-4-yl (III) *should read:* 5,6-dimethyl-2-methylaminopyrimidin-4-yl dimethylcarbamate (III)

On p. 107, nonuron, which appears in the legend of Fig. 13, *should read:* monuron

On p. vi, Aldicara, which appears in the title of Chapter 4, *should read:* Aldicarb

A 6
B 7
C 8
D 9
E 0
F 1
G 2
H 3
I 4
J 5